Research Reports in Physics

Research Reports in Physics

Nuclear Structure of the Zirconium Region
Editors: J. Eberth, R. A. Meyer, and K. Sistemich

Ecodynamics Contributions to Theoretical Ecology
Editors: W. Wolff, C.-J. Soeder, and F. R. Drepper

W. Wolff C.-J. Soeder F. R. Drepper (Eds.)

Ecodynamics

Contributions to Theoretical Ecology

Proceedings of an International Workshop,
Held at the Nuclear Research Centre,
Jülich, Fed. Rep. of Germany, 19-20 October 1987

With 116 Figures

Springer-Verlag Berlin Heidelberg New York
London Paris Tokyo

Dr. Wilfried Wolff
Professor Dr. Carl-Johannes Soeder
Dr. Friedhelm R. Drepper

Kernforschungsanlage Jülich, Postfach 1913,
D-5170 Jülich, Fed. Rep. of Germany

ISBN-13: 978-3-642-73955-2 e-ISBN-13: 978-3-642-73953-8
DOI: 10.1007/978-3-642-73953-8

Softcover reprint of the hardcover 1st edition 1988

2156/3150-543210 – Printed on acid-free paper

Preface

Given the problems of the environment in relation to the ever-growing civilization activities of man, it is not surprising that in recent years more and more related data have been collected, phenomena have been observed and concerns have been expressed. The ecological systems of man's environment react to the impact of these civilization activities. In fact, quite often these reactions occur as shocks and surprises. Acid rain and the related forest diebacks are but one example. Ecological systems consist of many components and their dynamics is thus complex.

The study of complexity is the underlying theme for work at KFA Jülich. A major part of this work concerns the physics of condensed matter, including polymers, spin glasses, surfaces and layers. Another part relates to microelectronics in the submicrometre range in its relation to the design of vector and parallel computers. Supercomputing characterizes this search for the understanding of complexity further. But equally the work on the chemistry of the atmosphere, radioagronomy, biotechnology and research on information processing on the molecular level must also be mentioned. To make the picture complete, at KFA Jülich work on reactor technology, fusion research and energy systems of course continues. These are complex systems too.

It is therefore quite natural that research work at KFA is taking an ever increasing interest in theoretical ecology as a new and quickly evolving field of science. KFA Jülich is a national laboratory. While research on theoretical ecology can and should be executed mostly in small but capable groups, the context of such research as described above points to places like KFA Jülich.

But what is the essence of theoretical ecology? The well-known and profound book "Theoretical Ecology" edited by R.M. May was probably the first of its kind and the first with this title. It is remarkable for a bystander that Robert May did not define what is meant by the term "theoretical ecology". Is it that self-evident?

Among those who laid the foundations of theoretical ecology, the Italian physicist Vito Volterra was certainly one of the most important. He offered the first mathematical solution to the classical problem how to operate a carp pond containing a certain number of pike in such a way that one would obtain an optimal harvest of good quality carp per unit pond surface and unit time. Since the problem goes to the roots of predator-prey relationships, Volterra also became one of the initiators of modern game theory.

Ernst Häckel who coined the term "ecology" defined it as follows: "Ecology is the science of interrelationships between living organisms among themselves and with their environment." From here it is easy to derive that theoretical ecology is the discipline that deals with theoretical analysis, simulation and modelling of ecological interrelationships. There are obviously two basic working modes in theoretical ecology; the deductive or so-called top down approach and the inductive or bottom up approach. Both modes are indispensable, each in its place, and interaction between the two is certainly as necessary as desirable.

Populations of living organisms are the biotic components of ecosystems such as a lake, a grassland, or a forest. Organismic productivity in an ecosystem relies on an abiotically limited and thereby defined carrying capacity consisting of climatic conditions, topology, and nutrient availability. Interaction of organisms with these physical and chemical primary conditions means, for example, that they determine whether and what trees can grow on a site. Once a forest has developed, the conditions are significantly different from what they had been before. Succession has taken place, i.e. short-term ecohistory. Interactions within a complex fraction of the biosphere consist in competition for food and space and in links in food chains, which are defined as patterns of eating and being eaten or, as recently expressed in a popular fashion, as patterns of "who eats whom". These trophic patterns are changing in time as ecological niches open and close in annual, multiannual or successional cycles, and as, deep underneath, evolution proceeds. As an analogue to an antique phrase one could say "you can never return to the same ecosystem" because succession and evolution are always on the move.

We have begun to perceive that there are common principles behind the functioning of rather different types of ecosystems such as oceans, savannas, and even the functioning of socio-economic systems like trade-commodity markets. At a higher level of abstraction, one views any entire complex of living organisms in an ecosystem as being a multiparticle-multispecies system, multiply interrelated. Because of the complexity of ecosystems, many, if not most, of the dynamic changes in the concentration of any species, or particles, or components, be they organisms or molecules, are bound to be of a non-linear nature. As far as this is the case, extreme computing power is required to cope with adequate simulation or modelling tasks. Fortunately enough, there are also basic laws and rules, which still appear to be linear.

Before we began to talk of environmental protection, we were taught that there were defined ecological equilibria or stabilities in ecosystems. This implied that an ecosystem, once disturbed, would swing back to its former state or pattern. C.S. Holling, however, was able to demonstrate a few years ago for Canadian forests how ecosystems do not in fact return exactly to their former state after sufficiently severe disturbance. The shift from one dynamic equilibrium to a homologous but non-identical next state was termed by Holling resilience.

Since resilience in forests is a non-linear phenomenon, it may turn out to be an impressive example of the universal validity of the rules and laws of non-linear dynamics in ecosystems, too.

By far the largest part of theoretical ecology consists in numerous attempts to interpret ecological field data by inductive modelling approaches, usually deterministic in nature. On purpose, this workshop was focussed strongly on deductive approaches to ecodynamics in order to promote this particular mode of theory generation. We consider it a necessity that future meetings be devoted to structured dialogues between the large inductionist and the small deductionist fraction of theoretical ecologists. Symbiosis of both fractions is essential to make our efforts in environment research successful in terms of enforcing the scientific basis of environment protection and in terms of taking to a greater extent the right measures at the right time, guided by theoretical ecology.

It is not sufficient to discover the ecodynamic pathways that lead to the development of a sustainable biosphere. The scientific messages, once we are sure of them, have to be translated into action. This is also a principal obligation of theoretical ecology.

It was due to these lines of thought that KFA Jülich organised this workshop. It is hoped that the proceedings contribute to the further advancement of this new field of science.

Jülich, *W. Häfele*
June 1988

Contents

Part V [illegible] Modelling Approaches

Part I

Evolution

Evolution: Why the Whole is Greater Than the Sum of the Parts

P.M. Allen

International Ecotechnology Research Centre,
Cranfield Institute of Technology, Bedford MK430AL, UK

ABSTRACT

The basis of systems' modelling has been the Newtonian paradigm. This states that the behaviour of a system can be understood, and anticipated by identifying its components and the causal links that act between them. This leads to a set of deterministic differential or difference equations which govern the behaviour of the system. However, this description is achieved by classifying elements into categories and by supposing that only the most probable events in fact occur. But real systems in fact evolve, that is they add and subtract mechanisms, components and interactions over time, whereas the deterministic model does not! Therefore, evolution itself must result from what has been removed from reality in order to arrive at the mechanistic model, and if we are ever to understand evolution better, and explore the real impacts of possible actions, then we must study the effects on system dynamics of non-average behaviour. From this come new insights into the nature of evolutionary processes, and better models which include the adaptive responses from within the system.

1. INTRODUCTION

Understanding the workings, and possible evolutions, of the complex systems which surround us has become a matter of urgent necessity. Global and local uncertainties seem to threaten us (Clark and Munn, 1986). For example at the global level we are faced with the increase in CO_2 and a possible change in climate and sea level, holes appearing in the ozone layer, desertification, uncertainty and volatility in the world's financial markets, the increasing world population and its exodus from the rural areas into the growing chaos of gigantic third world cities. In addition the ever growing interconnection between events at different points of the globe links their destinies and produces a still more complex system as responses and feedback forge new channels of information flow, and new subtleties of reaction.

Why has the incredible growth of knowledge and understanding realized by western science failed to anticipate and solve all these questions too?

The reason is simple. Science never 'solved' the problem of evolution. For example, what law of nature explains the diversity and connectivity of different ecosystems? The basic thought model, or paradigm, used by science did not encompass the phenomena which have created the living world (Prigogine and Stengers, 1983). Of course, if science only means a conscious attempt to learn and understand, then obviously nothing can ever fall outside its domain. But in reality, the great successes of Newton and of the ideas he espoused, led to the adoption of a particular view of the world as a mechanical system. And this, it turns out, while immensely powerful in certain domains, is not the way the living world works! (Rosen, 1984)

This idea is contained in the often repeated phrase' the whole is greater than the sum of its parts'. Despite its frequent use the exact meaning in never made absolutely clear. Indeed, what is the system if it is not the sum of its parts? And this is closely linked to the idea of modelling. Indeed, 'system dynamics' itself, in its 'traditional' form, is just such an attempt to find the 'sum', or collective effect, of the interacting components of a particular situation. It proposes to examine what a system will do, by building the machine that results from the sum of the observed mechanisms (Forrester, 1961, 1968). How can this be incomplete?

In what precise way can living systems be 'more than mechanical'? Is this a disguised form of 'vitalism', where divine intervention is invoked to explain the living world?

The answer proposed here is that the 'whole is greater than the sum of its parts' because of the evolutionary potential that is hidden within it. It is the fact that a system contains within itself the possibility of becoming something different, of 'adapting', of evolving, that makes the 'parts' less than the 'whole'. And this is also the reason why the changes that occur in complex systems are 'more than mechanical'. As we shall see, the evolutionary potential of a system is ultimately linked to the existence of microscopic freedom, represented as stochasticity, and non-average behaviour, resulting from the diversity and variability of its elements, which either make divine intervention unnecessary in the explanation of evolution, or perhaps is the means by which it occurs.

Of course, a partial answer to such questions was that offered by the brilliant insight of Charles Darwin (1859). But his ideas, although certainly correct as far as they go in biology, present evolution simply as the fruit of 'selective' forces acting on randomly occurring mutations. The theory is not really predictive, but instead is a plausible explanation of whatever is observed.

When translated into mathematical models, it represents a naturally evolved system as one in which each species has attained it 'optimal' behaviour as a result of some previous evolutionary exploration, and this is then simply 'maintained' by an appropriate mixture of nature (DNA) and nurture (Mayr, 1982; Wright, 1968, 1969, 1977, 1978).

While Darwin's theory represented a tremendous step forward for the biological sciences, its connection to physics was tenuous, and it was simply assumed that they applied to different, and separate domains.

2. THE NEWTONIAN PARADIGM

In Newtonian science, understanding of a system was to be obtained by identifying its 'parts' together with the causal connections between them. The resulting assemblage of mechanisms, written down as a set of differential or finite difference equations then constituted a 'model' of the system, and provided a tool for understanding observations and making predictions.

Two basic situations were found. First, in the absence of friction (planetary motion for example) the movement was unchecked, going on for ever. There was no 'net effect' from such movements, and there would be no way of telling whether a film of such events was being shown forwards or backwards. The movement was 'reversible'.

Second, with dissipative processes such a friction, any initial concerted motion would eventually be damped until the system reached thermodynamic equilibrium, and all its initially high grade energy had been dissipated into random, thermal motion. This was an irreversible, deterministic progression to equilibrium, and this final state could be predicted as the maximum of the appropriate thermodynamic potential. The image here is of a universe gradually 'winding down' as it uses its initial potential for creativity.

However, evolution in biology or the human sciences is about creative forces. What matters is how the system became what it is, and how its structure may evolve in the future. In other words, if the world is viewed as some kind of 'machine' made up of component parts which influence each other through causal connections, then instead of simply asking how it 'works', the real problem concerns how it got to be as it is. It is fundamentally about the origins of qualitative change in things, and how the 'parts' of a system came into being, and are maintained.

The real difference in approach between this Newtonian view and the new perspective today lies in whether we think of evolution as being over, or as still continuing. And the key issue is our understanding (or lack of understanding) of the passage between detailed microscopic complexity of the real world, which clearly can evolve, and any aggregate, macroscopic 'model' of this.

By using the analogy of a dissipative mechanical system, evolution in biological, economic and socio-cultural systems was (mistakenly) viewed as being similar to an approach to equilibrium. Behind this is the idea of evolution as an optimizing 'force', that has led to the retention of the individuals and organizations we see because of their functional superiority. In this way, the classical theories of economics, of evolutionary biology and of anthropological interpretation have been permeated by the materialist ideas of the mechanical paradigm of classical physics. Carried deep within this is the idea of change being necessarily 'progress', and that the long term evolution of a complex system must express some form of natural 'justice'. In other words, this leads to the idea, often unstated, that survival, and material success, is the ultimate measure of value.

Survival is of course a good idea, but in reality the new ideas concerning the evolution of complex systems show us that, in general, system survival will occur. But, this 'survival' can be accomplished in an almost infinite number of ways. Each carries its own values, structure, costs and rewards, distributed among its characteristic elements, and all these evolve over time as a result of the complex dialogue that is generated in the system.

3. MODELLING COMPLEXITY: THE PROBLEM OF REDUCTION

In order to even think about reality, to invent words and concepts with which to discuss it, we are forced to reduce its complexity. We cannot think of the trillions of molecules, living cells, organisms, individuals and events that surround us, each in its own place and with its own history. We must first make a taxonomic classification, and we must also make a spatial aggregation. This is shown in Fig. (1). On the left we represent the cloudy, confused complexity of the real world. Each part is special, each point unique. On the right, is a 'model' of that reality, in terms of 'typical elements' of the system, where classifications and spatial aggregation have been carried out. But the point is that however good the choice of variables, parameters and interaction mechanisms may be, these only concern **average behaviour**.

If we compare reality with the predictions of our model, then we shall necessarily find that variables and parameters 'fluctuate' around average values, and also that there is much greater **microscopic diversity** than that considered at the level of the macroscopic model.

By making the 'right' taxonomic and spatial aggregations we can model present reality by a system of boxes (components) and arrows (causal links). We can identify the 'parts' of the system, and also the factors that act to change them. However, because there is necessarily a loss of detail and precision, the description that must result is necessarily **probabilistic** in character. Clearly, also, if the exact identity and situation of each individual or element is not known with absolute precision, then there must exist an infinity of possible future evolutions, corresponding at least to the different trajectories emanating from each possible initial condition.

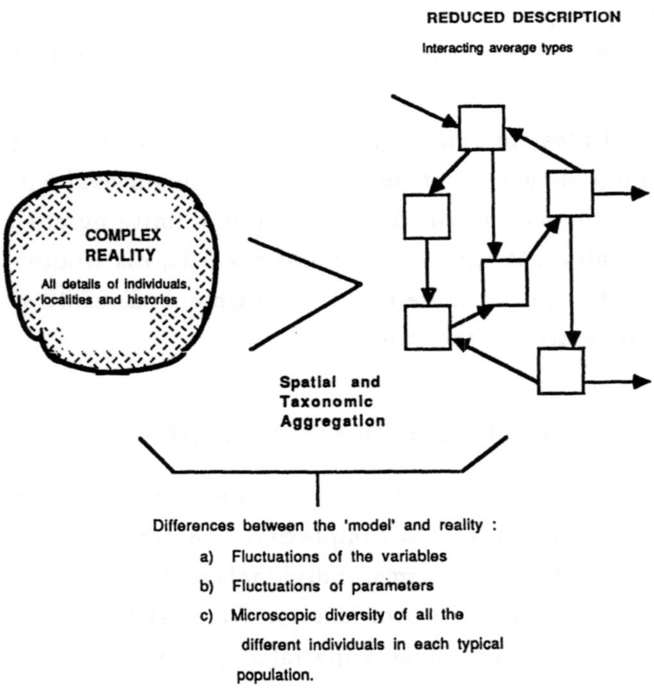

Fig. 1: Modelling, and even thinking about a complex system necessitates a simplification into categories, which constitute the system. We make 'mechanical' replica of present reality. But evolution concerns change in this structure – new boxes and arrows.

The real question of interest then becomes the extent to which these possible trajectories spread in the state space of the system. Obviously, a stable system will be characterized by only a limited divergence, while, unstable ones may display a dispersal of trajectories into areas corresponding to qualitatively different situations. Clearly then, the first 'model' that arises is therefore 'probabilistic' in nature, and takes into account all possible sequences of events into the future, from the most to the least probable.

In general, however, most realistic systems are much too difficult to handle mathematically at such a complex level of description, and further simplifications are therefore made. Also, such a description is not, it seems, psychologically satisfactory for most people. They require a single answer to the question 'What will happen?'. And in order to obtain a single answer the question is changed to – 'what will most probably happen?', or 'what would happen on average?'. By asking this question, we then can proceed to ignore all the non-average behaviour of the system, and find (or create) a satisfying (but misleading) single trajectory for the average behaviour of the system.

So if, in addition to our basic taxonomic and spatial aggregations, we assume that only average elements make up each category, and that only the most probable events actually occur, then our model reduces to a 'machine' which represents the system in terms of a set of differential equations governing its variables.

This is the 'Newtonian' vision of the world. But such a 'machine' is only capable of 'functioning', not of evolving. If cannot restructure itself or insert new cogs and wheels, while reality can! And this is because of the differences between the left and right hand side of Fig. (1), which must mean that the key to understanding evolution must lie in what has been taken out from complex reality in order to reduce it to the model on the right.

This is why the 'whole', the real thing, is in fact greater than the sum (or model) of its parts. We can always picture the system in from of us in terms of typical components, with typical interactions occurring between them. But our picture will lack the evolutionary potential that exists in the real system.

Clearly therefore, evolution is due to two things. First, to the effects of non–average values – fluctuations – of variables and parameters, and second, to change introduced by the microscopic diversity which underlies the 'taxonomic' classification of the model.

In principle then, we could capture all the effects of evolution in a complex system by listing every detail and circumstance of each element of the system.

But, of course, it would lead to a 'model' of the system as complicated as the original. It is not clear therefore, what 'understanding' would be obtained from such an exercise, and of course in practice it is impossibly onerous.

Instead of this, we must classify the system into characteristic populations, and consider the effect on them of the various processes and interactions present within the spatial zones specified. This means that we cannot know exactly when each event such as a birth, or a death, a scale or a purchase will occur, since we do not know the detailed condition, or precise location of each individual. All that we can say, therefore, is that there is a certain **probability** that an individual will be involved in an 'event'.

Furthermore, because conditions will vary locally as will the individuals actually involved in an event because of the spatial and taxonomic aggregations, even these probabilities themselves, expressed by the values taken by parameters such as the probability of birth or death, will be uncertain. In addition, the external conditions imposed on the system may also be changing in time, possibly with a consistent average, but fluctuations around this.

So the first level of description we arrive at is one in which the variables obey probabilistic equations for increase or decrease (e.g. birth or death) characterized by parameters which are themselves fluctuating.

This first approximation to the reality of a complex system can be simulated on a computer with lots of 'dice throwing' expressing the lack of information about where each element is and what its doing. The output of the model is an envelope of possibilities which includes **all possible sequences of events** – even the least likely! It is the view of science as being the search for a single, definite answer to questions concerning the future of the system, which turns out to exclude evolutionary processes.

4. EVOLUTIONARY VITALITY: THE ROLE OF THE NON – AVERAGE

In order to understand the difference that exists between a complex reality, and a deterministic model let us first describe briefly a simple example dealing with fluctuations of the variables in a problem. We shall suppose that we have a single population, X, and that there are only two processes operating in the system – the birth of new individuals, and the death of others. We shall suppose that for each individual, the probability of death during an interval of time dt, is mdt. Clearly, our model does not consider the aging of individuals, and m is simply an aggregate mortality rate.

The births we shall assume to occur at a certain rate per individual, but with an upper limit posed by the availability of some resource. So, in a short time interval dt, we shall assume that the probability of a new individual being born in $bX(1 - X/N)$. Clearly, this is kind of probability 'logistic' situation.

Let us suppose that initially the population was 10. We can perform a 'simulation' of the evolution, by using random numbers to decide on the course of events.

However, if we repeat the simulation starting from 10 again, then we find that the course of events can be considerably different. Depending on 'luck', we may have very few deaths during certain periods, or very many. Each simulation will be different as the sequences of random numbers changes. However, they all seem to give rise to a similar long term average. The problem facing us now is that of representing mathematically this process which gives many possible trajectories depending on 'luck'. We could run the system thousands of times, plot each result on the same graph as in Fig. (2) and examine the intensity of the 'ink' pattern that would result.

An alternative idea is that of the statistical ensemble. Here it is assumed that we have a large number of replicas of the initial system, and we study the evolution of the ensemble.

Let us assume that we have Ω (where Ω is a large number) replicas of our system, and that initially each one has a population of 10. As time goes on, different evolutions will occur in the different systems, and the population present in each will no longer be the same. If we ask what the probability of finding a particular value of the population is, at a given time, then it will be given by the fraction of the Ω replicas which have that population at that time.

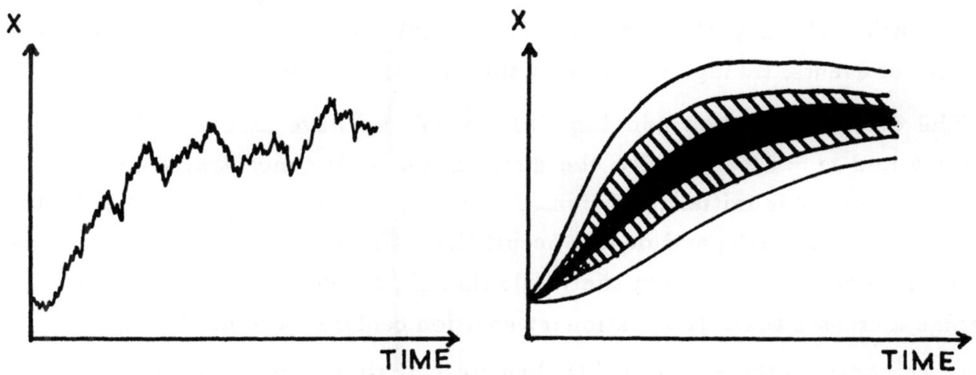

Fig. 2: a) a single simulation. b) the 'ink density' of repeated simulations.

We can construct an equation for the changing probability distribution function by saying that the change in the number of replicas with population X during a time interval dt, is due to the number that had $X-1$ before, and in which a birth occurred, minus those in which a birth occurred which had population X before. And also those that had $X+1$ before, and in which a death occurred, minus the number that had X before, but in which a death during the period dt reduced it to $X-1$.

If $\Omega(x,t)$ is the number of replicas with population X at time t, then we have:

$$\Omega(X,t+dt) = \Omega(X,t) + b(1-\frac{X-1}{N})(X-1)\Omega(X-1,t)dt - b(1-\frac{X}{N})X\Omega(X,t)dt$$

$$+m(X+1)\Omega(X+1,t)dt - mX\Omega(X,t)dt \qquad (1)$$

and if $\Delta\Omega(X) = \Omega(X,t+dt) - \Omega(X,t)$ then

$$\Delta\Omega(X,t)/dt = b(1-\frac{X-1}{N})(X-1)\Omega(X-1,t) - b(1-\frac{X}{N})X\Omega(X,t)$$

$$+m(X+1)\Omega(X+1,t) - mX\Omega(X,t) \quad . \qquad (2)$$

But as we have said the probability of finding a particular value of X, $P(X,t)$ is $=\Omega(X,t)/\Omega$, then taking the limit of small dt, we can write:

$$\frac{dP(X,t)}{dt} = b(1-\frac{X-1}{N})(X-1)P(X-1,t) - b(1-\frac{X}{N})XP(X,t)$$

$$+m(X+1)P(X+1,t) - mP(X,t) \quad . \qquad (3)$$

This is the so-called 'Master equation', also termed the Chapman – Kolmogorov equation, for the simple logistic process we have supposed. It describes the evolution of the probability of population X as a result of all possible sequences of events, taking into account their relative probability.

The evolution is shown in Fig. (3), where we have assumed that $b=1$, $N=100$ and $m=0.2$. Initially, the distribution is all concentrated in $X=10$, since we chose this initial condition. $P(10)=1$, and all the others $X \neq 10$ are zero. Over time, births and deaths occur, the distribution spreads out, and the mean value of X changes. On the whole though, births exceed deaths, and the average increases towards a stationary solution centered around $X=80$.

If we consider the equation (3) then we can study the stationary state by looking for the form of $P(X)$ which makes $dP(X)/dt = 0$.

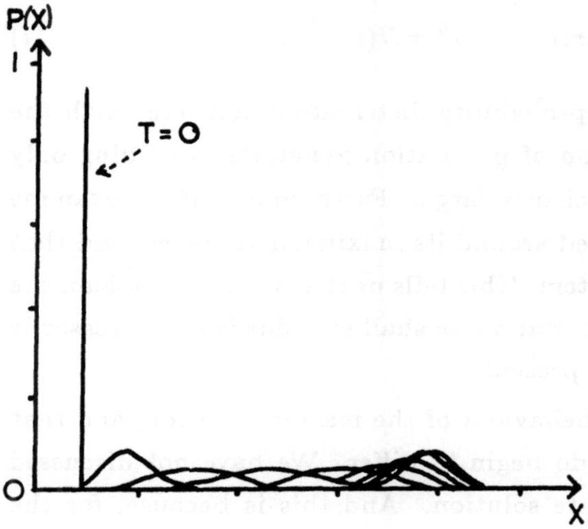

Fig. 3 For details compare text.

This will occur when:

$$b(1-\frac{X-1}{N})(X-1)P(X-1)-b(1-\frac{X}{N})XP(X)$$
$$+m(X+1)P(X+1)-mP(X)=0 \quad .$$

And this gives us the result:

$$P(X)=\frac{b}{m}(1-\frac{X-1}{N})(\frac{X-1}{X})P(X-1) \quad . \tag{4}$$

We therefore have a recurrence relation which can tell us what the probability of successive sized populations are. For example, the probability of finding a population of 49 is (for $b=1$, $m=0.2$, $N=100$) 2.037 times the probability of finding 48. Similarly, that of 85 is 0.6325 of finding 84. In other words we can find out what the most probable value of X is by asking when $P(X)=P(X-1)$. This occurs when

$$\frac{b}{m}(1-\frac{X-1}{N})\frac{(X-1)}{X}=1 \tag{5}$$

which is when:

$$X^2-(N(1-\frac{m}{b})+2))X+(N+1)=0 \quad .$$

Providing that $N>1$, there is only one positive root, and if in addition $N(1-m/b)\gg 2$, the root is:

$$X_{mostprobable}=N(1-\frac{m}{b}) \quad . \tag{6}$$

And this is interesting because it is the same solution as that of the deterministic logistic equation:

$$\frac{dx}{dt} = bx(1 - \frac{x}{N}) - mx : \qquad x^0 = N(1 - \frac{m}{b}). \tag{7}$$

In other words, the peak of the probability distribution coincides with the solution of the deterministic equation of population dynamics, providing only that the numbers involved are sufficiently large. Furthermore, if we examine how sharply the distribution is peaked around its maximum value, we find that this increases with the size of the system. This tells us that for this problem, the 'law of large number' is still working. But as we shall see, this is not necessarily the case if a stronger nonlinearity is present.

Of course, when X is small, the behaviour of the master equation, and that of the deterministic approximation do begin to differ. We have not discussed the value of $P(0)$ for the steady state solution. And this is because, for the master equation there is only one truly definitive stationary state. It is when $P(0) = 1$, and all the other $P(X) = 0$. This is because $X = 0$ is a 'sink'. There is a mechanism for going into it, but none for coming out! The population of 1 can die and fall to $X = 0$, but a population of 0 has no way of becoming a population of 1. In reality though, providing that the population is a sufficient size, the probability of extinction in a reasonable time is negligible, and we may study the form of the rest of the distribution even though, strictly speaking, this may only be 'transient'.

Now let us look at the behaviour of the master equation for a process with a little more nonlinearity in it. If we consider a problem of 'grazing' in a little detail, we find a kind of 'cooperative' effect at work among the prey population. As the numbers become large, the rate at which they are 'harvested' by a fixed number of predators, saturates. This means that per individual, the death rate actually drops as a result of the predator's incapacity to cope with such an abundant prey.

If we assume that the probability of a prey being caught by a predator will be proportional to the number of predators, y, and the fraction of the time that each one is able to search for prey, then we have:

$$\textit{Probability} \quad X \quad \longrightarrow \quad X - 1, \qquad \textit{grazing} = sxy/(1 + x\tau) \tag{10}$$

where y is the number of predators, and τ the average time taken to 'deal' with each unit of prey.

Now using these expressions, a master equation can be constructed in exactly the same way as for our previous example. By considering a statistical ensemble, we can write down an equation for the evolution of the probability of X, $P(X)$.

We have:

$$\frac{dP(X)}{dt} = P(X) + b(1 - \frac{X-1}{N})(X-1)P(X-1) - P(X) - b(1 - \frac{X}{N})XP(X)$$

$$+(m + \frac{sy}{1+s\tau(X+1)})(X+1)P(X+1) - (m + \frac{sy}{1+s\tau X})XP(X) \tag{11}$$

and this equation tells us how any initial distribution of population, $P_0(X)$ evolves over time, as a result of th probable and improbable sequences of events.

The equation is difficult to solve in general, to show the evolution of the distribution function. But, as before it can be simulated even on a very modest microcomputer. A typical simulation is shown in Fig. (4).

Here we see for an initial population of 12, the master equation gives rise to a spreading of the distribution function, until part of it falls into the 'sink' at $X = 0$, and the rest gradually accumulates at a higher population of about 40. In other words we are witnessing a bifurcation actually occurring! Some systems of the ensemble end up at $X = 0$, and others at $X = 40$! But the average of the distribution remains perfectly well behaved, and calculable but, of course, totally misleading.

In short, the idea of deterministic trajectory for a system can be seen to be inadequate, since even when we start from a well defined initial condition, $X = 12$, the probability distribution governing the system spreads out and flows to two different solutions!

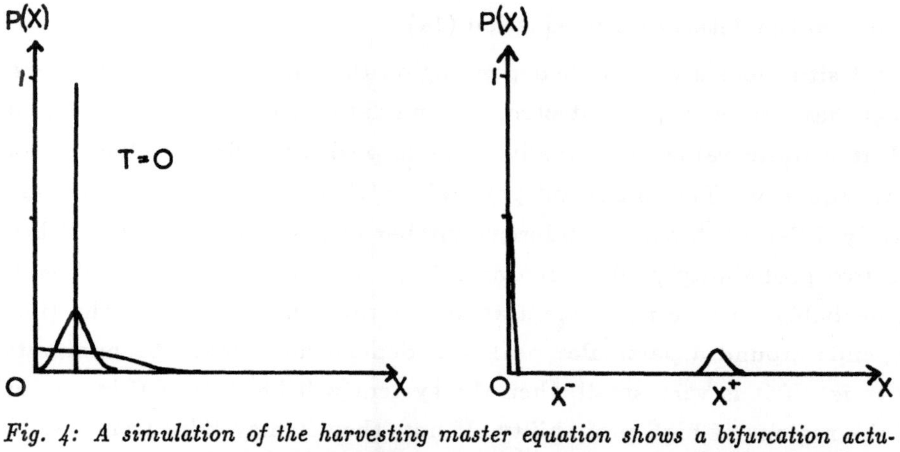

Fig. 4: A simulation of the harvesting master equation shows a bifurcation actually occurring. The distribution function spreads out, and until it has two maxima, one of which is at X=0, separated by a minimum.

Let us examine the stationary solution of the equation (11). Following a reasoning similar to that of the logistic equation, we find that:

$$b(1-\frac{X-1}{N})(X-1)P(X-1)-(m+\frac{sy}{1+s\tau X})XP(X)=0$$

and hence that the recurrence relation for $P(X)$ is:

$$P(X)=\frac{b(1-\frac{X-1}{N})(X-1)P(X-1)}{(m+\frac{sy}{1+s\tau X})X} \quad . \tag{12}$$

Once again this can be used to calculate the form of the stationary distribution function. In particular, we can once again ask when it passes through maxima or minima when the value of $P(X)$ and $P(X-1)$ are equal. This will occur when:

$$b(1-\frac{X-1}{N})(X-1)=(m+\frac{sy}{1+s\tau X})X \tag{13}$$

which, as we shall see, is exactly the same as that giving the solutions of the deterministic version of the problem:

$$\frac{dx}{dt}=bx(1-\frac{x}{N})-mx-\frac{syx}{1+s\tau x}=0 \tag{14}$$

$$x=0, \quad and$$

$$x^{\pm}=\frac{1}{2}[N-\frac{1}{s\tau}]\pm\frac{1}{2}\sqrt{[N-\frac{1}{s\tau}]^2-\frac{4Ny}{\tau b}} \quad .$$

In other words, providing that we are discussing populations which are considerably larger than 1, the location, and condition of existence of the maxima and minima of the distribution function $P(X)$ are exactly the same as those of the solutions of the deterministic equation (14).

Different situations are possible depending on the values of the parameters, and we can have either a peak at zero, a bi-modal distribution with peaks at zero and at a finite value, and finally a single peak at a finite value, as for the logistic equation. The important point about these results is that they are qualitatively different, having a different number of peaks and valleys. When there are two probability peaks, separated by a minimum, in an experiment, we shall probably observe the system at one or the other peak, and the time that it spends around a particular peak will depend how small the minimum probability is. If it is very small, then the system will be 'trapped' in one or other of the regions of high probability. If not, then it may 'flip' from one to the other at rather irregular intervals.

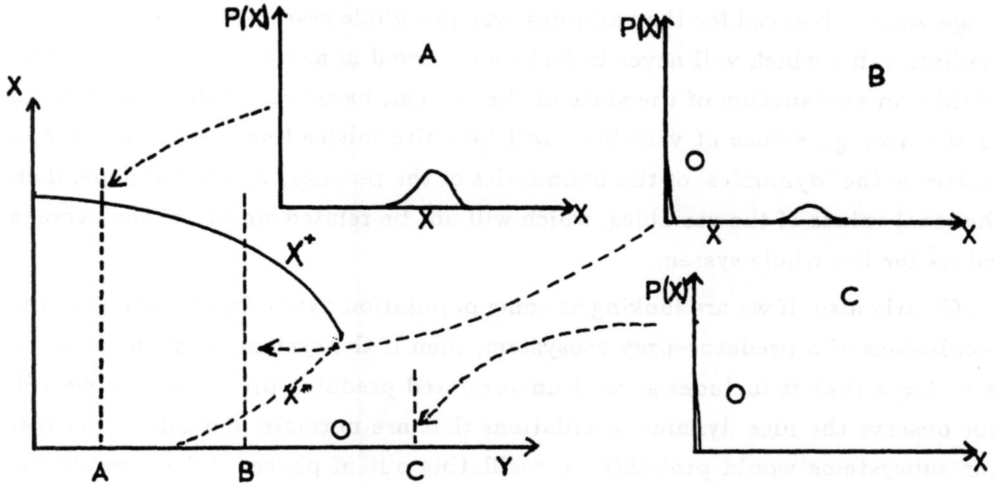

Fig. 5: The probability distribution has 3 possible forms: a peak at $x=0$; one peak at $x=0$ and another at x^+; and a single peak at x^+.

Having discussed and simulated the master equation, and the evolution of the probability distribution, we can also ask what behaviour this would predict on average.

This can be obtained from the master equation (11) by multiplying by X, and averaging over all possible values of X. Then, if we divide by the volume of the system in order to calculate the evolution of the density of X, x, we find:

$$\frac{dx}{dt} = bx(1 - \frac{x}{N}) - mx - \frac{sxy}{1 + sx} \tag{15}$$

which is, of course, the well known harvesting equation. And this is how the average of an infinite ensemble will apparently behave. But the problem is, what happens when there is more than one solution possible? The deterministic model tells us that we would observe one or the other solution ($x = 0$, or $x = x^+$). But, on the other hand, our master equation model tells us that as the system size increased beyond the size of any obvious mechanisms of spatial correlation, we would see the average of different subsystems some of which had $x = 0$ and others with $x = x^+$! In which case, we would really observe an intermediate average values, made of patches of dense population.

This is a very important point. It tells us that in order to obtain an understanding of a system, we must choose the right size for our model. If the system parameters are such that several different steady states are possible, and the system considered encompasses many virtually independent subsystems, the av-

erage values observed for the variables over the whole system will be some intermediate value which will never in fact be observed in any local region. Because of this, an explanation of the state of the system, based on mechanisms related to the average values of variables, will be quite misleading. What will really matter is the 'dynamics' of the boundaries of the patches, which will depend on the local values of the variables, which will not be related simply to the average values for the whole system.

Clearly also, if we are looking at some population dynamics, for example the oscillations of a predator–prey ecosystem, then if the system we wish to model is so large that it includes several uncorrelated predator prey systems, we will not observe the nice dynamic oscillations that are normally described, because our subsystems would probably be oscillating out of phase. All we would see would be some intermediate value of the variables, resulting from the average of the separate local dynamics. A non–spatial deterministic model is, it seems, incorrect for statistical reasons when applied to small systems, and could be misleading when applied to large ones.

Obviously, in nonlinear systems is it no longer sufficient to simply suppose that if there are many individuals involved, the law of large number will ensure that the average value will be a good representation of reality. Such systems generate spatial structure, and this can no longer be ignored.

The spreading of the distribution function is a dynamic property resulting from the nonlinear mechanisms. Because our state space has (for certain parameter values) two regions of accumulation of probability ($x = 0$, or $x = x^+$), the system acts as if these were attractors, and x^- as if it were the separation between the two basins of attraction. In the deterministic model, a system which is initially in a particular basin, on one side of the unstable solution x^- , will simply move towards that attractor. In the more realistic master equation approach however, the probability distribution may well straddle the separating value, in which case it is no longer certain that any single system will necessarily fall into the basin in which it starts.

From this analysis we begin to understand quite clearly the relation between the probabilistic 'master equation' and the deterministic kinetic equation. When we are in the situation where two stable solutions are possible, $x = 0$ and x^+, then the corresponding probability distribution for x is that of Fig. (5). And as we have seen the two peaks of probability occur at $x \simeq 0$ and at $x = x^+$. The minimum of probability, however, occurs at $x = x^-$.

Not only this, we can now see the meaning of the terms stability and instability. When we have a peak of probability, then we say that the system is at a stable solution of the kinetic equations. When, on the other hand, we have a probability minimum, then we say that the system is at an unstable solution of the kinetic equations. Clearly, the reason is that if the system is near a 'peak', then most probably it will move towards the peak – to stability. However, for a trough, the opposite is true and the system will most probably move away from that state.

The kinetic equations correspond to the most probable evolution of the system – but of course, in reality, improbable events must occur and when they do, they can take the system off in the opposite direction! The system can switch 'solutions' spontaneously because of the underlying stochastic behaviour.

The simple model above involved only a single variable, and in fact the nonlinearity led to a relatively simple situation where two stable stationary states could exist. With more variables and more complex nonlinearities, however, systems can display a wider range of possibilities, and periodic solutions are also possible, with various frequencies of oscillation around a so–called limit cycle solution. However, in addition to this, a system can also have a more complex solution in which it exhibits 'chaotic' motion (Lorentz, 1963).

This is a fascinating result, and has provoked a vast number of studies and publications. Deterministic equations can lead to random behaviour.

In the case of 'chaotic attractor', if we accept the description in terms of deterministic differential equations, we see that it corresponds to a situation in which neighbouring 'trajectories' diverge over time, although the overall state space (possible values of the variables) occupied by the system shrinks, and approaches that the surface of a 'strange attractor'.

But what does this mean if we consider the underlying stochastic picture? The sensitivity of the system to its precise initial condition really implies that probability distribution at any particular time is not sharply peaked for at least one variable, and therefore that if we performed repeated experiments on a real system in this situation, it would move along a new path each time. Probably, not only do 'neighbouring' trajectories diverge over time, but even repeated experiments from identical initial conditions would also. Another question which arises is that of the size of the system that would be correctly described by a set of non–spatial deterministic equations. Following the discussion above, we can see that a large system may well appear to be well behaved, with fairly stable global average values for variables, although each local region may be following a chaotic path. In modelling the real world, we must pay particular attention to the question of spatial structure, since only a certain size system can in fact be described correctly by a set of non-spatial dynamic equations. This size is related, in a chemical system, to the diffusion coefficient, but in biological, ecological and socio–cultural systems it is related to the flow of 'information' between the elements.

In most situations of interest, the generation and exchange of information will be a complex process which needs to be included in the dynamic model explicitly for a correct understanding. An example of such a model has been given recently in connection with fisheries (Allen and McGlade, 1986, 1987).

In addition to the discussion above, however, in the living world, we must examine the possibility that the internal structure of the individuals or elemental objects themselves could evolve in time. Indeed, these elements could themselves be dissipative structures in competition for the energy and matter that they need to maintain and transcend themselves.

In this connection then, it is the existence of microscopic diversity and modes of individuals liberty that must be discussed.

5. EVOLUTIONARY DRIVE: THE EFFECTS OF MICROSCOPIC DIVERSITY

In the previous sections we have discussed the effect of taking into account the non–average behaviour of variables which characterize the difference between the real world, which can evolve, and a mechanical 'model' of a system consisting of deterministic differential equations. Much work has been done on exploring the behaviour of these 'self– organizing' (Nicolis and Prigogine, 1977; Prigogine, Allen and Herman, 1977; Allen, 1986), synergetic (Haken, 1977) systems. Also, the effects of fluctuating parameters has also been investigated (May, 1977; Beddington and May, 1977; Horsthemke and Lefever, 1982) Here, however, we turn to the third aspect which has been somewhat neglected until now. It is that of the existence of microscopic diversity in the real world. In order to write down a model, it is always necessary to classify the elements of the system, and it is the changing numbers of these different types that becomes the 'variables' of the differential or difference equations which make up the model.

But, what about the difference that exist within each type? What of the variance of the characteristic parameters of individuals in each that will exist in any population? In this section we shall briefly describe some recent work that has examined this question. It discusses a simple problem of population dynamics, within a simple ecosystem. But clearly, the ideas and results can also be applied to economic and socio–cultural models.

We shall study the evolution of a particular population, as it feeds on resources, reproduces and dies, either of natural causes or because of some predator. In fact, this could be easily extended to other problems of socio–cultural evolution, such as evolving technology, the diffusion of ideas, economic activities etc. In that case it would concern the origin of different strategies, and their adoption or rejection by surrounding populations. But here we shall limit the discussion to this simple ecological situation. The important point is that 'evolution' implies some changes of form, character or behavioural strategy, which affects the manner in which individuals perform in capturing prey, reproducing and avoiding death. Clearly, there are many 'dimensions' of behaviour or morphology which can affect these mechanisms, but here we shall only consider a simple case of two dimensions.

In Fig. (6) we see a 2–dimensional space of possible morphologies or strategies for a population. Each point corresponds to a particular blend of the 2 dimensions of character. Below it, we may imagine a surface on which we represent the 'resources' (generally prey species) available to each particular character. In this way, a population with characteristic of the point i, say, can 'feed on' the resources below at and around the point i, depending on the degree of specialization. Above the 2–dimensional 'character space' of the population considered, we can also imagine a surface which expresses the 'dangers' which will result from a given morphology or strategy – that is from predator populations which consider the population we have chosen as a 'resource'.

For any particular 'character' of our population, there is a relief map of possible rewards and dangers which could result, given the prey and predator species that exist in the system at that time. This means that the 'character space' is spanned by a kind of 'contour map' of potential 'malthusian parameter' (birth rate – death rate), which will show the 'selective advantage' of any particular character choice, given the existing circumstances. In this paper we shall simply examine the simplest problem of how a evolution leads a population to 'climb' a hill in the evolutionary landscape.

Let us consider the simplest possible case of a single population x. According to its 'situation' in character space it will have a certain birth rate, and mortality, and we shall suppose that the limits of the resource on which if feeds will give rise to a logistic type population dynamics.

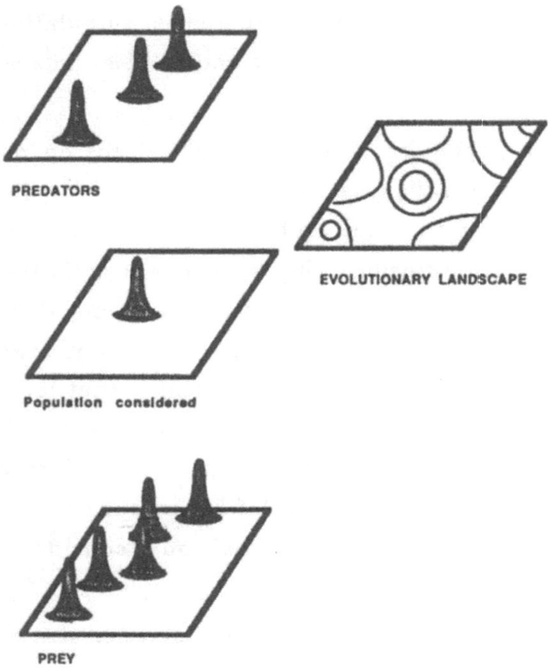

Fig. 6: A two dimensional space of morphology or character for a population feeding on prey and fed on by predators.

The equation will be:

$$\frac{dx}{dt} = b_1 x(1 - \frac{x}{n}) - m_1 x \tag{16}$$

where n is the limiting resource related to the prey species in the level below, and b_1 and m_1 the birth and death rates which correspond to the position in 'character space' – that is on the evolutionary landscape.

Now, what interests us here, is the question of 'evolution'. How does the 'character' of a population change over time in response to the 'rewards and dangers' of particular strategies? In order to examine this, we must consider two important aspects:

a) Imperfect reproduction leading to 'diffusion' in character space.

b) The operation of selection in amplifying 'favourable' mutations and suppressing unfavourable ones.

Usually, the models of population genetics or of neo–Darwinian evolution simply consider the **second term**. That is, they assume that all strategies are present initially, and examine the 'selection' of the 'fittest'. We shall attempt to retain both aspects.

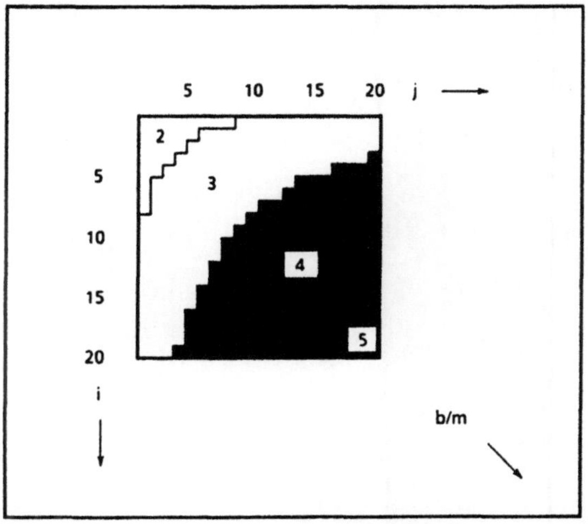

Fig. 7: The parameter space in which we can study co-evolution of x and y.

We can construct a parameter space of the effectiveness of the two species. For this, we have consider two indices, i and j, which characterize the effectiveness of b and m respectively. So, we have:

$$b(i) = b^0(1 + \alpha i)^\gamma, \qquad m(j) = m^0(\frac{1}{1 + \beta j})^\epsilon$$

and different values of b^0, m^0, α, β, γ and ϵ correspond to different 'evolutionary landscapes' which all show increasing effectiveness in a diagonal towards the bottom right hand corner, as in the Fig. (7).

In this simple example then, we are assuming that the characteristics corresponding to the two dimensions chosen for the 'evolutionary landscape' happen to be such that they affect reproduction and longevity separately.

For species which reproduce perfectly, coexistence is only possible if the two species are of exactly equal effectiveness, b/m. They are at the same 'altitude'.

But let us suppose that only a fraction (f) of the births of x are 'perfect' reproductions of the parents. On average, the would give rise to a negative effect. That is, the births of 'mutant' types would be biased towards the less effective.

We shall suppose, as a first assumption, that 3/4 of the mutants are characterized by a lower value of i and j, while only 1/4 have higher values as in Fig. (8).

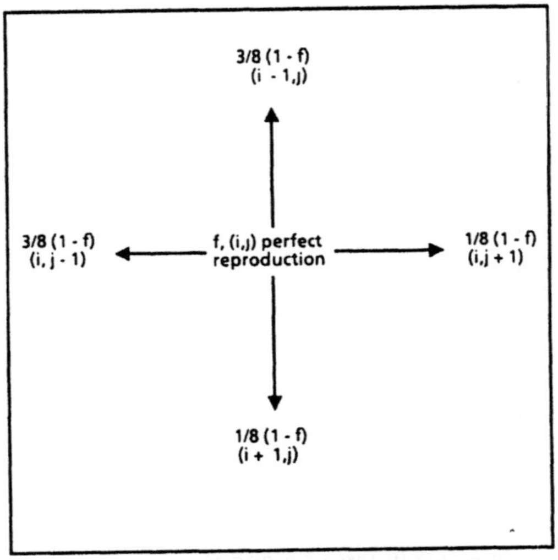

Fig. 8: The pattern of reproduction of each population x(i,j) and y(i,j), linking all points of the evolutionary landscape.

Each point in parameter space is connected to its vertical and horizontal neighbours through mutations or stochasticity of behaviour. Diagonal mutations are excluded because we assume that reproduction and longevity are independent.

The equations for the dynamics of a species is:

$$\frac{dx(i,j)}{dt} = [b(i)(fx(i,j) + \frac{3}{8}(1-f)x(i,j+1) + \frac{1}{8}(1-f)x(i,j-1)$$

$$+b(i+1)\frac{3}{8}(1-f)x(i+1,j) + b(i-1)\frac{1}{8}(1-f)x(i-1,j)]C$$

$$-m(j)x(i,j) \qquad (17)$$

where f is the fraction of perfect reproduction (e.g. 90 %), and C is the 'crowding' factor taking into account competition with other species.

Now let us examine first the possibility that a perfectly reproducing species ($f(x) = 1$) could be 'invaded' by an imperfectly reproducing one ($f(y) = 0.9$). In Fig. (9) we show the result of initially placing 200 units of x at position $10,10$, and then seeing if 2 units of y placed also at $10,10$ can invade.

Initially, as can be well imagined, the species x grows, while y remains at a low value. However, the errors in reproduction of y lead to a 'cloud' of small populations around its initial position $10,10$. These are mostly at lower values of i and j, but some are higher. As the selection operates, the least effective

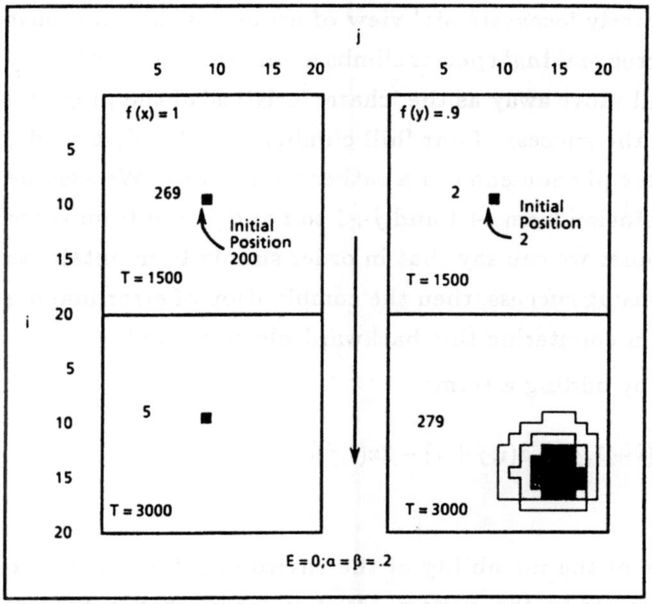

Fig. 9: The competition between a perfect reproducer (left) and an imperfect one (right) goes in favour of the imperfect one, given sufficient time. At time 3000 *population* $x = 5$ *and* $y = 269$.

populations of y are eliminated rapidly, but gradually small amounts of mutant populations 'invade' the more effective regions of the parameter space (7), and after some time they start to multiply at the expense of x, and then rapidly eliminate the perfect reproducer. His 'perfect reproduction' allows no possibility of an adaptive response.

The simulations above showed that even if a population of 'perfect reproducers' had ever evolved, it would have been invaded and replaced by error–makers. We can also examine the effect of 'error making' on a species' ability to maintain itself or evolve in its evolutionary landscape.

We find that if the slope is sufficiently large, then error making is advantageous, and we find a forward evolution whose rate depends on the slope of the landscape. Therefore if the slope decreases as b/m increases, then at the 'top of a hill', selection will operate in favour of lower error rates, because local exploration and discovery have yielded all there is to 'know' about that resource. But this situation would only exist if the 'resource' itself was not evolving! If it is another living population, then in fact the growing 'success' with which it is being eaten would in fact drive its own evolution faster.

If we are to adopt a properly 'ecosystemic' view of evolution then we must admit the fact that as our 'error making' species climbs a 'hill' in his evolutionary landscape, the 'hill' itself will move away as the 'characteristics' of the prey and predators change in view of the success of our 'hill climber'. In this first study, we can represent this complex phenomenon in a rather simple way. We assume that there is a backwards diffusion from $i+1$ and $j+1$ to i and j, and from there to $i-1$ and $j-1$. In this manner we can say that in order simply to maintain its position on an slope of increasing success, then the combination of error making and selection must succeed in countering this backward blowing wind.

We can account for this by adding a term:

$$E[x(i+1,j)+x(i,j+1)-2x(i,j)]$$

to the equation (17).

The term E is a measure of the instability of the environment and the rate of evolution of the other species in the system. If it is large, then a species which does not evolve in its behaviour will rapidly become less and less effective over time. This term sets the natural limit to evolutionary progress in terms of 'effectiveness'. An equilibrium distribution of populations is attained, and the larger the value of E, the lower the values of i and j for the centroid of the population.

The value of E which a species must counteract will result from the evolution of the other species in the system. In this way, there is a kind of 'competition' between the different levels which will ensure that E will not go to zero. This is why evolution can probably never be viewed as being 'over'. In reality the mutual linkages will ensure that each population maintains a variability as part of its strategy enabling survival. And this variability will be seen at each instant as being inefficient and suboptimal.

Whatever the precise fraction of superior and inferior 'mutants' is (and this is determined by the natural world not the species or the ecologist), there will exist an 'optimal' rate of mutation which will be selected for by evolution, and our model can be used to find this. The evolutionary process will therefore lead quite naturally to a rate of evolution which is itself regulated by the evolutionary process elsewhere in the system, so that the 'progress' being made in each level of the system is just counteracted by that in the others.

Perhaps evolution itself could set these rates. In order to explore this possibility, we can suppose that the 'fidelity' of reproduction of a population is an hereditary characteristic which could itself vary.

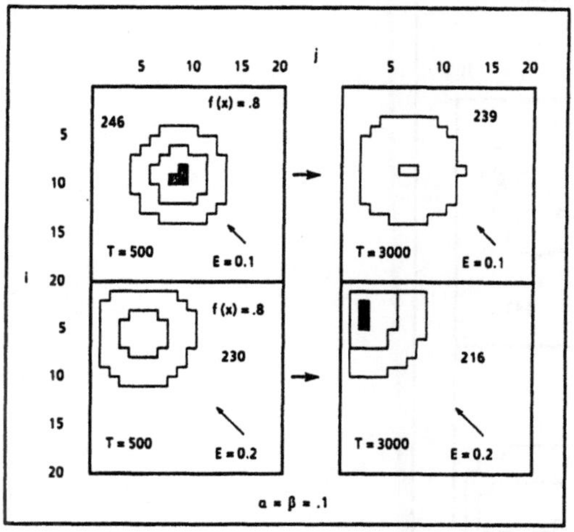

Fig. 10: With an environmental wind E of 0.1*, the equilibrium position of the population centroid is at* 8, 8*. With* $E = 0.2$ *it is at* 3, 3.

For these simulations the vertical, i–axis will still represent the rate of reproduction of any sub-population, and will still serve to represent the 'hill' up which evolution is struggling to climb. Clearly, the higher the value of i attained, the higher the birth rate. However, the value of j will indicate how 'true' replication is, ($j = 0$, $f = 0\%$: $j = 19$, $f = 100\%$). In this way selection itself will act on the system to decide the degree of 'variability' which is the most effective in climbing the hill.

In Fig. (11) with $\alpha = \beta = 0.1$ for species with initially equal values of b, but different values of f (for x, 95%: for y, 85%) we see that the species with initially the lower fidelity 'wins'. On its way to 'winning', however, the species goes through an 'identity crisis' and plunges to very low average fidelity, but then gains until it has only a distribution of fidelity that ranges from 85% to 100%.

From these simulations we see that if a certain amount of 'random variability' is a property of a particular population, then selection can operate on it to regulate the variability which is 'necessary' for 'hill climbing' or countering the evolution of other species represented by E.

We see that variability itself is part of a species' strategy. In other words what we see as the result of evolution are not populations with optimal behaviour but rather populations which can learn!

This idea is similar to that used by Eigen and Schuster (1979) for describing the evolution of competing polymers in their theory of the Hypercycle. They

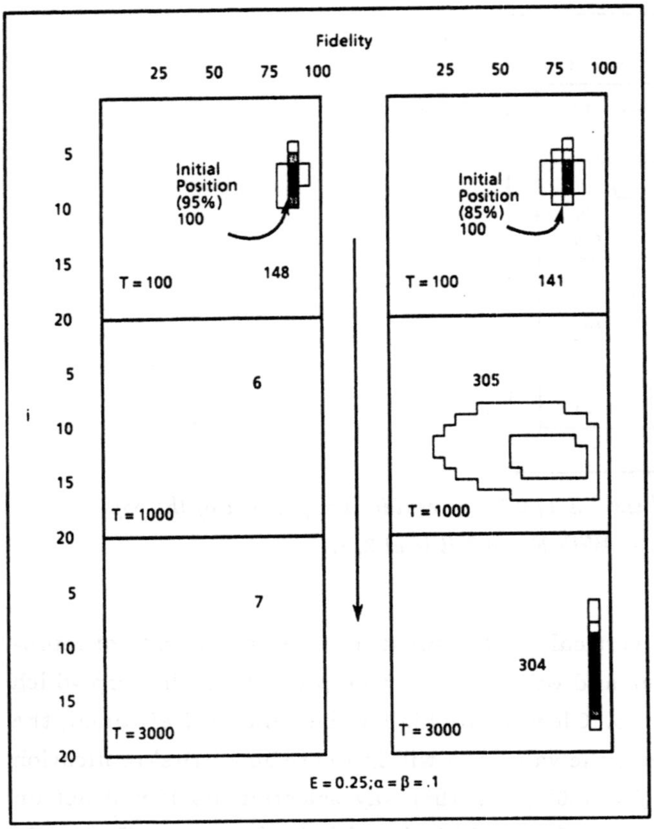

Fig. 11: Evolution of fidelity in an evolutionary landscape with a steep slope and strong environmental wind. Here in order to move forward 'variability' is strongly selected for, then high effectiveness is maintained by high fidelity.

retain both 'mutations' and 'selection', and obtain a similar kind of result to that which we show here. They show that evolution should not be looked at in terms of single population types, but instead in terms of a successful quasi-species, whose average behaviour is the most effective. In the system which they discuss, a chemical reactor into which flow carefully controlled fluxes of monomers, there is in fact only a single 'evolutionary hill' to climb, and the polymers duly perform this task. The hill itself does not move, because there are no other levels of population involved, and in consequence it is certainly correct to say that evolution will come to and end, when the quasi-species is at the top of the adaptive landscape.

Very recently, Ebeling and Engel (1986) have also proposed a model which retains both variability and selection. They discuss their application to the solving of optimization problems, and fundamentally they adopt the view that evolution leads to 'hill–climbing'.

However we feel that when we consider the natural evolution of ecosystems, then the need to adapt and to change will be a permanent feature, owing to the evolution of other levels (prey, predators) of the system, and that therefore the ability to adapt will remain as part of the evolutionary strategy.

We have presented here only the very simple problem of climbing a particular 'hill', and have explored how small mutations or errors in reproduction could achieve this. Obviously, other landscapes and mutation patterns can and will be modelled. We find that variability is necessary to success in hill–climbing, even though this necessarily implies a certain level of 'waste' or 'sub–optimality' at any given instant. Furthermore, we show how the mechanism of variability itself could be adjusted by the evolutionary process itself, leading to the idea that evolution is driven by the noise to which it leads. It is driven from within, each species being part of the environment of others, and strongly connected to a changing physical environment as well. Evolution is no longer simply the selection of 'optimal behaviour', but the selection of species which both produce, and can cope with change.

Co–evolution within an ecosystem will lead to behaviour on the part of individuals and species which is not optimal in terms of the simple rationality of maximal exploitation of known resources, or maximal avoidance of death. However, the variability in behaviour and phenotype will ensure the longer term survival of the species.

One way of ensuring variability in behaviour of single individuals would be to include internal mechanisms giving rise to 'deterministic chaos'. This would then give rise to exploration, and hence to the ability to learn and to adapt. In a sense than, the origin of 'intelligence', which in some species has replaced genetics as the principle way of adapting behaviour, could well be 'chaotic attractors'.

Because behaviour is sub–optimal at any given instant, with loss making parts, selection is relaxed and the species remaining in competition are not operating a 'knife–edge' selection. In this way, partly, the rigours of the mechanical models of population dynamics are relaxed, and we can better imagine the emergence and maintainance of the complex and diverse systems of reality.

6. DISCUSSION

The fundamental question raised in this paper is that of the origin and nature of evolutionary processes. These, as we have seen, reside in the non-average behaviour of the elements: either as fluctuations of parameters or variables, or in the actual diversity of individuals. It is this microscopic variability and randomness in the system which both drives evolution, and which is selected for by evolution. This is because it confers on a system the ability to 'learn' and hence to 'adapt'.

The real message of the new concepts in science are that change and disequilibria are probably more 'natural' than equilibrium and stasis. Those who can adapt and learn will survive. And this will depend on their 'creativity'. And it is here that we see a new and exciting role for 'deterministic chaos' in the evolutionary process. Clearly, one way that organisms and systems can actively maintain an internal mechanism of exploration and discovery is if they contain subsystems characterized by chaotic behaviour. Not only that, but in the evolutionary process, behaviour that is not predictable according to any simple rule will confer an advantage on its possessor, perplexing both its predators and its prey.

What we can see from the discussion above is that the ultimate source of 'creativity', and hence of evolutionary change, is the underlying 'random' or 'chaotic' behaviour and nature of elements of a complex system. Even though such randomness may on the whole seem wasteful at any given time, nonetheless it is this microscopic freedom from any single selection criterion that confers adaptive potential on the system, and hence allows it to survive.

ACKNOWLEDGEMENTS

We wish to thank Guy Engelen for help with the computer program, and the Solvay Institute for Physics and Chemistry for financial support. This work resulted in part from a project sponsored by the United Nations University, Global Learning Division, and also Fisheries and Oceans, Canada.

REFERENCES

Allen P.M., 1986, 'Towards a New Science of Complex System', In, 'The Praxis and Management of Complexity', *United Nations University Press, Tokyo*

Allen P.M. and J.M. McGlade, 1987, 'Evolutionary Drive: The Effect of Microscopic Diversity, Error Making and Noise', in *Foundations of Physics, Special Edition, July, 1987*

Allen P.M. and J.M. McGlade, 1986, 'Dynamics of Discovery and Exploitation: The Case of the Scotian Shelf Fisheries', *Canadian Journal of Fisheries and Aquatic Sciences, Vol. 43*

Allen P.M. and J.M. McGlade, 1987, 'The Modelling of Human Systems: A Fisheries Example', *European Journal of Operational Research, June.*

Beddington J.R. and May R.M., 1977, 'Harvesting natural populations in a randomly fluctuating environment', *Science,* **197**, *p463-465*

Clark W.C. and R.E. Munn, 1986, 'Sustainable Development of the Biosphere', *IIASA and Cambridge University Press, Cambridge, England.*

Cyert R. and J. March, 1963, 'A Behavioural Theory of the Firm', *Englewood Cliffs, N.J., Prentice Hall.*

Darwin, C., 1859, 'The Origin of Species', *John Murray, London*

Ebeling W. and A. Engel, 1987, *Systems Analysis Modelisation and Simulation. Vol. 3 No. 5 p377-385*

Eigen M. and P. Schuster, 1979, 'The Hypercycle', *Springer Verlag, Berlin.*

Fisher R.A., 1958, 'The Genetical Theory of Natural Selection', *Dover, New York*

Ford B.B., 1964, 'Ecological Genetics', *Methuen, London*

Forrester J.W., 1961, 'Industrial Dynamics', *MIT Press*

Haken H., 1977, 'Synergetics: An Introduction', *Springer Series in Synergetics, Volume 1, Berlin*

Horsthemke W. and Lefever, R., 1984, 'Noise Induced Transitions. Theory and Applications in Physics, Chemistry, and Biology', *Springer Series in Synergetics, Vol. 15, Spinger New York, 318p*

May R.M., 1977, 'Stability and Complexity in Model Ecosystems', *Princeton University Press, Princeton, 1973*

Mayr E., 1982, 'The Growth of Biological Thought. Diversity, Evolution and Inheritance', *Belknap Press, Harvard University Press, Cambridge, Mass. USA*

Morecroft, J., 1983, 'System Dynamics: Portraying Bounded Rationality', *Omega,* **11**, *131-142*

Morecroft, J., 1985, 'Rationality in the Analysis of Behavioural Simulation Models', *Management Science, 31 (7), p 900 - 916*

Nicolis G. and I. Prigogine, 1977, 'Self-Organization in Non-Equilibrium Systems', *Wiley Interscience, New York*

Prigogine I. and I. Stengers, 1985, 'Order Out of Chaos', *Bantam Books, New York*

Prigogine I., P.M. Allen and R. Hermann, 1977, 'The Laws of Nature and Complex Systems', in 'Goals for a Global Community', Ed. E. Laszlo, *Pergamon Press, New York*

Rosen S., 1985, 'Information and Complexity', in 'Ecosystems Theory for Biological Oceanography', Eds. Ulanowicz and Platt. *Canadian Bulletin of Fisheries and Aquatic Sciences, Vol. 213. Fisheries and Oceans, Ottawa, Canada*

Sterman, J., 1986, 'The Economic Long Wave: Theory and Evidence', *System Dynamics Review, 2 (2), p 87-125*

Wright S., 1968, 1969, 1977, 1978, 'Evolution and the Genetics of Populations', Volumes I, II, III, IV, *University of Chicago Press*

The Evolution of Spatially Inhomogeneous Populations

J.M. McGlade

Darwin College, Cambridge, CB39EU, UK, and
International Ecotechnology Research Centre,
Cranfield Institute of Technology, Bedford, MK430AL, UK

INTRODUCTION

Essentially in this paper I shall argue that any undertaking in ecology requires an understanding of the spatial properties of organisms and systems, and that such an understanding comes from a knowledge of (i) historical and evolutionary events and processes, (ii) ecological, genetic and behavioural processes and (iii) scaling phenomena within and between systems. Two forms of knowledge are required, one a context-free, formal and abstract kind -- or theory, and the other a more situated, contextual and narrative mode -- or acccount. These must in turn be used to recognise the separate forms of space: indeed, the generative source of interpretation of spatial inhomogeneities is the identification of physical space, mental space relating to cognition and representation, and of the social space linking groups and individuals to themselves and to material nature (Soja 1985). It is this plurality of spatial and processual attributes that abrogates much of current ecological theory. In this paper I look specifically at the evidence from many aspects of research that will allow a reconstruction and a better understanding of the evolution of spatial inhomogeneities in lower vertebrates as represented by fishes.

HIERARCHICAL REPRESENTATIONS OF ECOSYSTEMS

Ecosystems are generally shown as food webs or graphs in which organisms are linked through Lindeman type trophic relationships, or through an Eltonian pyramid of numbers or biomass. The dynamics of such hierarchical systems was set aside until more recently, when May and others began to look at the stability and structure of such systems with respect to perturbations, colonization and patterns of diversity (May 1974; MacArthur and Wilson 1967; Gilroy 1975; O'Neill et al. 1986). One interesting feature that emerged was the limit on food-chain length to four or five levels, which Pimm and Lawton (1977) argued had to do with the decreasing resilience of a system rather than the more generally held view of decreasing available energy. These ideas about ecosystems have since been explored analytically and in more general terms by Hogg et al. (1988), who show that ecologies with unstructured cooperative interactions between arbitrary species are less stable than previously suggested by Gardner and Ashby (1970) and May (1974), and that hierarchical ecologies are more robust with respect to their stability.

This has important consequences for understanding spatial patterning, because there has been a considerable effort made to provide a theoretical representation of the practical ability of an organism to "go on" in the absence of any knowledge about the system or its infrastructure, implying that living is simply putting theory into practice. Theories however are not formulated in ways appropriate to their application to the situations they are supposed to depict (Wittgenstein 1980). Rather living activities are found in the flow of the functions they constitute: animals rather than living in space and time, account for it. The behavioural aspects of the life of an individual can thus function as insurance that living activities are perceptually-rational and reportable i.e. accountable to the other members of a group. Unfortunately there is often a tacit assumption in ecology that organisms and hence populations and systems are adapted to their environment in the sense of being fitted (*adaptare*) to (*ad*) it. This formulation of a situation from an uninvolved third-person has given rise to a dogmatism that berates the contextual account of a system aimed at understanding those within the system better able to appreciate the significance of events and processes. The question is whether or not the negation of insider or pluralistic views is a valid attitude for ecologists to hold.

EVOLUTION AND ECOLOGY

The next issue concerns the aspects of time and evolution in understanding spatial patterns. Darwin (1837:139) stated that

Time [is] no element in making change, only in fixing it:
only circumstances [are] a contingency of time.

And most ichthyologists do in fact believe that historical events have provided a large part of the brief for the spatial and life-history characteristics of most ichthyofaunas derived during the post-glacial period some 10,000 yrs BP. However if we wish to consider the current discreteness of pelagic and benthic ichthyofaunal regions, we find that the present distribution of fishes throughout the worlds oceans is well documented in the case of the temperate less speciose communities, but less so for the tropics. Nevertheless, it is still possible to understand the large scale distributional features by considering past histories in terms of vicariance, taxon pulses and dispersion. In the Atlantic Ocean vicariance in the post-glacial period was largely concerned with changes in sea-level and the effect on the continental shelf. Brain (1985), argued that for the continent of Africa, changes in sea-level as a result of alterations in temperature probably had a radical effect on the continental shelf area available for habitation by marine organisms. Such alterations in global temperatures have been well documented in the oxygen isotope determinations of foraminiferal shells of deep-sea sediments from Australia, New Zealand, the Pacific and Atlantic (Shackelton 1975; Donk 1976) and are continuously being extended (Mercer 1983). From these and other stratigraphic data it has been possible to reconstruct the appearance and disappearance of faunal barriers throughout the Pleistocene. It has also been possible to identify phylogenetic and distributional anomalies of single genera. For example the species in the Atlantic from the genus *Sebastes* are not only phylogenetic outliers in an otherwise low latitude, diverse shallow water, tropical family (Scorpaenidae) but probabaly have a North Pacific origin (Eschmeyer 1969).

In a recent study of demersal fishes on the continental slopes of Nova Scotia and Newfoundland (Fowler and McGlade 1988), both primary species (those that had evolved endemically within the area) and secondary species (those that had evolved outside the sampling area) were contributing to a highly diverse assemblage: in particular there was a disjunct zone of co-dominance among the two types between 600 - 900m, below which primary species dominated. This physical notation is important because changes in sea-level during the Pleistocene would have meant a disruption for colonization and habitation of the continental shelf at these depths. Invasion by a faunistic outlier would clearly have been possible in a perturbed environment, and the fact that *Sebastes* is ovoviparous and a benthopelagic feeder probably mitigated in its favour as a colonizing species. However, the Laurentian Channel that runs out from the Gulf of St. Lawrence and which separates the continental shelf off Nova Scotia from that of Newfoundland is still considered to be a true biogeographic boundary. Such distinctions are clearly important in unravelling spatial patterns, and underline the major problem with this approach -- that of a dependancy on adequate sampling and quantitative analysis.

McKelvie (1985) on the other hand has showed that the association of organisms with water masses as hypothesised by Sverdrup and others in the 1940's is not nearly so well defined for pelagic faunal regions in the same oceanographic area. He demonstrated that there was more of a graduation between faunas rather than any abrupt change, and few species were regularly associated with each other: he therefore concluded that individual pelagic species probably responded to environmental gradients. Yet the formation of distinct faunas are known from other dynamic areas such as the coastal waters off California, the Gulf Stream and the Sargasso Sea; thus the question must then be one of areal definition. McKelvie queries the existence of structured communities in mesopelagic fishes as suggested by many other workers and criticizes the use of certain statistical techniques to identify patterns, and whilst agreeing with this latter critique, it is important to note that a number of pelagic fishes such as *Clupea harengus* spawn in some of the most energetic parts of the ocean (Iles and Sinclair 1983). What is important then is that faunal assemblages reflect not only ontogenetic positioning and temporal movements (Haedrich and Judkins 1979) but also historical events.

If we consider that the presence of one or another species in a system can be explained in terms of its past biogeographically inferred history, we are in a somewhat better position to understand extant spatial patterns. For example, the carp *Cyprinus carpio* was presumed to have arisen in Asia and been introduced by way of Europe to Britain during mediaeval times for pond culture. Recent excavations of a Cromerian site at Little Oakley in Essex have revealed the presence of this species, suggesting not only the type of ecologocal regime that was present at the time, but also a very obvious link with the European river system more than 0.5 mya (Lister et al. 1988). In reconstructing present distributions of the cyprinids, it may be that this and other more recent arrivals should in fact be considered endemics. This calls to mind the problem of sampling the fossil and extant world, because it was this issue that led to the controversy between the theoretical aspirations of advocates of punctuated equilibrium, and those gradualists who deemed the gaps filled but missed by sampling. As far as fishes are concerned, however, macroevolutionary processes and extinctions are not at all well-documented, so that traditionally only larger biogeographic and speciation issues have been addressed.

COMMUNITY ECOLOGY

Accepting that there are scales in the spatial patterning that we are interested in, most discussions of biogeographic zones are generally posited for distances greater than 100 km. Below this we generally enter into the domain of community ecology: Schlosser (1987) states that the goals of this latter area are to record patterns occurring in nature, elucidate causal processe and generalize these explanations. But as we have seen above, the structure that exists in a system may in fact be open to invasion, and so it is just as important to define the inmates of a community as well as those excluded but spatially adjacent. Whether or not biological or physical features determine the exclusion is then a case in point which requires detailed examination. For example there have been a number of studies on coral reefs that have examined spatial heterogeneity, and results generally support the idea that above 10km, differences cannot be detected in aggregate form, but that below this at the level of individual reefs there are distinct differences due in part to recruitment of the larval forms

There are many physical and chemical attributes that can be measured, and to which fishes apparently respond: they include light, soluble gases, temperature, salinity and turbidity. Photoperiodic-induced shifts for example have been observed in the behaviour of fishes, although how light influences these changes is not known for most species (MacFarland 1986). Generally classifications such as diurnal or nocturnal are usually defined by the time of active feeding, although this may extend to patterns of light activity during the twighlight hours. The reasons for this may have to do with predator avoidance, prey movements or changes in retinal adaptation, but solid answers can only be arrived at through experimentation and observation at cellular, organismal and environmental levels. Similarly we see in the literature references to salinity (Loneragan et al. 1987), physical morphology and other aspects of the physical environment lifted out for examination as causal influences of spatial patterns. But of course these are only one type of spatial problem.

If we examine those aspects of social space, we find in particular work relating to interspecific competition, foraging theory and predator-prey dynamics. Some ecologists (e.g. Drury and Nisbit 1973) maintain that competition never occurs. Others argue that it is difficult to demonstrate. Similar arguments exist for foraging theory and predator-prey dynamics. And yet if we look carefully at experimental studies such as those by Norman and Jones (1984) for territorial fishes we find that intra-specific interactions constrain terrritory size well below optimum in terms of abundance of "preferred" food algal species. Further examples of this can be seen in the size-structuring that often occurs within fish species that are notable cannibals, with respect to their territories. This concept of social space is extended further into a motivational system (Toates 1986) and cognitive space in the work of Bence (1986) and Marcotte and Browman (1986) who suggest from experimental work on *Gambusia affinis* and *Salmo salar* that cognition and learning with respect to predation and thus extended to other forms of "adaptation" are commonplace in determining feeding behaviours and spatial dynamics. Often prey items are selected because an individual has previously been successful in capturing one of a similar type. The optimal diet imperative is thus open to serious challenge, because learning results in a form of manifested ignorance of the third-party theory. Other results of work by Ebeling and Laur (1986) and Mittelbach (1986) show that the consequences of foraging may indeed be the result of co-evolution of mutually interested species or that fishes may even build up perceptions of predator-mediated habitat use.

The non-lethal effects of these types of spatial processes can be as important in pattern formation as the death of prey items themselves.

CONCLUSIONS

In addressing the various levels and processes involved in the spatial dynamics of fishes, we see that there are scaling phenomena that should be taken account of; for example biogeographic domains are often of the order of 1000 - 100km; water masses of 100 - 10km; fixed territorial communities 100 - <1km and individual fixed territories often less than 10m. Migration patterns cut across these, but it is important to link these types of biological scales with those spatial dimensions discussed above.

Finally, fishes have a life-history that must be completed successfully both spatially and temporally. The scale of processes that bring an organism back to its point of birth for spawning is one of the mysteries of life, but the recogintion that even lower vertebrates at all stages of their life-cycle can learn, be aware of and control to some extent their surroundings has been put forward. Whether it is through the interception of biological molecules or a recognition of the presence of spatial scale in a group of similar organisms is not clear; but what is needed is a method of introspection such that any individual can unambiguously recognise members of its own group. This single factor becomes the wild card in our understanding of the evolution of spatial inhomogeneties, and yet without it our ability to proceed will remain severely curtailed.

REFERENCES

Bence, J.R. 1986. Feeding rate and attack specialization: the roles of predator experience and energetic tradeoffs. Env. Biol. Fishes 16: 113–121.

Brain, C.K. 1985. Temperature – induced environmental changes in Africa as evolutionary stimuli. pp 45–52. In E.S. Vrba [ed.] Species and Speciation. Transvaal Museum Monograph 4.

Darwin, C. 1837. Notebook B. pp. 139. University Library, Cambridge University.

Donk van J. 1976. O^{18} record of the Atlantic Ocean for the entire Pleistocene Epoch. Geol. Soc. Am. Memoirs 145: 147–163.

Drury, W.H. and C.T. Nisbit. 1973. Succession. J. Arnold Arboretum 54: 331–368

Ebeling, A.W. and D.R. Laur. 1986. Foraging in surfperches: resource partitioning or individualistic responses ? Env. Biol. Fishes 16: 123–133.

Eschmeyer, W.N. 1969. A systematic review of the scorpionfishes of the Atlantic Ocean (Pisces: Scorpaenidae). Occ. Papers Calif. Acad. Sci. 79: 1–130.

Fowler, M. and J.M. McGlade. 1988. Bathymetrc and geographic distribution of demersal fishes in the continental slopes of Nova Scotia and Newfoundland. (MS Submitted Deep Sea Res.).

Gardner, M.R. amd W.R. Ashby. 1970. Connectance of large dynamical (cybernetic) systems: critical values for stability. Nature 228: 784.

Gilroy, D. 1975. The determination of the rate constants of island colonization. Ecology 56: 915–923.

Haedrich, R. and D.C. Judkins. 1979. Macrozooplankton and its environment. pp. 4–28. In S. van der Spoel and A.C. Pierrot–Bults [eds.] Zoogeography and diversity of plankton. Bunge Sci. Pub., Utrecht.

Hogg, T., B.A. Huberman, and J.M. McGlade. 1988. The stability of ecosystems. (MS Submitted Nature).

Iles, T.D. and M. Sinclair. 1982. Atlantic herring: stock discreteness and abundance. Science 215: 627–633.

Lister, A.M., J.M. McGlade and A.J. Stuart. 1988. The early Pleistocene vertebrate fauna of Little Oakley, Essex. (MS Submitted Trans. Roy. Phil. Soc.).

Loneragan, N.R., I.C. Potter, R.C.J. Lenanton and N. Caputi. 1987. Influence of environmental variables on the fish fauna of the deeper waters of a large australian estuary. Mar. Biol. (Berl.) 94: 631–642.

MacArthur, R.H. and E.O. Wilson. 1967. The theory of island biogeography. Princeton Univ. Press, New Jersey.

MacFarland, W.N. 1986. Light in the sea – correlations with behaviours of fishes and invertebrates. Amer. Zool. 26: 389–401.

Marcotte, B.M. and H.I. Browman. 1986. Foraging behaviour in fishes: perspectives on variance. Env. Biol. Fishes 16: 25–33.

May, R.M. 1974. Ecosystem patterns in randomly fluctuating environments. Progress in Theor. Biol. 3: 1–50.

McKelvie, D.S. 1985. Discreteness of pelagic faunal regions. Mar. Biol. 88: 125–133.

Mercer, J.H. 1983. Cenozoic glaciation in the southern Hemisphere. Ann Rev. Earth Planetary Sci. 11: 99–132.

Mittlebach, G. 1986. Predator–mediated habitat use: some consequences for species interactions. Env. Biol. Fishes 16: 1–3.

Norman, M.D. and G.P. Jones. 1984. Determinants of territory size in the pomacentrid reef fish, *Parma victoriae*. Oecologia 61: 60–69.

O'Neill, R.V., D.L. DeAngelis, J.B. Waide and T.F.H. Allen. 1986. A hierarchical concept of ecosystems. Princeton Univ. Press, New Jersey.

Pimm, S.L. and J.H. Lawton. 1977. Number of trophic levels in ecological communities. Nature 268: 329–331.

Schackleton, N.J. 1975. The stratigraphic record of deep–sea cores and its implications for the assessment of glacial, interglacials, stadials and interstadials in the Mid–Pleistocene. pp. 1–24. In K.W. Butzer and G.L. Isaac [eds.] After the Australopithecines. Mouton, The Hague.

Schlosser, I.J. 1987. The role of predation in age and size–related habitat use by stream fishes. Ecology 68: 651–659.

Soja, E.W. 1985. The spatiality of social life: towards a transformative retheorisation. pp. 90–127. In D. Gregory and J. Urry [eds.] Social Relations and Spatial Structures. MacMillan Education Ltd., London.

Toates, F. 1986. Motivational Systems. Cambridge Univ. Press., Cambridge.

Wittgenstein, L. 1980. Remarks on the Philosophy of Psychology. Vols. I and II. Basil Blackwell, Oxford.

Intraspecific Competition and Evolution

F. Bugge Christiansen

Department of Ecology and Genetics, University of Aarhus,
Ny Munkegade, DK-8000 Aarhus C, Denmark

1. INTRODUCTION

Adaptation through natural selection is one of the key elements of the Darwinian theory of evolution. This process has been the focus of much attention, partly because of its importance in the understanding of biological evolution, and partly because it appears as a nice directed deterministic process. The annoying aspect of Darwin's theory of natural selection is, however, that the raw material of the process, the phenotypic variation, originates as random errors in the hereditary transmission.

The process of natural selection is deterministic, in the sense that hereditary variation among individuals in their adaptation to the environment leads to natural selection and subsequently to evolutionary change. The process, however, has significant stochastic elements. The dynamics of the genetic variation has random elements and the environment varies in an unpredictable fashion that makes natural selection stochastic. Even the property of natural selection that it depends upon individual survival and reproduction makes it stochastic. These erratic properties of the process should be accounted for in theories of natural selection. However, natural selection is still a deterministic force given the variation in the population. The dynamics of the variation is, on the other hand, very unpredictable and governed by inherently stochastic processes.

Variation originates from mutations which are essentially random errors in the transmission of genetic information. On the phenotypic scale mutations may usually be considered either as noise around a norm or as clearly deleterious aberrations. For evolutionary arguments we need only consider the variations on the existing phenotypes. In theoretical discussions this may lead to considerations of the process of perfection of a given phenotype, but this adaptionist attitude neglects a fundamental property of biological evolution. The process of evolution leads generally to change if more fit variants exist in the population and it leads, by and large, to constancy if variants generally are less fit than than the normal individuals in the population. The genetic details may complicate this simple picture, but nothing in the process points to a strive towards perfection. The evolutionary process is opportunistic, much in the same way as the amateur carpenter's principle of the available nails. A closer inspection of the product in both cases hardly allude to notions of perfection, but it works.

With respect to the process of anagenetic evolution ecology contributes to the understanding of many aspects of natural selection. It is therefore natural that evolutionary ecology usually discusses phenomena of adaptation given a collection of species and given some hereditary phenotypic variation in each species. The process of cladogenetic evolution is considerably more complicated and less well understood. The process of speciation is not primarily a process influenced by ecological factors, rather ecological processes are determinants in the persistence and spread of a new species. On the other hand, species extinction is, almost by definition, an ecological process. Evolutionary ecology is dominated, however, by the most deterministic aspects of the evolutionary process, and the thorough understanding of the deterministic atoms of the process contribute to the comprehension of the larger evolutionary scenario.

2. MODELS IN EVOLUTIONARY ECOLOGY

Theoretical considerations in evolutionary ecology may at times be faced with problems of interpretation of results, problems which have caused discussions and controversy. The difficulty is that ecological models are made to describe the dynamics of populations, whereas evolutionary models need specifications of individual fitness. This conflict is not trivial, and it may be illustrated by a discussion of the theoretical result of evolutionary ecology that is most well known: the r- and K-selection dichotomy.

The classical result referred to in this distinction is Fisher's fundamental theorem of natural selection from 1930: "The rate of increase in fitness of any organism at any time is equal to its genetic variance in fitness at that time" [1]. This theorem is based on a simple description of natural selection in terms of the demographic model of unlimited population growth, that is, the individual fitness is given by a Malthusian parameter, r. Further, the theorem explores simplifying genetic assumptions [2], in that genes can be considered as independent with small effects. The interpretation of this result with respect to a particular character has been that evolution of the character will proceed in a direction that increases fitness and that given enough heritable variation the character will evolve towards the value that maximizes fitness, if such a value exists. Evolution in Fisher's model occurs in the direction of steepest increase in fitness [3].

MacArthur [4] considered a similar model with density dependent selection based on the logistic model of population growth:

$$\frac{dx}{dt} = rx \frac{K - x}{K}, \tag{1}$$

where x is the population size, r is the Malthusian parameter describing the unlimited growth of the population at low densities, and K is the carrying capacity which is the population size where growth ceases. Here evolution will proceed so K increases, with r playing a role only in the initial development of the population. Thus density dependent selection is qualitatively different from density independent selection. MacArthur and Wilson [5] hypothesized that in an unpredictable environment natural selection will depend more on r, because the population often finds itself in a phase of exponential increase, so r is maximized according to Fisher's theorem. The two extremes were named r- and K-selection.

MacArthur's calculations were preceeded in the 1930'ies by Kostitzin [6], who extended Fisher's model to the study of ecological phenomena, density dependent selection in particular. However, Kostitzin formulated the logistic model in the form used by Pearl, Lotka and Volterra. The birth and death rates of the Malthusian model are assumed to be linear functions of the population density:

$$\frac{dx}{dt} = x[b - (d+cx)], \tag{2}$$

where b and d are the per capita birth and death rates, and where c is a coefficient describing the effect of intraspecific competition (included here as a density dependent death rate coefficient). Kostitzin concluded that evolution proceeds in much the same way as in the density independent case, only modified by differences among genotypes in their susceptibility to density effects. Thus, the Malthusian parameter r = b-d is still a major determinant of fitness in the density dependent case.

The mathematical models (1) and (2) are exactly the same if we identify the Malthusian parameters, r = b-d, and if the equilibrium population size of Kostitzin's model is identified with the carrying capacity parameter of MacArthur's model, K = (b-d)/c. The difference between the models is in the biological interpretation of the results. In both models evolution maximizes the equilibrium population size, K, but in (2) K is proprotional to r. Kostitzin's model, like the Malthusian model, is formulated in terms of birth and death rates of the individuals. MacArthurs model, on the other hand, uses Gause's [7] version of the logistic model made for describing population data.

The r- and K-selection concepts are in this sense founded on a particular model. The dichotomy is a product of the structural instability of model (1), an instability which produces the discrepancy between a constant and a varying environment. The concepts, however, has led to the interpretation of life-history data in terms of r- and K-stragegies. For instance, the production of many and small offspring has been viewed as an r adaptation, and the production of few large offspring has been viewed as a K adaptation.

2.1. Reproduction in Marine Invertebrates

This dichotomy in reproductive characteristics is especially pronounced in marine invertebrates. Within many taxonomic groups two modes of reproduction exists: planktotrophic development, where the offspring are small and feed as planktonic larvae for some time ending with a metamorphosis to the adult form, and direct (or lecithotrophic) development, where the offspring are produced in a size close to the metamorphosis size of closely related species (Figs. 1 and 2). The evolutionary cause of this variation, however, may be understood from simple properties of the process of reproduction [8,9], and as we shall see in the next section these results are valid without reference to whether selection is density dependent or density independent.

From considerations of the energy budget of the organism we expect a rough inverse proportionality between offspring size, s, and number, B, in that E_g = Bs is a measure of the total biomass devoted to reproduc-

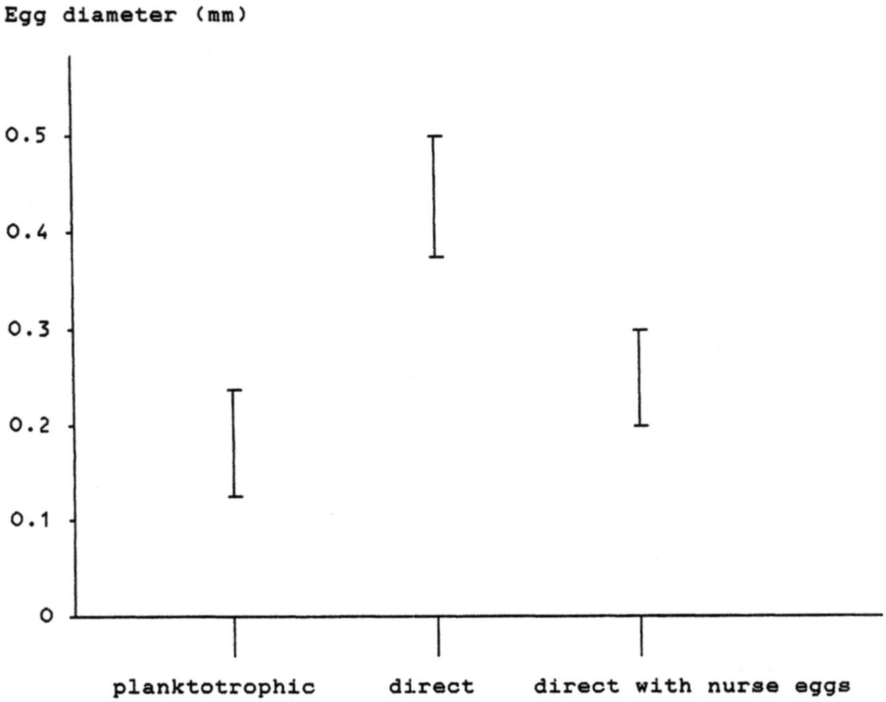

Figure 1. Egg diameters in marine snails of the genus <u>Crepidula</u> for three different modes of reproduction. The figure is based on data from 15 American species [10,11]: eight with planktotrophic development, five with direct development and two with direct development, where food for the developing larvae are supplied by supernumerous nurse eggs. The figure show the range of variation in egg sizes within the three categories of development. The egg plus nurse eggs that go into the production of one larvae in the third category adds up to a biomass which is within the range of biomasses of the eggs with direct development. Simplified after Christiansen and Fenchel [8].

tion. Further, the robustness of the offspring is expected to increase as a function of their size, in that the death rate is a decreasing function of s. Thus, higher fecundity means lower offsping survival. This relationship between a component of selection working on the parents and a component working on the offspring has the consequence that evolution will either be towards a very small or a large offspring size [8]. In marine invertebrates the small offspring size correspond to the planktotrophic development, the large to direct development. The either-or nature of the process means that closely related species living in similar environments may evolve widely different offspring sizes.

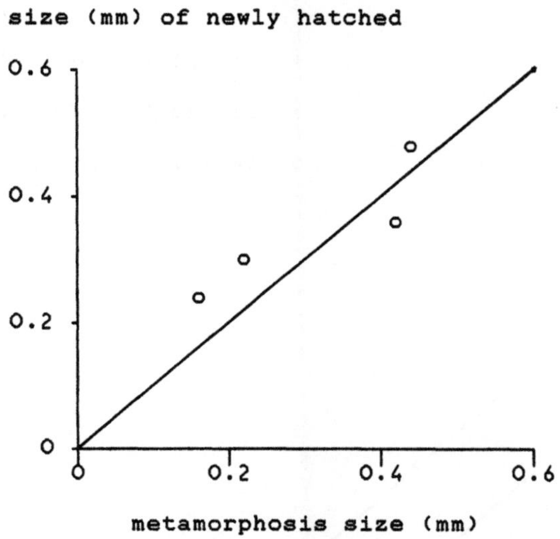

Figure 2. The size relationship between metamorphosis size of planktotrophic larvae (abscissa) and the hatching size of forms with direct development in four genera of marine snails where both modes of development are found. After Christiansen and Fenchel [8].

The impact of reproduction on the parent is not in terms of the fecundity, B, but in terms of the total energy expenditure, the reproductive effort, E_g. Again from considerations of the energy budget of the parent we expect a high reproductive effort to result in a low survival of the individual prior to reproduction. This relationship between two components of selection working on the same individuals means that reproductive effort does not exhibit the evolutionary instability shown by the offspring size [12], and closely related species living in similar environments are expected to show similar reproductive efforts (Fig. 3).

3. ENVIRONMENTAL VARIATION

The dichotomy between constant and variable environments in evolutionary ecology is in contrast to the very relaxed attitude to environmental variation in population genetics. Haldane and Jayakar showed that for evolutionary arguments a variable environment, by and large, is represented by a constant environment where the individual fitnesses are

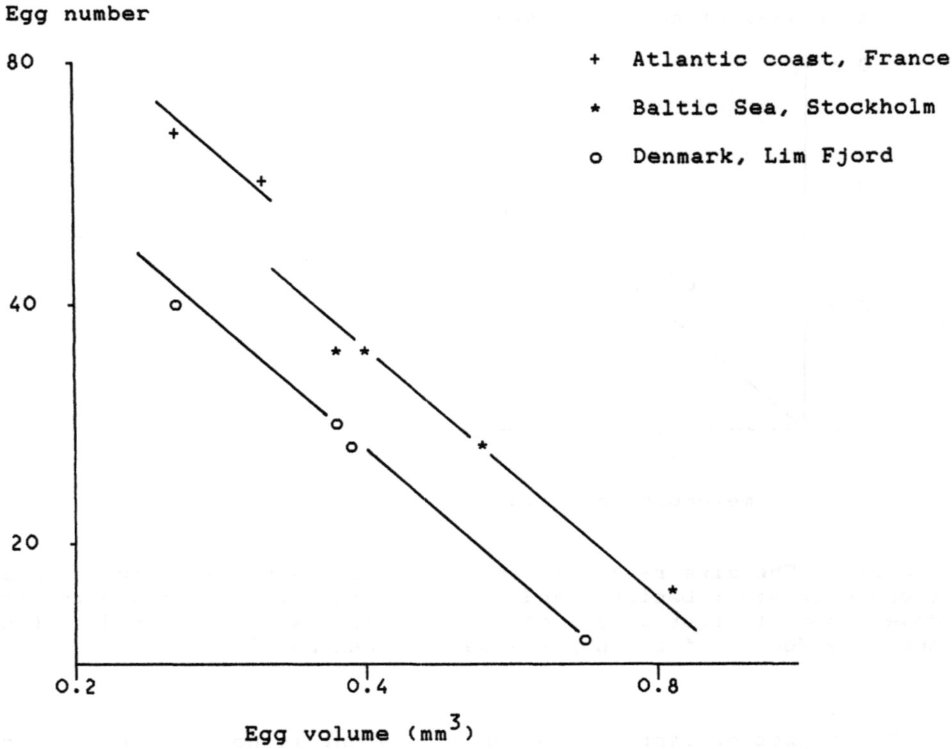

Figure 3. The variation in the egg volume (abscissa) and the egg number per clutch (ordinate) in four amphipod species of the genus Gammarus in three geographical areas. The reproductive effort is measured as the product of egg volume and number. The lines of constant product (with slope -1) connect the points corresponding to species in the same area (only two species were investigated in France). Simplified after Fenchel and Kolding [13].

the geometric average fitnesses through time [14,15]. The same holds if (2) instead of (1) is used as a basis for the definition of fitness.

The discrete non-overlapping generation version of Kostitzin's model due to Poulsen [16] is more convenient for general population genetic arguments. Let x be the population size among adults in a given generation, and assume that each adult produce B offspring. In total $u(0) = Bx$ offspring are produced. The number of surviving offspring of age t is designated $u(t)$, and offspring survival is described by

$$\frac{du}{dt} = -(d+cu)u. \tag{3}$$

The offspring matures at age T, so the number of adults is the surviving offspring at age T, $x' = u(T)$. The dynamics can be summarized in the recurrence equation

$$x' = \frac{DBx}{1 + (c/d)(1-D)Bx}, \tag{4}$$

where $D = e^{-dT}$ is the density independent survival probability.

At low densities the population size multiplies by the factor DB every generation, so the population grows at low densities when $BD > 1$. If so it settles eventually on an equilibrium population size given by $x = (BD-1)/[(c/d)(1-D)B]$.

Natural selection is studied in this model by allowing variation in the fecundity B and in the death rate parameters d and c. We will consider a simple genetic situation: an autosomal locus with two alleles A and a in a diploid organism reproducing by random mating. Let x_1, x_2 and x_3 be the numbers of adults of genotype AA, Aa and aa, respectively ($x = x_1+x_2+x_3$). The frequency of allele A among the genes in the population is $p = (2x_1+x_2)/(2x)$ and the frequency of allele a is $q = 1-p = (2x_3+x_2)/(2x)$. With equal fecundity the genotypic frequencies among offspring are in the Hardy-Weinberg proportions, i.e., the number of offspring are $u_1(0) = p^2Bx$, $u_2(0) = 2pqBx$ and $u_3(0) = q^2Bx$. We will also consider genetic variation in fecundity, and this produce genotypic frequencies which are a bit more complicated [17]. However, for the simple genetic arguments that we produce here, we need not go into these complications. We will assume, however, that the fecundity only depends on the female genotype, i.e., B_1, B_2 and B_3 are the fecundities of females of genotype AA, Aa and aa, respectively. Genetic variation in survival during development is described by the process (3), so equation (3) is replaced by

$$\frac{du_i}{dt} = -(d_i + c\sum_{j=1}^{3} \alpha_{ij}u_j)u_i, \tag{5}$$

where α_{ij} is the intraspecific competition coefficient, that is, $c\alpha_{ij}$ is the increased death rate of genotype i per individual of genotype j.

Consider a population monomorphic aa ($p = 0$) and introduce allele A in low frequency, then the new allele will increase in frequency when

$$\frac{\log(D_2B_2)}{\alpha_{23}} > \frac{\log(D_3B_3)}{\alpha_{33}} \tag{6}$$

and A will vanish from the population if the opposite is true [16]. The numerators, log(DB), are closely related to the Malthusian parameter, $r = b-d$, so condition (6) is closely related to the condition for maximization of $(b-d)/c$ of the continuous time model.

With stochastic variation (of a reasonably stationary nature) in B_i, D_i ($i = 1,2,3$) and c the condition for initial increase of allele A is

$$\frac{E[\log(D_2B_2)]}{\alpha_{23}} > \frac{E[\log(D_3B_3)]}{\alpha_{33}}, \tag{7}$$

where E designate expected value [18]. The numerators are the geometric average density independent fitnesses DB, so this result is closely related to the principle of Haldane and Jayakar [14,15]. The modification due to intraspecific competition is in a varying environment exactly the same as in a constant environment (6).

These results may be extended to allow for age dependent death rates and more complicated density dependence including time-lag in the density responce [18]. With complete lag in the density responce we get

$$\frac{du}{dt} = -[d+cu(0)]u(t) \tag{8}$$

which provide the widely used exponential model of discrete generation population growth with the recurrence equation: $y' = yB\exp[-(d+cy)T]$, where $y = u(0)$. The important assumption employed to reach (7) is that the intraspecific competition coefficients are constant with only the density dependent death rate coefficient, c, varying through time.

Therefore, in a varying environment density dependent selection is not qualitatively different from density independent selection. The study of evolution in a varying environment does produce insight into new phenomena, but the results from models of density independent selection carry to models of density dependent selection. These simple results use a description of individual fitness in terms death rates and fecundity. The results obtained using the population descriptive parametrization of Gause and MacArthur uses fitness defined in terms of the parameter K which is really a compound parameter describing population and not individual behaviour [5]. The focus on K has also been viewed as an assumption of proportionality of r and c, the susceptibility to population density. Indeed, the various aspects of fitness cannot be viewed as independent, but correlations among fitness parameters should be specifically modeled as dynamical pleiotropy [12].

4. EXPLOITATIVE COMPETITION

The descriptive approach of MacArthur is usefull for the study of specific ecological phenomena and the related characters. The ecological description of the relation between an individual and its environment may be used as a basis for the specification of natural selection on phenotypic variation of relevance to the individual adaptation.

The above considered discrete model is realistic, but rather complicated in its description of natural selection as a process occurring throughout the life of an organism [2]. A simpler approach was devised by Anderson [19] and Roughgarden [20] based on a discrete difference equation analog of the model (1). We use the Anderson-Roughgarden model in the form $\Delta x = x'-x = V(K-x)x$ (the constant V is assumed small). The recurrence equation may be written as $x' = wx$, where $w = 1 + V(K-x)$ may serve as a basis for the definition of fitnesses corresponding to the fitnesses of classical discrete generation population genetic models. The linear form of w allude to model (1), and equation (4) may as well serve as the basis for the fitness definition [21], but this procedure leads to a fitness description which is fundamentally different from the one used in the previous section.

A model for the study of intraspecific competition can be made by specifying the genotypic fitnesses at a locus with alleles A and a in an organism reproducing by random mating as $w_i = 1 + V(K_i - x\gamma_{io})$, $i = 1,2,3$, where the index o signifies average: $\gamma_{io} = p^2\gamma_{i1} + 2pq\gamma_{i2} + q^2\gamma_{i3}$, and γ_{ij} is the intraspecific competition coefficient between genotypes i and j. We will be concerned with the model of symmetric competition, where $\gamma_{ij} = \gamma_{ji}$, $i,j = 1,2,3$, which covers most models of exploitative competition. MacArthur's maximization result [4] was generalized to this model by Matessi and Jayakar [22] as the maximization during evolution of the quantity $\psi = K_o^2/\gamma_{oo}$, where $K_o = p_2K_1 + 2pqK_2 + q^2K_3$ and $\gamma_{oo} = p^2\gamma_{o1} + 2pq\gamma_{o2} + q^2\gamma_{o3}$. Thus, evolution tends to maximize the average carrying capacity, K_o, and minimize the average intraspecific competition, γ_{oo}, with the "trade-off" between the two aspects given by the ratio ψ.

4.1. Competition for Resources

One particularly interesting application of this formulation is the study of exploitative competition using the resource competition model of MacArthur and Levins [23]. The resource renewal and consumption in this model are based on a Lotka-Volterra predator-prey model with a very high prey turnover, so the prey abundance at any time is directly related to the utilization of the consumer species of interest, the predator [24]. A well studied example where this model applies is the grasing of mudsnails, Hydrobia, on the microflora on marine sediments [25]. The snail utilizes food particles with a characteristic size distribution with a mean that depends on the size of the snail (Fig. 4).

In the model the abundance of food particles is described by their size distribution, the resource spectrum S, in a situation where no exploitation by the snails occur. This distribution will be assumed normal with mean zero (an arbitrary normalization) and variance σ^2. The exploitation of the resources by the snails modify the size distribution of the food particles. Let R denote this modified resource distribution, i.e., $R(\rho)$ is the actual available ammount of resources of quality ρ. The consumption of resources by the snails is described by the utilization function, U, which is the size distribution of ingested particles with unlimited resources. The ammount of food that a snail gets is proportional to $\int R(\rho)U(\rho)\,d\rho$, so we assume that the choice of the snail is independent of the distribution of available resources. This is a reasonable assumption for the mudsnails, because they seems to ingest sediment particles rather indiscriminantly. The utilization function also is assumed to be Gaussian.

The ammount of food that a snail of a given size gets is specified as a function of the environment in terms of the potential availability of resource and in terms of the exploitation by other snails. The model therefore specifies the fitness variation related to the size variation among snails. An evolutionary model is now built from a specification of the hereditary basis of the variation in this character. A reasonable assumption is that the character shows polygenic inheritance, so using the standard Fisherian model we may consider a locus with two alleles,

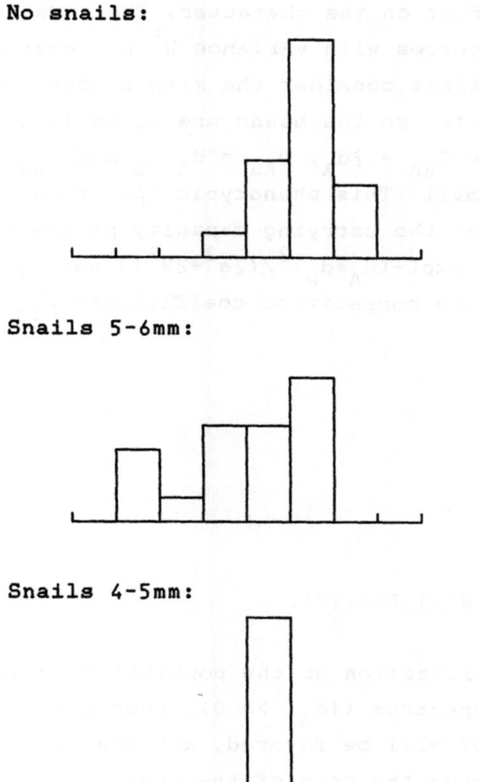

Figure 4. The size distribution of diatoms in experiments, where different size groups of Hydrobia ulvae have grased sediments which undisturbed show a diatom distribution shown in the top figure. After Fenchel and Kofoed [26]

and a, that has a small effect on the character. The individuals of a given genotype utilize resources with variance W^2 and means D_{AA}, D_{Aa} and D_{aa}, respectively. Let us first consider the simple model that the alleles have additive effects, so the means are given in terms of the allele effects d_A and d_a as $D_{AA} = 2d_A$, $D_{Aa} = d_A+d_a$ and $D_{aa} = 2d_a$, where $|d_A-d_a|$ is assumed to be small. This phenotypic specification provides the fitnesses w_i in terms of the carrying capacity parameters, $K_1 = \exp[-4d_A^2/(2\sigma^2+2W^2)]$, $K_2 = \exp[-(d_A+d_a)^2/(2\sigma^2+2W^2)]$ and $K_3 = \exp[-4d_a^2/(2\sigma^2+2W^2)]$, and the competition coefficients γ_{ij} as the symmetric matrix

$$\begin{bmatrix} 1 & \gamma & \gamma^4 \\ \gamma & 1 & \gamma \\ \gamma^4 & \gamma & 1 \end{bmatrix}, \tag{9}$$

where $\gamma = \exp[-(d_A-d_a)^2/(4W^2)]$ [24,27].

If the mean resource utilization of the population is displaced from the mean of the resource spectrum ($|d_A| \gg 0$), then the allele with the numerically smallest effect will be favored, and the population mean utilization will move towards the mean of the resource spectrum (zero). As the population mean approaches zero a stable polymorphism is maintained provided

$$\begin{aligned} &(d_A-d_a)[(\kappa^2-1)d_A - (\kappa^2+7)d_a] > 0 \quad \text{and} \\ &(d_A-d_a)[(\kappa^2+7)d_A - (\kappa^2-1)d_a] > 0, \end{aligned} \tag{10}$$

where $\kappa^2 = \sigma^2/W^2$, and otherwize the allele with the smallest effect is favored [27]. When condition (10) is fulfilled, then the population converges to a unique and globally stable polymorphic equilibrium. Further, a maximum of two allele may be maintained in the population.

The properties of condition (10) depends on whether the resource spectrum is wide ($\sigma^2 > W^2$ and therefore $\kappa^2 > 1$) or narrow ($\sigma^2 < W^2$ and $\kappa^2 > 1$). Polymorphism is likely for a wide resource spectrum, and at a polymorphic equilibrium a third allele can increase in the population if its effect lies outside the interval formed by d_A and d_a [27]. The successful invasion of a new allele causes one of the old alleles to disappear and a new two allele polymorphism is established. Therefore, in a wide resource the exploitation of the resources by the species is expected to broaden due to the buildup of genetic variation. In a narrow resource spectrum ($\sigma^2 < W^2$), on the other hand, the population is likely to become monomorphic. Polymorphism is only possible when d_A and d_a are of opposite sign. If the population is initially polymorphic a third allele can increase if its effect lies inside the interval formed by d_A and d_a, and one or both of the old alleles are lost from the population.

4.2. Multi-Dimensional Resources

These results generalize to resources described by m qualities [28]. The resource spectrum is again assumed Gaussian and described by the variance-covariance matrix Ω. The utilization functions are Gaussian with mean D and variance-covariance matrix Γ. We consider the evolution at a multi-allelic locus with additive contributions d_k of allele A_k, $k = 1,2,\ldots,n$, so the mean utilization of genotype A_kA_l is $D_{kl} = d_k + d_l$. Again we assume a small effect of the locus, i.e., $|d_k + d_l|$ is small for all $k,l = 1,2,\ldots,n$. An overview of the process may be obtained from the quadratic forms

$$M_k = (d_k-Cd_o)^T\Gamma^{-1}(\Omega-\Gamma)(\Omega+\Gamma)^{-1}(d_k-Cd_o), \tag{11}$$

where $C = (\Omega+3\Gamma)(\Omega-\Gamma)^{-1}$ and $d_o = p_1d_1 + p_2d_2 +\ldots + p_nd_n$ with p_k being the gene frequency of allele A_k, $k = 1,2,\ldots,n$. Allele A_k will increase in frequency if and only if $M_k > M_o$, where $M_o = p_1M_1 + p_2M_2 +\ldots + p_nM_n$. In particular, the population is at a genetic equilibrium when $M_k = M_o$ for all k where $p_k \neq 0$. Thus, a genetic equilibrium is characterized by the allelic contributions positioned on an (m-1)-dimensional manifold, where each two-dimensional slice is a cone section curve, and this almost certainly restricts the number of alleles to m+1 in a polymorphic population.

An everywhere wide resource is characterized by $\Omega-\Gamma$ being positive definite, and then the condition for the initial increase of a rare allele is that its contribution is outside the elipsoidal equilibrium manifold. When $\Omega-\Gamma$ is negative definite then the resource is everywhere narrow and a new allele increases if its contribution is inside the elipsoidal equilibrium manifold. In general, the equilibrium manifold is not ellipsoidal, but natural selection has a diversifying tendency in directions where the resource spectrum is wider than the utilization function and a normalizing tendency where the resource is narrow [28].

4.3. Deviations from additivity

The results so far have been obtained with the assumption of additive effects of the alleles on the genotypic mean value. Deviations from additivity, referred to as dominance, can be modeled by specifying the genotypic mean values as $D_{AA} = 2d_A$, $D_{Aa} = d_A+d_a - \alpha(d_A-d_a)$ and $D_{aa} = 2d_a$, where the dominance parameter α is usually considered to be in the interval [-1,1]. We will still assume a small contribution of the locus to the character, i.e., $|d_A-d_a|$ is small, and we may for simplicity in the arguments assume that $d_A > |d_a|$, i.e., the genotype aa is always closest to the resource optimum. The additive model correspond to $\alpha = 0$, and complete dominance corresponds to $|\alpha| = 1$; $\alpha = -1$ means that $D_{Aa} = D_{aa}$, so the heterozygote is equal to the homozygote closest to the resource optimum, and $\alpha = 1$ means that $D_{Aa} = D_{AA}$, so the allele with the largest effect is dominant and allele a is recessive.

For intermediate dominance, $-1 < \alpha < 1$, a single globally stable polymorphism exists when

$$\frac{d_a}{d_A} < \frac{(1-\alpha)(\kappa^2-1)}{(\kappa^2+7) - \alpha(\kappa^2-1)} \tag{12}$$

which becomes condition (10) for $\alpha = 0$ [29]. For complete dominance, $|\alpha| = 1$, the corresponding condition is

$$\frac{d_a}{d_A} < \frac{(\kappa^2-1)}{(\kappa^2+3)} \; . \tag{13}$$

This immidiately points to a discontinuity when $\alpha = 1$, and indeed for $(\kappa^2+3)/(3\kappa^2+1) < \alpha < 1$ there is an interval of positive d_a/d_A values where a stable polymorphic equilibrium exist which is not globally stable: the allele a outcompetes allele A if its initial frequency is sufficiently high, and otherwize a stable polymorphism results. In any other case, where condition (12) or (13) is not fulfilled, the population will eventually become monomorphic aa [29].

The indeterminate behaviour of the model occurs when all genotypes utilize resources with a mean to the same side of the resource optimum and when, in addition, the mean of the heterozygote is rather far from the optimum. In this case the variation at the locus in question is kep in a polymorphic state. In a population in an environment where the resource optimum is initially to the other side of the population mean, the heterozygote is rather close to the optimum, and in a similar situation the variation at the locus may be lost.

4.4 Interspecific competition

The evolution of interspecific exploitative competition may be studied by similar models [22,24,30,31]. The results are rather parallel in that each species evolves towards an optimum location determined by the resource spectrum and the position of the other species. With two species this process may lead to character displacement as observed in coexisting populations of Hydrobia ulvae and H. ventrosa [25]. The process occurs within each species and leads eventually to a coevolved stable state where each species is at an optimum location. At this state the guild of competitors is well protected from invasion by other species [32]. With a given collection of potential species the composition of a natural community therefore will depend heavily on the order in which the species are introduced.

In a coevolved community the species has a tendency to become polymorphic as a result of selection by intraspecific competition [27,28]. The resource spectrum available to the single species tends to be "wide" whenever the species may coexist. Invasion into this polymorphic guild of species may cause the residents to evolve rather rapidly away from the newcomer under the reasonable assumption that the newcomer exhibit less genetic variance than the residents [24]. This process may eventually cause one or more of the resident species to move to a marginal position with respect to the resource spectrum, so it may go extinct due to low population size or fluctuations in the resource availability.

5. DISCUSSION

Theoretical consideration in evolutionary ecology may suggest ways to understand observed patterns of variation within and among species in nature. The evolutionary explanations of phenomena which usually are considered as parts of the prerequisites of ecological descriptions of natural ecosystems may ease the widespread deterministic tendencies in ecological explanations of the dynamics of ecosystems. Determinism in ecological explanations of natural systems is the assumption that given sufficient knowledge about the environment the dynamics of the biota may in principle be specified. This determinism may be hidden by the consideration of models that allow for stochastic environmental variation, or models that display seemingly indeterministic behavior because of dynamical chaos. The basic indeterminism of any biological system is that it depends critically on the history of the system, both the long term evolutionary history of its biological components and the short term history of, e.g., environmental variations and biological invasions.

The models considered here are extremely simplified in their genetics. The assumptions are basically the simplifying assumptions used by Fisher [1]: one locus with a small effect on a character assuming that such small effects adds up over many loci. The complications of genetic linkage and recombination are ignored, even though it is known that the principle of fitness maximalization breaks down when these effects are incorporated into the models. In addition, the evolution at a gene locus may be highly influenced by the variation a linked loci due to accidental associations of alleles at different loci. These effects are aggravated by the realization that any natural population is finite with random genetic drift disturbing the action of natural selection [33]. The emphasis on randomness and on accidental phenomena in evolution is due to Wright [34].

The other assumption in the simplified Fisherian model is that natural selection may be described by individual fitness values. In Fisher's fundamental theorem fitness is the Malthusian parameter, r, calculated from the individual demographic data. In Section 2 the proof of MacArthur's principle again rest on an individual fitness parameter, viz., $r(x-K)/K$. In Section 3 natural selection is described in terms of the survival and reproduction of the individuals in the population. This reflects the action of selection in nature, and even in the simplest circumstances the maximization principles break down. However, the models based on individual fitness parameters are considerably simpler than the descriptive models of Section 3, and the study of the evolutionary process has advanced by this simplification. The principle of fitness maximalization, in Fisher's, MacArthur's or in Matessi and Jayakar's formulation, should be viewed as a usefull guide in the analysis of simple evolutionary models, and as a fruitfull intuitive concept. The interpretation of these principles, however, should be founded on models describing the connection between individual variation and natural selection.

The modern realization of indeterminism in evolution is probably best described by the summary of Darwin's view as expressed in his autobiography: "There seems to be no more design in the variability of organic beings and in the action of natural selection, than in the course which the wind blows" [35]

Acknowledgement: Comments on a draft manuscript by Dr. Volker Loeschcke were very helpful.

[1] Fisher, R. A.: The Genetical Theory of Natural Selection. Claredon Press, Oxford 1930.

[2] Christiansen, F. B.: The definition and measurement of fitness. In: Evolutionary Ecology. Shorrocks, B., (Ed.). Blackwell Sci. Publ., Oxford 1984, p. 65-71.

[3] Kimura, M.: On the change of population fitness by natural selection. Heredity 12 (1958) 145-167.

[4] MacArthur, R. H.: Some generalized theorems of natural selection. Proc. Natl. Acad. Sci. USA 48 (1962) 1893-1897.

[5] MacArthur, R. H., and Wilson, E. O.: The theory of island biogeography. Princeton Univ. Press, Princeton NJ 1967.

[6] Kostitzin, V. A.: Equation différentielles générales du problème de sélection naturelle. C. R. Acad. Sci. Paris 203 (1936) 156-157.

[7] Gause, G. F.: The Struggle for Existence. Williams and Wilkins, Baltimore 1934.

[8] Christiansen, F. B., and Fenchel, T. M.: Evolution of marine invertebrate reproductive patterns. Theor. Popul. Biol. 16 (1979) 267-282.

[9] Vance, R. R.: On reproductive strategies in marine benthic invertebrates. Amer. Natur. 107 (1973) 339-352.

[10] Coe, W. R.: Divergent methods of development in morphologically similar species of prosobranch gastropods. J. Morphol. 84 (1949) 383-399.

[11] Gallardo, C. S.: Two modes of development in the morphospecies Crepidula dilatata (Gastropoda: Calyptraeidae). Mar. Biol. 39 (1977) 217-224.

[12] Christiansen, F. B.: Pleiotropy in dynamical parameters of models on the evolution of simple phenotypes. In: Genetic Constraints on Adaptive Evolution. Loeschcke, V., (Ed.). Springer-Verlag, Berlin 1987, p. 108-124.

[13] Kolding, S., and Fenchel, T. M.: Patterns of reproduction in different populations of five species of the amphipod genus Gammarus. Oikos 37 (1981) 167-172.

[14] Haldane, J. B. S., and Jayakar, S. D.: Polymorphism due to selection of varying direction. J. Genet. 58 (1963) 237-242.

[15] Gillespie, J.: Polymorphism in random environments. Theor. Popul. Biol. 4 (1973) 193-195.

[16] Poulsen, E. T.: A model for population regulation with density- and frequency-dependent selection. J. Math. Biol. 8 (1979) 325-343.

[17] Feldman, M. W., Christiansen, F. B., and Liberman, U.: On some models of fertility selection, Genetics 105 (1983) 1003-1010.

[18] Christiansen, F. B.: Evolution in a temporally varying environment: Density and composition dependent genotypic fitnesse. In: Population Biology and Evolution. Wöhrmann, K., and Loeschcke, V. (Eds.). Springer-Verlag, Berlin 1984, p. 115-124.

[19] Anderson, W. W.: Genetic equilibrium and population growth under density-regulated selection. Amer. Natur. 105 (1991) 489-498.

[20] Roughgarden, J.: Density-dependent natural selection. Evolution 52 (1971) 453-468.

[21] Clarke, B.: Density-dependent selection. Amer. Natur. 106 (1972) 1-13.

[22] Matessi, C., and Jayakar, S. D.: Coevolution of species in competition: A theoretical study. Proc. Natl. Acad. Sci., USA, 78 (1981) 1081-1084.

[23] Macarthur, R. H., and Levins, R.: The limiting similarity convergence and divergence of coexisting species. Amer. Natur. 101 (1967) 377-385.

[24] Christiansen, F. B., and Fenchel, T. M.: Theories of Populations in Biological Communities. Springer-Verlag, Berlin 1977.

[25] Fenchel, T.: Character displacement and coexistence in mud snails (Hydrobiidae). Oecologia 20 (1975) 19-32.

[26] Fenchel, T., and Kofoed, L. H.: Evidence for exploitative interspecific competition in mud snails (Hydrobiidae). Oikos 27 (1976) 367-376.

[27] Christiansen, F. B., and Loeschcke, V.: Evolution and intraspecific exploitative competition I. One-locus theory for small additive gene effects. Theor. Popul. Biol. 18 (1980) 297-313.

[28] Christiansen, F. B., and Loeschcke, V.: Evolution and intraspecific exploitative competition III. One-locus theory for small additive gene effects and multidimentional resource qualites. Theor. Popul. Biol. 31 (1987) 33-46.

[29] Christiansen, F. B.: Natural selection related to the biotic environment. In: Evolution and the Genetics of Populations. Jayakar, S. D., and Zonta, L., (Eds.). Suppl. Atti Ass. Genet. Ital. XXIX (1984) 85-102.

[30] Roughgarden, J.: Resource partitioning among competing species - A coevolutionary approach. Theor. Popul. Biol. 9 (1976) 388-424.

[31] Fenchel, T. M., and Christiansen, F. B.: Selection and interspecific competition. In: Measuring Selection in Natural Populations. Christiansen, F. B., and Fenchel, T. M., (Eds.). Springer-Verlag, Berlin 1977, p. 477-498.

[32] Loeschcke, V.: Coevolution and invasion in competitive guilds. Amer. Natur. 126 (1985) 505-520.

[33] Kimura, M.: The Neutral Theory of Molecular Evolution. Cambridge Univ. Press, Cambridge 1983.

[34] Wright, S.: The Shifting Balance Theory and Macroevolution. Ann. Rev. Genet. 16 (1982) 1-19.

[35] Darwin, C.: The Autobiography of Charles Darwin. W. W. Norton & Co., London 1876 (republished 1969).

Fundamental Components in Ecology and Evolution: Hierarchy, Concepts and Descriptions

S.H. Cousins

Energy and Environment Research Unit, The Open University, Milton Keynes, MK76AA, UK

In a review /1/ of two recent and seminal books on the theory of evolution /2,3/ the evolutionist Ghiselin points to limitations at the root of ecological science, "Ecologists are most unclear about the nature of their fundamental units, and about what such units do". This is a useful challenge to ecologists and one that can be answered at least in part. It is also interesting from a further aspect as it illustrates the value of interdisciplinary contact which may hold some lessons for the future collaboration of biologists and physicists under the banner of Ecodynamics. Thus in this paper I shall try to address the question of fundamental components in particularly in ecology and where possible make connections to concepts in physics. If this makes for a rather strangely structured paper then at least the reason for it is clear.

You might ask why Ghiselin's criticism of ecology should be found in a review of two books on evolution. It arises because Eldredge and Salthe have attemped to define the fundamental components involved in evolution while expanding the conventional synthesis (see /2/) of evolutionary theory to include ecology more explicitly. The stimulus to re-examine the synthesis came from observations /4/ suggesting that evolution occurs in rapid bursts which follow (punctuate) long periods of stasis or equilibrium. This punctuated equilibrium theory has become a focus of evolutionary debate in the 70's and early 80's. In the search for a mechanism for punctuated equilibrium Eldredge and Salthe /5/ have proposed two interacting hierarchies of fundamental components. These are hierarchies of replication (genes make more genes, species make more species etc) and a hierarchy of energy and matter transfer (enzymes, ecosystems etc). We first consider hierarchy itself.

Hierarchy

The pursuit of fundamental biological components has often been associated with the construction of hierarchies of these components. Rowe /6/ traces the history of the use of hierarchy in ecology. He acknowledges Comte as classifying the sciences in a hierarchy according to levels of inclusiveness and Kant's distinction between logical classification and physical classification. Combining these approaches, Rowe sees a logical classification of biological sciences in a hierarchy: Ecology - morphology - anatomy - physiology, which parallels a physical hierarchy of objects: Ecosystem - organism - organ - cell. However in the search for mechanism we are interested in the physical hierarchy of objects or fundamental components. Although this distinction between the conceptual hierarchy of biological sciences and the hierarchy of objects appears clear in this case (possibly with the exception of ecosystem as an object) many of the important debates concerning evolution are predicated on whether something is a concept or an object; for example, treating species as classes (concepts) or as individuals (objects) /7-9/.

Although I am mainly concerned with Eldredge and Salthe's proposals for a dual hierarchy, another group of workers, O'Neill et al /10/ have adopted a dual hierarchy of ecological phenomena first suggested by MacMahon et al /11/. Each of these views has significant implications for ecology; Salthe's /3/ deduction that species are not entities in ecosystems has far reaching effects on current debates in ecology. Resolving the dual hierarchy of O'Neill et al may require a change in how ecosystem is defined. Rejecting the case for Eldredge and Salthe's dual hierarchy rests on showing how important taxonomic categories (eg phyla) are as classes of functional organisms in food webs /12/. As a result a single hierarchy can be constructed in which evolution may be studied and suggesting new priorities for research.

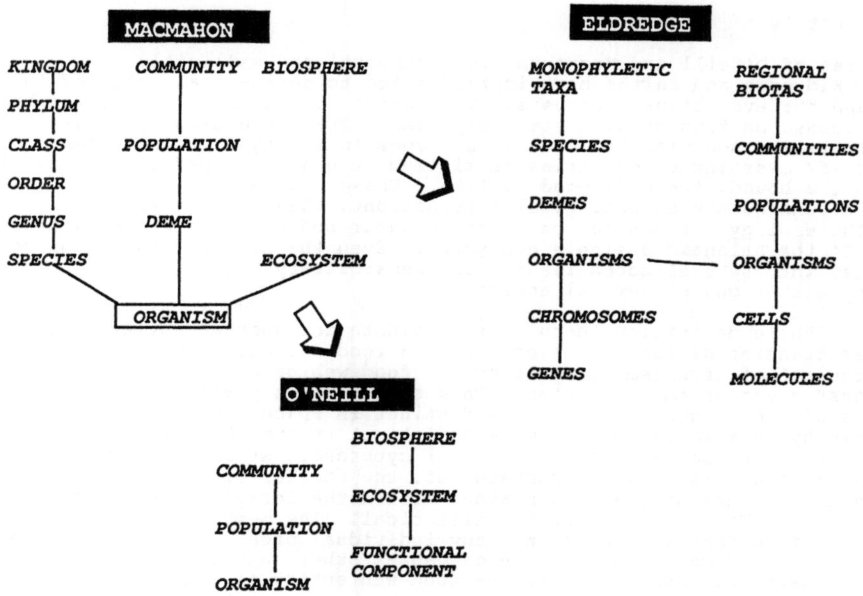

Fig. 1 Recent hierarchical approaches to ecology and evolution.

We will trace the origin of and examine these dual hierarchy proposals to see if a single hierarchy of biological objects is possible. The following terms are used i) Class - a group or set which is not itself an entity or object, ii)Entity - a thing defined in space and time with a cybernetic system of parts.

In recent times the origins of these hierarchical approaches stem from MacMahon et al /11/ figure 1. They proposed four hierarchies involoving the individual organism, with the organism given a special status or importance. One hierarchy is contained in the organism, organ - cell - molecule etc while the others contain the organism. Somewhat ironically in my view, the phylogenic hierarchy is separate from two ecological hierarchies. Ironic because phylogeny can describe classes of biological machine active in the 'ecosystem', the basis for this lies in the ecological importance of body size and with these allometric relationships holding most precisely within taxonomic groups /12/. As MacMahon et al note their greatest departure from tradition comes in removing community and population from the biosphere and ecosystem hierarchy. This is required by their adoption of the definition of ecosystem as 'a set of organisms and inanimate entities connected by exchanges of energy and matter'. Thus since the community - population - deme hierarchy only includes organisms a separate hierarchy is required by those authors.

MacMahon et al acknowledge that the terms community and ecosystem do not represent bounded objects . The boundaries must be supplied, they say, by the observer. O'Neill et al take a similar approach "the ecosystem comes to look more like an abstraction with spatiotemporal properties that can be specified only within the context of an observation set. The ecosystem as an independent discrete entity begins to look less tenable." We are left with ecosystem as a viable concept but not as a fundamental component or entity.

Before proceeding further we need to attend to the problem of entity in the ecological hierarchy. What would such an ecological hierarchy look like? Certainly organisms and the biosphere qualify as entities which also provide the lower and upper boundaries to the hierarchy. However, candidate intermediate levels such as populations, communities, ecosystems, trophic levels, and food webs all fail the test as ecological entities because they are unbounded in space and time.

An ecosystem entity

The hierarchies of O'Neill and MacMahon are conceptual and explicitly subjective. Eldredge and Salthe have instead tried to define real entities in ecological and the evolutionary process. The search for ecological entities has typically focussed on finding discrete ecosystems. These are usually cases where some spatial discontinuity leads to a change in the type of ecosystem found. Rowe /6/ sees these boundaries as minimum interaction surfaces as may characterise the boundaries of a pond or lake. These boundaries are not produced as a consequence of ecological interactions, they are externally imposed on the ecology. Ponds can be discrete water bodies but are they single ecosystems? Is the Atlantic a single ecosystem? Even the pond ecosystem can be differentiated into several zones leading to the arbitrary definition of the pond as being either one or several ecosystems.

By adopting Salthe's definition there is a candidate for such an entity in the energy-matter transfer hierarchy. I call this a food-web entity which is defined as the set of organisms comprising the food web of a single individual of the largest predator in a location. This entity has a given size, the foraging area of the largest predator; is distinct in space from similar foraging areas by behavioural mechanims and distinct in time by birth and death of the predator; the food web has a dynamical cybernetic structure. This definition also meets the test of containment, the smaller organisms and their feeding interactions are spatially contained within the foraging area of the largest creatures. This containment is statistical. The membership of the food-web entity is probabilistic in that any individual organism within the area enclosed by the entity may emigrate, die of causes other than predation, or be predated. The fact that membership of the food-web entity is statistical does not alter the possibility of calling it a real object or entity. Similarly, an organism is made up of protein molecules etc. which have a much shorter membership of the organism's body than the lifetime of the organism.

It remains an important question to ask if there are any other entities between the food-web entity and the biosphere. There are interactions between top predators for example when pack hunting is undertaken or when young are raised co-operatively by parents. Clearly there are also flows of organisms from one food-web entity to another. Do these flows themselves create some larger entity which is not the biosphere? How indeed does the effect of human industrial and social activity interact with the biosphere or the food-web entity?

A clue to the solution of this problem may be provided by how species are to be treated in the proposed hierarchy. Sexually reproducing species have well differentiated parts eg male and female. These parts interact and it is argued /7,8/ that these interactions and the offspring produced mean that species are real entities which can therefore act as causal agents in evolution. However, while the reproductive pair constitutes an entity, the set of pairs which constitute the species can be seen as additive and not constituting an interacting set of parts of an entity. While I have used the word pair I wish this social entity to cover any social grouping such that a species is a set or class which is the sum of social groups (entities) of organisms which share the same specific-mate recognition system /13/.

It is now possible to propose a single hierarchy of evolutionary phenomena. We can define the largest naturally occuring ecological entity as the ecosystem trophic module (ecotrophic module) which is the food web of the social group of largest predators in a location. This entity has a given size, the foraging area of the social group; is distinct in space from other such groups by behavioural means and is distinct in time due to the origination and extinction of the social group; the food-web has a dynamical cybernetic structure. Note that the ecotrophic module is a lineage which is therefore capable of evolution. This definition of the ecotrophic module supercedes the exploratory definition /14/. Figure 2 shows the spatial boundaries of such ETMs, the clans of hyaenas in the Ngorongoro crater, each clan contains between 30 and 80 hyaenas /15/. Although spatially well defined these clans are lineages and have a lifetimes longer than the individual hyaenas which compose them.

The suggestion that the reproductive or social unit is a feature of the ecotrophic module could at least in principle be extended to human society. However in present day ecology observability is to me of vital importance and so the impact of human society on ecotrophic modules is perhaps the best measure we have of the existence of some human ecological object. This impact could be measured by size and numbers of predators active in a human impacted ecosystem

Fig. 2 Ecotrophic modules formed by clans of Hyaenas in the Ngorongoro Crater, Tanzania /15/.

relative to a reference size and density typical of that biome. Human intervention in the ecosystem tends to eliminate top predators and therefore fragments the largest biome ETMs into numerous smaller ETMs based on the predators present in the human impacted ecosystem. A large human population in the Ngorongoro crater would almost certainly radically alter the observed ETM structure.

Other flows between ecotrophic modules such as migrations can be considered to be caused by some climatic effect which is properly considered to be part of the biospheric entity.

Single hierarchy

To further explore the evolutionary process and in the interests of parsimony I propose the following single hierarchy of levels, figure 3.

There are many kinds of entity at each level in this proposed single hierarchy; many kinds of molecule - proteins, DNA and so on; many kinds of organism - different castes, sexes, different species, different taxonomic sets of organisms. It is a condition of hierarchy therory that entities engage in interactions at the same hierarchical level /16/ and in so doing create the (emergent) properties of the entities at the next level and are constrained by the interactions at that level also. Thus molecular interactions create the metabolic patterns of the cell organelles, DNA together with other molecules in the metabolic and transcription pathways, replicate the DNA, sub-cellular structures, which include genetic structures and organelles which form the cell. Perhaps the most controversial aspect of the proposed hierarchy is the omission of species as entities and their treatment as sets of organisms. Species as sets are here treated as names we give to similar organisms. Species are not treated as causal agents in their own right. I take the view that species are types of organisms which share the same mate recogniton system /13/ and thus sets of organisms, the species members, have a potential to breed at some future time. Although species are observable (countable) at any one time, they are not countable through time except by convention.

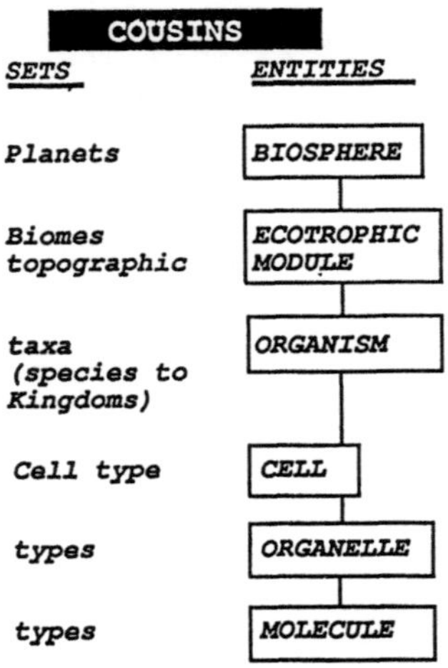

Fig. 3 The proposed single hierarchy of biological entities.

My reluctance to accept the proposed separate genealogical and ecological hierarchies is that below the level of organism the entities are components interacting at the molecular organelle and cell levels; the existence of different types of entity being a necessary condition for hierarchy theory and not reason for the creation of separate hierarchies. Above the level of organism I follow Grene /17/. Grene excludes taxa above the species level as being entities in the present even though they are linked through history by descent. I extend her view to the species itself while placing the interactions of species parts (eg the male female interactions) into the framework of the ecotrophic modlule.

There are certain consequences of the definition of ETM for the definition of ecosystem itself. First the geographical or topographical descriptions of ecosystem (pond, ocean, prairie, forest) become classes of trophic modules. That is, they become the names we give to different types of real objects the ecosystem trophic modules. Second, the definition does not mention the interaction with environment or the cyling of nutrients. These are essential prerequisites for the operation of ecosystems which may determine the type of ecosystem trophic module found at a location but are not a necessary part of the definition of an ecological entity. In a similar way, sociologists may define social systems without specifying the air needed to breathe or other essential aspects of bio-geological flows needed for society's operation. Flows of nutrients are from the hierarchy viewpoint too low in the hierarchy to occur in the definition of ecosystem or society.

Concepts

Concept formation is the step in science said to precede the formation of theories and laws. Ecology is often perceived as a young science and the view taken as suggested by physicists at the Ecodynamics conference, that concept formation and exploration is particularly appropriate in ecology. There is no doubting the much longer experimental and mathematical tradition of physics which does indeed make ecology look young by comparison. However it can be shown /18/ that aspects of ecological science date from earlier than Haekel's definition of ecology in 1869. In particular making species distribution atlases /19/ follows a clear line of descent from 17th century descriptions of national or local flora and fauna /20/.

Certainly there is a perception by some ecologists that ecology is coming out of its youth into early middle age. Youth in science is characterised by a proliferation of concepts which are not coherently interrelated. Concepts are indeed welcome but they must be justified by their contribution to the development of an integrated theory of ecology. Concepts should be clearly identified as ideas rather than as pieces of reality. Progress in science obviously also occurs by rejecting concepts in the search for theories and laws or better concepts. The rapid rise and the slow decline /12/ of one of ecologies most cherished concepts, the trophic level /22/ is perhaps an interesting example.

Lindeman and Hutchinson /22/ viewed the population dynamics of lake organisms as a problem in energy transfer. Beginning with photosynthesis in phytoplankton energy would be transferred to herbivorous zooplankton and from there to carnivorous organisms. These stages, S, renamed by Lindeman trophic levels, were energy categories. Thus energy passes from S1 the plant to S2,...S3..Sn. The concept of a feeding system in which organisms of one species were dependent on another for energy can be traced to Wallace /21/ and Semper /23/. Semper explicitly described a 10% rule in which an ecosystem comprised a mass of carnivores ten times less than the mass of herbivores which in turn weighed ten times less than the weight of plants. Thus the basic ingredients of the trophic level paradigm were present before Lindeman's classic paper in 1942. What Lindeman added was the hypothesis that succession in animal and plant communities is driven by improvement in energy transfer efficiency between trophic levels. In short, Lindeman invoked the second law of thermodynamics and suggested that the improvement of the efficiency of energy transfers may provide the ultimate cause of ecosystem organisation.

But there were problems with Lindemans's view. Energy transfer efficiencies were not uniformly improved during succession as is illustrated by the inclusion of homiothermic predators on heterothermic prey. One important problem with the concept and one which can be related to thermodynamic concepts is the confusion between body size and trophic level concepts. Elton /24/ had earlier described the animals in an ecosystem as constituting a pyramid of numbers in which there were very many small organisms and progressively fewer larger organisms. Lindeman wrote " The Eltonian Pyramid may also be expressed in terms of biomass ". This is very simply done by multiplying the number of individuals in a size class by their average weight, however, Lindeman did not propose this. Instead he erected the new classification of "progressive energy relationships of food levels" and asks us to imagine these as the pyramid of numbers. However feeding level (trophic level) and size class are mutually exclusive. A size class in Elton's pyramid contains organisms of different trophic levels while a trophic level contains organisms of different sizes. The claim that the second law of thermodynamics was being applied when the trophic level methodology was being used was convincing enough to almost totally displace the Eltonian method of ecosystem analysis. It was not for 50 years that the energetics of the Eltonian pyramid was worked out /25/ or the principles of body sized based mechanisms of predation, reproduction, movement and animal abundance more comprehensively described /26/. Physicists (now marine biologists) have added to this subject area papers on spectral analysis of ecosystems /27/ and dimensional analysis in ecology /28/.

It was natural that Lindeman should have introduced his concept against the background of the then dominant model of Elton. However Lindeman and Hutchinson had structured their model upon historical energy classes when energy transfer is dependent on the present state and not previous states. In short the trophic level model is a non-Markovian model of a Markovian process whereas the Eltonian pyramid does provide a Markovian description.

In Lindeman's model energy transfers are limited to respiration, egestion and ingestion. Respiration of some of the energy ultimately collected by the plant is used for growth either in the plant or animal or other kingdoms. However, we may however consider energy transfer as a thermodynamic problem which is posed in a way consistent with Elton's model.

Entropy and food particle size

Suppose we take a uniform food item and break it into smaller parts and scatter them over an ecosystem. This act produces a calculable change in the entropy of that food item.

Let the food item be broken into n particles of equal size and scattered over an ecosystem which is itself partitioned into m boxes each of the size of the food particles. Then the change in entropy, ΔS is given by,

$$\Delta S = -k \cdot lnW \tag{1.1}$$

where k is the Boltzmann constant and W the number of ways of rearranging n particles in m boxes.

$$W = \binom{m}{n} \tag{1.2}$$

Thus, taking 10^2 particles and scattering them over an ecosystem of 10^6 boxes of the same particle size then,

$$\Delta S = -k\frac{10^6!}{(10^6-10^2)!\ \ 10^2!} \tag{1.3}$$

$$\Delta S = 1.4 \cdot 10^{-20} J/K \tag{1.4}$$

We may in general consider the ecosystem to be populated by statistically unlikely occurances of local high energy dense states which are the organisms themselves. These arise in a uniform distribution of energy in the form of sunlight incident on the ecosystem. Thus different sized organisms can be seen as differently likely states while growth in size occurs as a result of predation (ecological collision). In this model the plant can also be considered as a variety of energy states, ranging from simple to complex molecules present in different densities ranging from soluable sugars found in leaf drip to energy dense nuts and seeds /29/.

Descriptions

Ulanowicz /30/ and O'Neill et al /10/ stress that any system boundaries can be drawn on a system and measurements made on the parts that have been created. Ulanowicz's defense of the trophic level is that it provides a quantitative description of an ecosystem rather in the same way as a statistical description may summarize certain data. The system boundaries implied by the trophic level concept cannot in that sense be right or wrong but merely more or less efficient than other methods for a purpose. But there is the danger that we are therefore left to evaluate concepts without the prospect of refutation.

An abstract a concept has to be operationalised by making measurements on, or, by counting observable entities. In the trophic level case a measure of the biomass, B, of the number Ni of creatures weight i provided by records of feeding behaviour such as stomach contents.

$$B = \sum_i n_i W_i \tag{2.1}$$

But we know from the entropy discussion that different particle sizes are not equivalent and so the same biomass can have different entropy values. Thus instead of B, we need to keep the data in the form of a distribution, n,

$$n(w) \tag{2.2}$$

If we wish this description to have a predictive value then we need to take into account that at t=1 two particles may have the same mass while at t=2 one particle may have grown (eg a juvenile organism) while the other which was an adult of a different species had stayed the same weight. For predictive purposes therefore, we need,

$$n(w, w_\infty, t) \tag{2.3}$$

where, w_∞ is the assymptotic weight to which the particle is growing, w is its present weight at time t. The w_∞ variable is also characteristic of each species; each animal species has a characteristic adult weight.

Different taxa have different bauplans which give them different properties in the ecosytem; warm-bloodedness, flight, photosysnthesis etc. Thus a whole ecosystem model can be created to any degree of complexity by subdividing all biota into taxonomic number distibutions, ni, giving for all heterotrophs,

$$\sum_i n(w_i, w_\infty, t) \tag{2.4}$$

By adopting number distributions, ni, for each species rather than each higher taxon, and since each species has a characteristic w_∞ , then the biota is described by

$$\sum_i n(w_i, t) \tag{2.5}$$

The time dependent change of this distibution due to internal causes of predation and respiration can be found from a set of partial differential equations for predation, starvation, growth and reproduction /31,32/.The complexity of the latter model can be reduced by setting delay terms (caused by prey handling times) to zero and by treating the model as a series of single species interactions. Then distributions of type 2.5 rather than type 2.4 can be used.

Conclusion

The case for a single hierarchy of biological objects has been made against a prevailing tendency towards dual or multiple hierarchies. In an attempt to find an ecological object intermediate between individual and biosphere a new ecological entity has been defined, the Ecosystem Trophic Module. The ETM is located within the pyramid of number and not the trophic level approach. A possible framework for the application of thermodynamics to the Eltonian pyramid is proposed. Finally the role of concepts in ecology has been highlighted by the difficulty of appraising concepts once they become widely used. The trophic continuum model of the Eltonian pyramid can be applied at a variety of levels of complexity showing that simple versus complex model debate need not be polarised into models with different general structures.

REFERENCES

/1/ Ghiselin, M. T. : Hierarchies and their components, Paleobiol. 13 (1987) 108-111.

/2/ Eldredge, N. : Unfinished Synthesis. Oxford University Press, Oxford 1985.

/3/ Salthe, S. N. : Evolving hierarchial systems. Columbia University Press, New York 1985.

/4/ Eldredge, N and Gould S. J. : Punctuated equilibria: an alternative to phyletic gradualism. In: Models in paleobiology. Schopf, T.J.M. (Ed.). Freeman, San Francisco 1972 p. 82-115.

/5/ Eldredge, N. and Salthe, S. N. : Hierarchy and evolution, Oxford Surv.Evol.Biol. 1 (1985) 184-208.

/6/ Rowe, J. S. : The level-of-integration concept and ecology, Ecology 42 (1961) 420-427.

/7/ Ghiselin, M.T. : A radical solution to the species problem, Syst. zool. 23 (1974) 536-544.

/8/ Hull, D. L. : Individuality and selection, Ann. Rev. Ecol. Syst. 11 (1980) 311-332.

/9/ Mayr, E. : The ontological status of species: scientific progress and philosophical terminology, Biol. Philos. 2 (1987) 145-166.

/10/ O'Neill, R. V., D. L. DeAngelis, J. B. Wade and T. F. H. Allen. A hierarchical concept of ecosystems. Princeton University Press, Princeton 1986.

/11/ MacMahon, J. A., D. L. Phillips, J. V. Robinson, and D. J. Schimpf. : Levels of biological organisation: an organism centered approach, Bioscience 28 (1978) 700-704.

/12/ Cousins, S. H. : The decline of the trophic level concept, TREE 2 (1987) 312-316.

/13/ Paterson, H. E. H. : The recogniton concept of species. In Species and specieation. Vrba, E. S., (Ed.). Transvaal museum, Pretoria 1978.

/14/ Cousins, S. H. : Can we count natural ecosystems? Bull. Brit. Ecol. Soc. 18 (1987) 156-158.

/15/ Kruuk, H. : Hyaena. Oxford Univ. Press, Oxford 1975.

/16/ Simon H. A. : The organisation of complex systems. In: Hierarchy theory. Pattee, H. H. (Ed.). Braziller, New York 1973 p. 1-27.

/17/ Grene, M. : Hierarchies in Biology, Amer. Scient. 75 (1987) 504-510.

/18/ Cousins, S. H. : An alignment of energy and diversity models of ecosystems. Phd Thesis, The Open University, Milton Keynes 1983.

/19/ Sharrock, J. T. R. : The atlas of breeding birds in Britain and Ireland. The British trust for ornithology, Tring 1976.

/20/ Carew, I. : A survey of Cornwall. 1603.

/21/ Wallace, A. R. : On the tendency of species to form varieties. In Darwin C and Wallace A. R., Proc. Linn. Soc. Lond. 3 (1858) 53-62.

/22/ Lindeman, R. L. : The trophic-dynamic aspect of ecosystems, Ecology 23 (1942) 399-418.

/23/ Semper, K. : Animal life as affected by the natural conditions of existence. Appleton, New York 1881.

/24/ Elton, C. S. : Animal Ecology. Sidgewick and Jackson, London 1927.

/25/ Platt, T. : Structure of the marine ecosystem: its allometric basis. In : Ecosystem theory for biological oceanography. Ulanowicz, R. E. and Platt, T. (Eds.). Fisheries and Oceans, Ottawa 1985 p 55-64.

/26/ Peters, R. H. : The ecological implications of body size. Cambridge University press, Cambridge 1983.

/27/ Platt, T. and Denman K. L. : Spectral analysis in ecology. Ann. Rev. Ecol. Syst. 6 (1975) 189-210.

/28/ Platt, T. and Silvert, W. : Ecology, physiology, allometry and dimensionality. J. theor. Biol. 83 (1981) 855-860.

/29/ Cousins, S. H. : The trophic continuum derived from plant structure, animal size and a detritus cascade. J. theor. Biol. 82 (1980) 607-618.

/30/ Ulanowicz, R. E. : Lindeman exonerata. In : Progress in systems ecology. Patten, B. C. and Jorgensen, S. E. (Eds.). Elsevier, New York in press.

/31/ Parkin, H. and Cousins, S. H. : Towards a global model of large ecosystems: equations for the trophic continuum. Energy Research Group Report 041, The Open University, Milton Keynes 1981.

/32/ Cousins, S. H. : The trophic continuum in marine ecosystems: structure and equations for a predictive model. In: Ecosystem theory for biological oceanography. Fisheries and Oceans, Ottawa 1985 p. 76-93.

Simulated Evolution of Primitive Organisms

G. Duchateau[1], *J.A. Meyer*[1], *P. Tarroux*[1], *P. Vincens*[1], *and G. Weisbuch*[2]

[1]Département de biologie, Ecole Normale Superieure,
46 rue d'Ulm, F- 75230 Paris Cedex 05, France

[2]Département de physique, Ecole Normale Superieure,
46 rue d'Ulm, F- 75230 Paris Cedex 05, France

ABSTRACT

Computer simulations have been developed for describing the evolution of primitive organisms whose phenotypic changes (acquisition of food reserves, reproduction and death) are assumed to depend on the dynamic properties of a network of threshold automata. Such a network models the interactions among the genes of a living cell. Simulated organisms compete for environmental resources and undergo genetic mutations which concern the intensity of their connections. This approach allows the correlation of phenotypic characteristics with the properties of the whole genome. During the simulations, the variations of selective and neutral characters have been monitored.

INTRODUCTION

Models of evolution based on organisms necessarily rely upon a description of the relationship between genotype and phenotype. A simple assumption is that fitness - a particular phenotypic character - is determined uniquely by one gene. For example, Fischer's Fundamental Theorem of Natural Selection is based on such an assumption (/1/).

Since such a model is highly oversimplified, it was deemed necessary to simulate organisms whose genomes are modeled by a network of automata (according to a proposition of Kauffman /2/) where an automaton represents a gene and in which the collective dynamic properties of the network are interpreted as phenotypic properties. In a previous paper (/3/), one of us has shown that this approach leads to a smooth fitness landscape, with neighboring networks having similar fitnesses. The model, based on a differential system describing population dynamics, exhibits punctuated equilibria as a generic behavior.

This paper is based on similar assumptions. By simulating organisms, rather than populations, it is hoped that complex mutation mechanisms and complex genomic structures can be described.

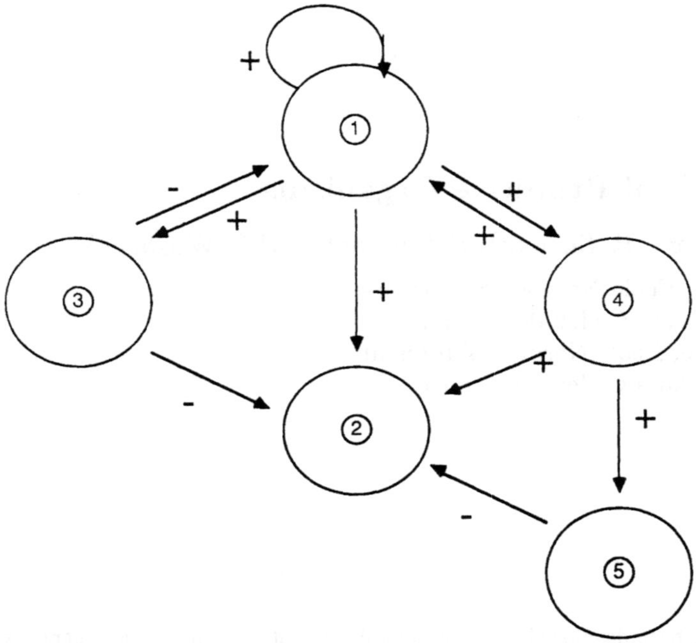

Figure 1
The five automata are represented by circles and their connections by arrows, "+'s" indicate connection of strength +1 and "-'s", -1.

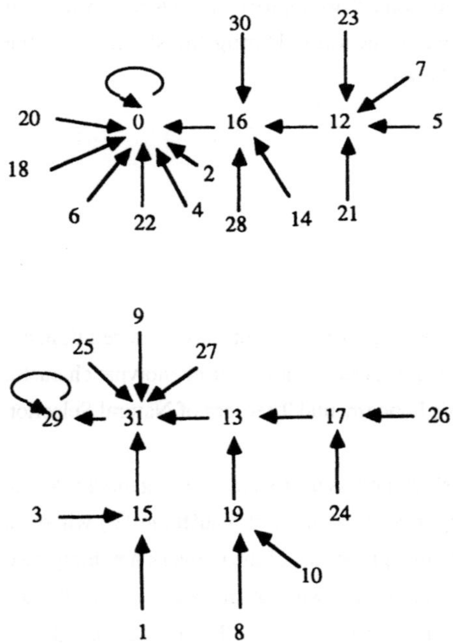

Figure 2
Iteration graph of the network of figure 1, exhibiting 2 attractors of period 1 and their basins of attraction.The network configurations are coded by decimal transcription of the five binary states of the automata. For instance 5 stands for 00101 where the automata 1 to 5 are respectively in states 1, 0, 1, 0, 0.

The automata can be in two states:1 (active) or 0 (inactive), depending upon interactions with other genes. The configuration of the genome (set of the states of the genes), which changes at each time step, determines the activities of the organism, i.e. nutrition, reproduction and death.

An organism is specified by its genome which is composed of n interacting genes (n is between 16 and 20), and by the set of the interactions between these genes.

The general aim of this model is to study the evolution of successful genomes and the emergence of various adaptative strategies. The preliminary results that are reported here take into account only the mutations that concern the connections between automata.

DESCRIPTION OF THE MODEL

At the beginning of the computer simulation, about 20-40 organisms are randomly generated . The evolution of these organisms, is described as a function of the dynamics of their genomes and their interactions with the environment. Limited energy resources are available from the environment. New organisms are generated by duplication. Then mutations are randomly applied to some of them. At each time step, a random sampling is performed in order to provoke the death of certain organisms.

Dynamics of the network of automata.

The connections between automata can be equal to -1, 0 or +1. The number and the value of the connections determine the dynamics of the network.

The state $S_{i(t+1)}$ of the gene i, at time t +1 depends on the states $S_{j(t)}$, at time t, of all the automata j which are connected to i, and is computed according to the following algorithm:

We first compute $G_{i(t)}$

$$G_{i(t)} = \Sigma_j T_{ij} S_{j(t)}.$$

where T_{ij} is the connection between i and j, and $S_{j(t)}$ the state of automaton j.

S_j is then determined by the following rules :

$$\begin{array}{llll} \text{if} & G_{i(t)} > 0 & \text{then} & S_{i(t+1)} = 1 \\ \text{if} & G_{i(t)} = 0 & \text{then} & S_{i(t+1)} = S_{i(t)} \\ \text{if} & G_{i(t)} < 0 & \text{then} & S_{i(t+1)} = 0 \end{array}$$

All the automata of a given organism are updated simultaneously (parallel iteration). Starting from any initial configuration, these rules allow the determination of the set of configurations successively reached by each genome during the simulation. As soon as a given configuration has been reached twice on a trajectory, the subset of intermediate configurations is visited indefinitely. This subset is called an attractor. Obviously, if a configuration is its own successor, it is an attractor of period 1.

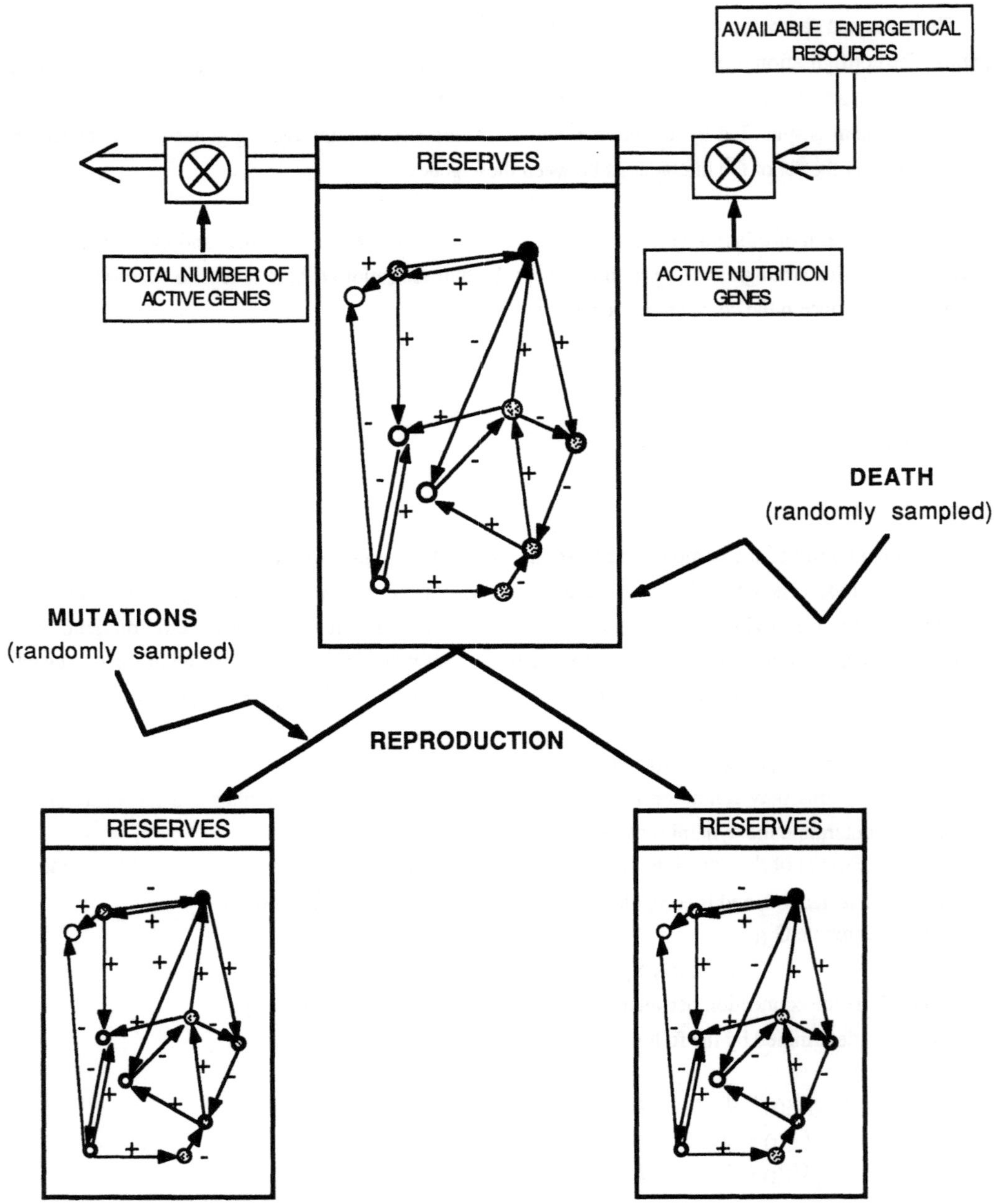

Figure 3 : Scheme of the biological activity of an organism during one time step

The different circles illustrate the different types of genes (reproduction , nutrition or neutral genes). The arrows and the associated "+'s" and "-'s" represent the connections between genes.

Figure 1 is an example of a network with five automata, which can occupy 32 different configurations. Figure 2 represents the time succession of the different network configurations,i.e. the iteration graph of the network described in figure 1.

A network made of n binary automata can be in 2^n different configurations and the number of possible realizations of such a net is $3^{(n*n)}$. For n = 10, there are already 1024 configurations and 9^{50} different nets. We have chosen the numerical simulation approach in order to avoid dealing with such a large number of equations.

Biological functions

Some genes have a biological function, such as nutrition or reproduction. Some other genes, without any biological function, are said to be neutral. The dynamics of a network activates or inhibits certain genes which determine the activities of the organism (see figure 3). For example, reproduction only occurs if a "reproduction" gene is active.

Population dynamics depends on three processes that can be performed by the organisms at each time step: energy uptake or consumption, reproduction and death.

a- *Energy balance*

At each time step a constant food supply E is shared by all organisms.The energy available per organism, $\Delta E_{(t)}$, is then limited by the total population, $Pt_{(t-1)}$, according to:

$$\Delta E_{(t)} = E / Pt_{(t-1)}.$$

Only a fraction of the energy, proportional to the number ($na_{(t)}$) of its nutrition genes which are active, is in fact absorbed. Energy intake, $\Delta_1 R_{(t)}$, by an organism during a time step is then :

$$\Delta_1 R_{(t)} = \Delta E_{(t)} * na_{(t)}.$$

The energy loss by each organism during each time step is denoted $\Delta_1 R_{(t)}$, and depends upon the total number ($nt_{(t)}$) of its active genes.

$$\Delta_2 R_{(t)} = C * nt_{(t)}.$$

where C is a constant

Therefore $R_{(t)}$, the energy reserves of each organism, varies in time according to:

$$R_{(t)} = R_{(t-1)} + \Delta E_{(t)} * na_{(t)} - C * nt_{(t)}.$$

b- *Reproduction process*

At each time step and for each organism, if the reserves are above a given threshold (S_R) and if at least one reproduction gene is active, duplication occurs. The duplicated organisms are identical, except for the connections among genes which can be changed by mutations. Mutations involve at most one gene with a probability (P_M) of 0.2 .

c- *Death process*

If reserves are insufficient, i.e. less than S_D, the organism dies from starvation. If they are above S_D, accidental death occurs with a probability (P_D) of 0.08.

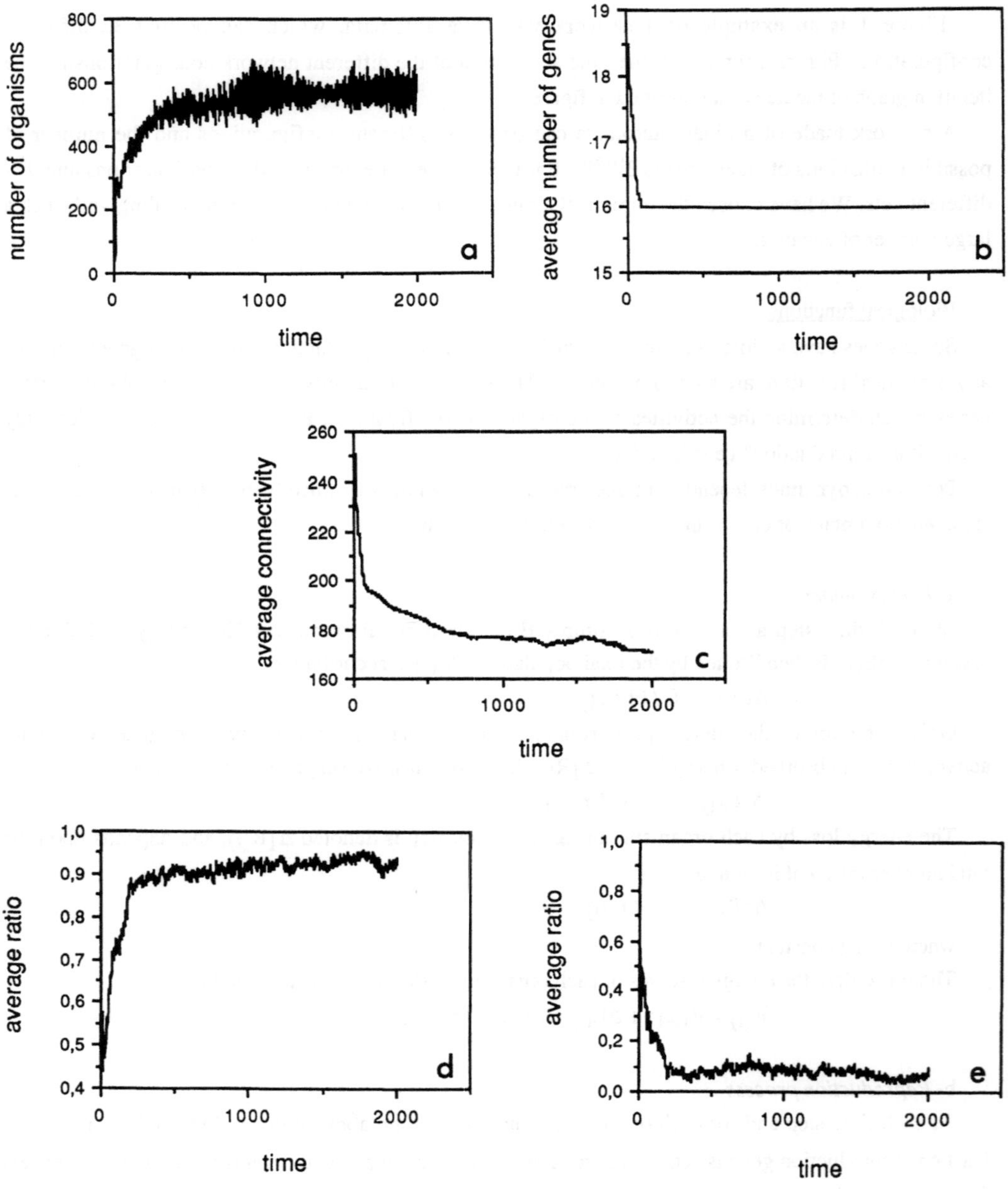

Figure 4
a - evolution over time of the total population.
b - evolution over time of the average number of genes.
c - evolution over time of the average number of non-zero connections per genome.
d - evolution over time of the ratio (active nutrition genes)/(total nutrition genes).
e - evolution over time of the ratio (active neutral genes)/(total neutral genes).

We have described the evolution of the population for 3000 cycles, which takes roughly 12 hours. The program is written in Ada® and runs on an IBM PC AT.

RESULTS

The results of the first simulations are reported here. The organisms are competing for energy resources and their mutations concern only the connections between the genes. The following figures have been used in the numerical simulation :

$$E = 500 \;-\; C = 0.45 \;-\; P_M = 0.2 \;-\; P_D = 0.08 \;-\; S_R = 35 \;-\; S_D = 4.$$

Total population (fig.4a)

After an initial increase, the population stabilizes around 550, because of the limited energetical resources in the environment. Since the energy resources must increase in order for reproduction to balance death at equilibrium the following inequality is verified :

$$Pt_{(t)} < (E/C) * (na_{(t)} / nt_{(t)}) < (E/C).$$

Number of genes (fig. 4b).

Starting from a population of 10 organisms with 16 to 20 genes, it can be seen that the number of genes is not changed by mutations. The fact that energy consumption is proportionnal to the number of active genes benefits those organisms with a small number of genes. The organisms with the minimum number of genes, 16, are selected in less than 100 time steps.

Average connectivity (fig. 4c).

The degree of connectivity contributes to the dynamics of the network which in turn determines which genes are active. Its monotonic variation with time is an indication of its functional role but we cannot at this time interpret these data. Two relaxation times for the variation of the connectivity can be observed on the graph. The short one is equal to roughly 30 time steps and the longer one to 1000 time steps.

Nutrition and reproductive genes (fig.4d).

Figure 4d represents the time evolution of the ratio of active nutrition genes to total nutrition genes. This ratio is directly responsible for the energy balance of the organism. This ratio grows from an initial value of 0.4 towards 1. The ratio of active reproduction genes to total reproduction genes follows the same behavior. We see that selection favors maximal nutrition of the organisms and increases the chances of reproduction.

Neutral genes (fig.4e).

Starting from a random population of organisms with an initial average proportion of 0.5 active neutral genes to total number of neutral genes, it can be seen that these genes have a negative effect on fitness, because they are only responsible for energy consumption and nothing else. Therefore, an organism with a minimal proportion of active neutral genes loses a minimum of energy. Selection acts by decreasing their proportion to roughly 0.1.

CONCLUSIONS

The model described here takes into account the fact that the phenotype depends on collective properties of the genome, which are determined by the interactions among whole set of genes.

The preliminary simulations reported here took into account only mutations of the connections among genes. Other mutations will be considered in future studies, for example, mutations of biological functions associated with each gene or mutations of the number of genes.

Quantities monitored in future simulations will be:

-the genetic distance between organisms alive at a given time,

-various characteristics of the selected genomes (number of genes, degree of connectivity, number of attractors and width of the attraction basins).

The consequences of environmental fluctuations also need to be studied.

REFERENCES.

/ 1 / Ewens, W. J. : Remarks on the evolutionary effect of natural selection, Genetics. 83 (1976) 601-607.

/ 2 / Kauffman, S. A. : Metabolic stability and epigenesis in randomly constructed genetic nets, J. Theor. Biol. 22 (1969) 437-467.

/ 3 / Weisbuch, G. : Un modèle de l'évolution des espèces à trois niveaux basé sur les propriétés globales des réseaux booléens, C. R. Acad. Sci. 298 (1984) 375-378.

Dynamics of Developmentally Constrained Populations

M. Kerszberg

IFF der KFA, P.O.Box 1913, D-5170 Jülich, Fed. Rep. of Germany

1. Introduction

The structure of biological organisms is known to be subject to developmental constraints [1]. The possible effects of such constraints on the evolutionary fate of species have been the center of much debate lately [2]. In the present work, I investigate, by computer simulation, the dynamics of populations in which each individual is endowed with a developmental program responsible for expressing in a constrained fashion its genotype into a set of phenotypic characters. I find that survival of species may be enhanced by certain types of developmental mechanisms: these are nonlinear and involve three basic ingredients, namely

– error correction (many small mutations have no phenotypic effect);

– pleiotropism (one mutation may affect more than a single phenotypic character);

– polygenic inheritance (a given phenotypic character is under the dependence of many genes).

Because of the error correction feature (*i.e.*, canalization), many small mutations which would affect adversely the fitness of the population remain hidden; while pleiotropy ensures that, in time of need, a few mutations suffice to generate a large phenotypic shift. Thus, we have a coherent explanation of phylogenetic stasis: canalized development improves fitness, *and* it results in punctuated patterns of evolution.

In what follows I describe a simple developmental model embodying the three features just introduced, the "majority" model [3]. I then study the evolution of mixed populations involving this and other developmental programs: it is the majority model which fares best. I also show how the majority mechanism sometimes enables a population to overcome major environmental upheavals, but can at the same time be responsible for evolutionary stasis.

2. The Majority Model

The model consists in a discrete, multilayer nonlinear developmental network [4]. It is assumed that information content at a given level is embodied in a string of binary digits, and these are determined entirely from information at the previous level by the so-called *majority rule* which sets a bit to "1" if and only if the majority of its three upper neighbors are themselves set to "1" (Fig.1a).

The top layer represents the "genotype," the last layer the "phenotype." The biological relevance of the model has been discussed in [3]. It is clear from the figure that the rule implies pleiotropism and polygenic inheritance; from the fact that four different possibilities at level n do generate a "1" at level $n+1$, it is also evident that local error correction is implemented.

Remarkably, even such a simple nonlinear mapping leads to surprising behavior. Small changes in genotypic content usually affect the phenotype either weakly or not at all (canalization); but, occasionally, small mutations can lead to large phenotypic effects (nonlinearity, see Fig. 1b).

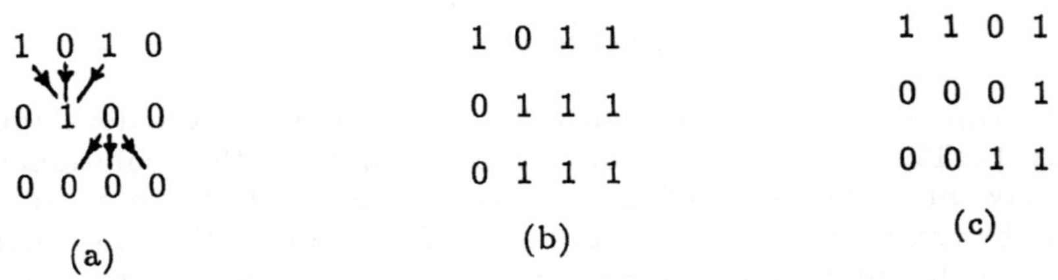

Fig.1 : Developmental rules. a) The majority rule: a bit on level n is set to one if and only if the sum of the three bits on level $n-1$ is greater than or equal to 2. "0"s are supplied on the sides of the array. The model involves pleiotropy and polygenic inheritance (arrows), and possesses error correction properties. b) Nonlinear error amplification by the majority rule: a one bit mutation leads to a large phenotypic effect (all bits on the bottom level changed, except one). c) Processing according to the "one-only" rule: a given bit at level n is set to one if and only if a single of its three neighbors on level $n-1$ is equal to one. There is pleiotropy and polygenic inheritance, but very weak error correction.

In our simulations, we compare populations of individuals with majority-rule as well as two other types of processing. The first of these is the "one-only" rule (Fig.1c), where a given binary digit is set to "1" *if and only if* one only of its three upper neighbors is itself set to "1;" this rule has the features of pleiotropy as well as polygenic inheritance, but error correction is weak – a pair or a triplet of ones is always lost. We also use the identity rule, where the bits at any level are simply copied from the previous level (this rule lacks all three features of error correction, pleiotropy or polygenic inheritance).

3. Population Dynamics

The general principle of the simulations is as follows. A "fitness" is associated with each individual's phenotype and determines his probability of a) leaving offspring, and b) living or dying. "Leaving offspring" means that a new network is produced, possessing a genotype nearly identical with that of the parent. Reproduction is asexual and takes place with a constant mutation rate per generation, *i.e.*, each time reproduction occurs, a fixed fraction of genotypic "errors" is introduced. After birth, "expression" of the offspring's genetic message into phenotype is instantaneous.

The "fitness" of an individual is essentially one minus the Hamming distance of its phenotype to some "preferred" phenotype. The Hamming distance corresponds to counting the fraction of bits which are different in two bit strings: thus if the preferred phenotype is 000111 and some individual has phenotype 111111 then its fitness will be $1 - 0.5 = 0.5$, while if its phenotype reads 111000 it will have fitness $1 - 1 = 0$. The definition of "fitness" just given will be completed in Section 5 when we introduce time-dependent preferred phenotypes.

In terms of the "fitness", the birth and death probabilities can now be introduced; at any given moment, these probabilities are functions of the individual's "fitness", which depends on the external conditions, and of the "fitnesses" of all other phenotypes with wich the individual is competing. In particular, the total number of individuals present influences the fate of each: when few are alive, births increase as there is less competition for fixed resources, and on the other hand births will decrease when the population gets too numerous. The details of how the simulations achieve this will not concern us here; let us simply mention that the the average net number of births per generation is, for individuals with fitness f,

$$N(f-a) - \frac{N^2}{N_i}\bar{f}, \tag{1}$$

where $\bar{f}$ is the average fitness of the population, N_i the carrying capacity, and a the rate of death from old age. Eq.(1) is consistent with the standard logistic form, but for the fact that the coefficients of N and N^2 are now dependent on the fitnesses of the individuals currently alive [5].

The individuals in our population have a size of 20 (width) times 10 (depth). The total number of possible genotypic configurations is thus 2^{20}, or roughly one million. As opposed to this, the total number N of individuals is about 100: thus, a very small fraction of the total genotypic space can ever be explored, even when allowing for the accumulated effect of mutations over the entire length of the simulation (a thousand generations maybe, with a mutation rate of 0.001 per bit and per generation). We have thus placed ourselves in the conditions under which macroevolution operates.

4. Coexistence of Species with Different Canalization Properties

In this Section I demonstrate the sort of selective advantage which can be conferred through developmental canalization, by first simulating mixed populations where individuals may exhibit one or another form of developmental mechanism (this form being of course inherited from their parent). The preferred phenotype is assumed to consist entirely of "0"s: the more "1"s are present, the higher the penalty.

Majority versus Identity. The coexistence of majority- and identity-rule populations always ends with the survival of the majority-rule "species", although we may have started with a greater number of identity-rule networks. We have assumed here that the efficiency, in terms of, say, food requirements, are similar for the two species: in other words, canalization is assumed here not to entail any additional energy consumption. The survival of "majority-rule", however, occurs with such overwhelmingly large probability that it is unlikely that a small disadvantage with respect to energy consumption should influence the results in any essential way.

The origin of this predominance of "majority rule" becomes clear when the simulations are analyzed in some detail: it stems from the absence, in the corresponding subpopulation, of the low-fitness tails with which the "identity-rule" species must live. The difference can in turn be traced to the canalization properties of the majority rule: it suppresses the phenotypic effects of many mutations, leaving the individual perfectly adapted. The tails afflicting the "identity-rule" group could be made to dwindle by diminishing the mutation rate for this group; such improvement, however, would be at the cost of a reduced probability of its ever surviving any major environmental change, and this is already dismally low, as we shall see in Section 5.

Majority versus One-Only. The majority rule also displaces a population of one-only individuals. This is now due to the better error-correction properties of the majority-rule mapping. Small mutations too often result, for the one-only case, in large phenotypic effects, and this again contributes to the appearance of tails of unfit individuals.

Identity versus One-Only. Although the one-only rule certainly does not perform too well against the majority processing, its canalization properties are still sufficient for it to beat the identity rule. The only way to ensure the victory of the identity rule is by imposing a starting condition where all individuals have an all-zero genotype. Then the bad stability of "one-only" plays against it: in general, several mutations can, through a cascade effect, multiply their nefarious effects when processed in a one-only fashion, while such overbalancing is impossible with the identity rule. This example illustrates how subtle the effects of developmental mechanisms on the relative success of species groups can be; ours has only been a preliminary investigation of this rich field.

5. Adaptive Shifts and Developmental Canalization

In the rest of this paper, we shall be concerned with phenotypic shifts. I let a population adapt to a given "environment", and observe what happens when a gradual shift occurs in the position of the adaptive peak. I also consider the case where a new peak appears in the adaptive landscape and finally supplants the existing one. In both cases, it is seen that extinction is extremely frequent; that the environmental transformation is best survived by the majority-rule populations; and that phylogenetic stasis occurs as a result of the majority type of developmental dynamics.

5.1. With No Adaptive Gap

We now study the occurrence of a shift of adaptive peak between phenotypes p_1 and p_2 separated by a Hamming distance d. Usually, canalization ensures that many genotypes, taken from the sets g_1 and g_2, are expressed as p_1 or p_2. The fitness of a phenotype p is evaluated in the following way. Call d_1, d_2 the Hamming distances between p and p_1, p_2 respectively. Define then

$$f = (1-t)|1-d_1| + t|1-d_2|,$$

where t is the reduced time, a variable running smoothly between 0 at time T_1 to 1 at time T_2. We see that at $t = 0$, fitness has its maximum value of 1 when p corresponds to p_1, while at $t = 1$, p_2 is favored. Note also that, at $t = 0.5$, all p's for which $d_1 + d_2 = d$ have the same fitness, and there is thus no adaptation gap between p_1 and p_2.

The simulations proceed in the following way. We start with random genotypes and wait some time. When the population has adapted, we start changing the conditions, i.e. the parameter t. We observe the variation of several selected parameters in the population as a function of time during this process. This will be a function of the mutation rate, the population size, the sizes of the individuals, the magnitude and rapidity of the change between p_1 and p_2, and so on. We consider here the influence of a few combinations of these parameters, namely magnitude and rapidity of change in external conditions, these quantities being of course defined for a given mutation rate.

We choose p_1 to consist of "0"s only, and p_2 of "1"s only. Such a shift requires that the population perform a large phenotypic transition, and represents, therefore, a difficult hurdle. This explains why, when the adaptive peaks are shifted very quickly (in, say, a few tens of generations), extinction is *always* observed. Only when the transition is imposed sufficiently slowly (over hundreds of generations), do more interesting phenomena occur. In the following I therefore limit myself to transitions generated by slow variations in the external conditions.

Identity Processing. Gradual shift of the phenotype towards p_2 is sometimes observed, but this is rather rare: most of the time, extinction occurs. When a successful jump to the new peak occurs, the change in phenotype is smooth and progressive, its rate being essentially determined by the mutation rate, together with the rate at which the "slanting" in adaptive landscape appears (this sets the selective advantage enjoyed by favorable mutants.)

Majority Processing. Much more interesting behavior can be observed in the case of majority-rule processing. Not only is survival much more frequent, but, depending on the random initial conditions, a variety of situations occur, exhibiting from essentially no stasis, all the way to a characteristic "punctuated equilibria" pattern (see Fig. 2.) In the present case, the occurence of changes in phenotypic appearance, as well as their amplitudes, are determined strictly by the details of the developmental network.

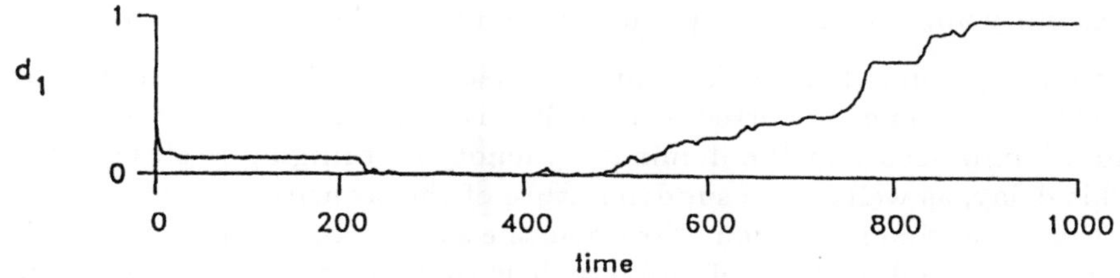

Fig.2 : Phenotypic shift in a population of majority-rule networks, in the absence of an adaptive gap. Time is in generations; the abscissa denotes the average fraction of "1"s in the phenotypes of the population. The transition takes place smoothly between $T_1 = 100$ *and* $T_2 = 800$. *Stasis episodes are seen, which are of purely developmental origin.*

Here then is the major result of our simulations: the emergence of punctuated evolution patterns generated, not by the *ad hoc* introduction of adaptation gaps, but through the necessary existence of error damping, pleiotropism and polygenic inheritance in the mechanisms which ensure the expression of genetic information. It is worth stressing that the arguments from classical genetics which have been invoked in order to understand stasis in the presence of adaptive gaps [6] rely entirely on the existence of such gaps. The type of stasis here depicted is of a completely different nature. It is due solely to the existence of internal constraints on the type of phenotypic adaptation which an organism is capable of effecting. These constraints stem from the very nature of the developmental processes at work in this organism. Such processes improve fitness and enable the organism to survive certain drastic environmental perturbations; but at the same time they limit its internal dynamics, thus reducing the repertoire of possible adaptations. There is no continuum of possible phenotypes, and certain transitions must therefore occur saltationally. These *internal* factors in evolution interact in complex ways with the external ones; we have seen in Section 4 that they are probably *themselves* under selective pressure.

One-Only Processing. The case of the one-only rule is somewhat similar to that of the majority-rule. However, we observe that successful transitions are a relatively less frequent event. This seems due again to the rather poor error-correction properties of the rule.

5.2. When Adaptive Gaps Exist

We introduce an adaptive gap between p_1 and p_2 through the following formula:

$$f = (1-t)(1-d_1)^2 + t(1-d_2)^2.$$

It is now easy to see that the fitness of p equidistant of p_1, p_2 is *less* that that of p_1, p_2 themselves when $t = 0.5$.

Identity Processing. Not very surprisingly, we never observe a successful transition, even when the shift in preferred phenotype is extremely slow.

Majority Processing. Extinction is also the usual fate of the population when majority-rule processing is substituted for the identity rule. However, here, another possibility appears in some simulations, namely adaptation and survival.

Figure 3 displays one of those few "lucky" cases. Note that in principle, at generation 450, p_2 becomes preferred over p_1; but it is not before generation 650 that a sudden jump occurs in the dominant phenotype, bringing it close to the p_2 form. This delay, as well as the sudden nature of the transition, are recurring features in this type of computation. Note also the severe population bottleneck. Remarkably, a very small number of individuals is thus responsible for the jump: the population which performs the switch seems always to be effectively small. This is due, as we shall see, to the fact that a successful transition apparently requires a rare combination of appropriate genetic make up, together with the occurence of very specific mutations in the offspring. The phenomenon is worthy of more detailed discussion. At generation 625, the population has dwindled to

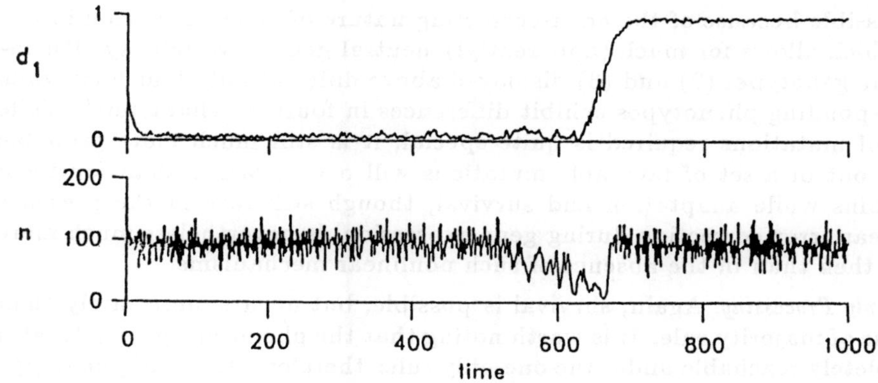

Fig.3 : Phenotypic shift in a population of majority-rule networks, in the presence of an adaptive gap. Time is in generations; the abscissa in the top diagram denotes the average fraction of "1"s in the phenotypes of the population. The bottom graph shows the population number. The transition in external conditions takes place smoothly between T_1 = 100 and T_2 = 800, but the phenotype shifts in a sudden, delayed manner (see text.)

18; all individuals have, up to one or two mutations, the same genotype. Only one individual out of the 18 presents any 1's in its phenotype (namely, three). At generation 650, the population numbers 29. All of the 29 share a common ancestor at generation 625, one which had an all-0 phenotype, and genotype:

$$10000101000010101000. \qquad (2)$$

The population of 29 displays much more genetic variability than that of its ancestor, since one may now observe individuals differing in up to seven loci, for instance

10000101100011101000 and

11101101000010111100.

Many of the individuals are now over the adaptive gap, and have more 1's than 0's in their phenotype (for example, the last genotype displayed above has 14 ones in the corresponding phenotype); selective pressure alone can thus complete the transition: and, indeed, at generation 675, we find a population of 97, all originating from one of the 29, which possessed genotype:

$$11101101000010101100. \qquad (3)$$

There is still a large amount of genetic variability, but this will gradually diminish as the population settles in its new niche.

Thus, in our model at least, a population which succeeds in effecting a large adaptive movement does so through the agency of one or a few "founder" members. These successful individuals seem to have, in most simulations, a rather interesting genetical status, in the sense that, while their phenotype is p_1 or almost p_1, their genotype sits very close to the "boundary" between g_1 and g_2:

this is possible because of the error-correcting nature of gene expression in our model, which allows for much (apparently!) neutral genetic variability. For instance, the genotypes (2) and (3) displayed above differ at only four loci; while the corresponding phenotypes exhibit differences in fourteen characters! While the sort of mutations required is quite special, it is still much more probable that four out of a set of favorable mutations will occur, rather than fourteen: this explains while adaptation and survival, though still rare in the presence of nonlinear error correction during gene expression, is nevertheless much more probable then than in the absence of such nonlinear mechanisms.

One-Only Processing. Again, survival is possible, but occurs more rarely than in the case of majority rule. It is worth noting that the phenotype p_2 (all "1"s) is not completely reachable under the one-only rule: therefore, the final phenotype exhibited by the population is not p_2, but some "optimal" phenotype with a large proportion of 1's, but also large fluctuations in their frequency at any given position in the phenotypic string. Thus we have here an example where the existence of developmental mechanisms, while making survival possible, also hinder the subsequent adaptation from becoming perfect.

6. Perspectives

Our work has demonstrated the benefits of error damping ("canalization"), pleiotropism and polygenic inheritance for species survival, assuming these nonlinear features can be acquired at a small energetic cost. In fact the majority rule, for instance, creates *bridges* over certain adaptive valleys. Thus, an environmental fluctuation leading from a preferred phenotype

000000000 to

000111000

requires, in order to be surmounted, three mutations with joint probability a^3 when the identity rule is used, but, for *some* majority rule individuals at least, just *one* mutation with probability a.

We also showed how such developmental nonlinearities lead sometimes to punctuated equilibrium patterns, even in the absence of adaptive valleys.

Ultimately, since nonlinear developmental mechanisms are of such potential advantage, they must have been selected for, which may be part of why we are more than just a genotype. This opens the perspective that not only the phenotype itself, but the mechanisms leading to its realization starting from a given genetic message in a given (possibly time dependent) environment, are all under selective pressure, albeit on what may be very different time scales. Under this hypothesis, one may conceive that mass extinctions could take place when large, novel environmental transformations appear – such as shifts from one climate alternation pattern into a new one: the basic developmental processes of whole groups of species might then become inappropriate. While a certain amount of canalization can be advantageous, it is also clear, however, that too much of it, or canalization into "wrong" phenotypes, can be detrimental. An optimum must be found in this respect, which can only be maintained *in status quo* so long as certain basic external patterns do not alter too drastically.

It would certainly be interesting to extend our simulations to include the effects of life history, which have been ignored here; in addition, although they certainly are among the most important factors in speciation, geographic (*i.e.*, space-dependent) effects, as well as the effects of genetic recombination and sexual selection mechanisms, have not been included at all in this work. It is certainly to be expected that mating and recombination statistics, for instance, will strongly influence the detailed processes affecting the small "founder" subpopulations observed in Section 5.2. Nevertheless, we believe that our simulations are sufficiently suggestive as to provide some insight into the process of adaptation under changing conditions. Our aim has only been to open an exploratory path leading to these fascinating questions.

References

[1] Alberch, P. (1982) *Developmental Constraints in Evolutionary Processes*, pp. 313-332 in J.T. Bonner, Ed., *Evolution and Development*, Dahlem Konferenzen, Springer Verlag, Berlin.

[2] Gould, S.J. and N. Eldredge (1977) *Punctuated Equilibria: the Tempo and Mode of Evolution Reconsidered*, Paleobiology 3, pp. 115-151.

[3] Agur, Z. and M. Kerszberg (1987) *The Emergence of Phenotypic Novelties Through Progressive Genetic Change*, Amer. Nat. 129, pp. 862-875.

[4] Sander, K. (1983) *The Evolution of Patterning Mechanisms: Gleanings from Insect Embryogenesis and Spermatogenesis*, pp. 137-160 in B.C. Goodwin, N. Holder and C.C. Wylie, Eds., *Evolution and Development*, Cambridge University Press.

[5] Roughgarden, J. (1979) *Theory of Population Genetics and Evolutionary Ecology: an Introduction*, Macmillan, New York.

[6] Lande, R. (1985) *Expected Time for Random Genetic Drift of a Population Between Stable Phenotypic States*, Proc. Nat. Acad. Sci. USA 82, pp. 7641-7645; Newman, C.M., J.E. Cohen and C. Kipnis (1985) *Neo-Darwinian Evolution Implies Punctuated Equilibria*, Nature 315, pp. 400-401.

Differential Evolution of Pesticide Resistance in Predators and Prey

J.D. van der Laan and P. Hogeweg

Bioinformatics Group, University of Utrecht, Padualaan 8,
NL-3584 CH Utrecht, The Netherlands

SUMMARY

The differences between the development of pesticide resistance in insect predators and prey are investigated with a simple discrete-time predator-prey model.

The results, which hold for a large class of models and parameter values, suggest that the lack of resistant predators might not be caused by retardation of their resistance development but might be a stable phenomenon.

INTRODUCTION

A major problem in insect and mite pest control is the evolution of resistance to chemical pesticides. More than 447 species of insects and mites have now developed resistance. It is remarkable that only 3% of them are predators and parasites of pests (Georghiou, 1986; Croft & Strickler, 1982). Following the elimination of their natural enemies (the predators) the resistant insects are often present in much higher numbers than before the pesticide applications.

Resistance in this context is a practical concept which not only refers to a high frequency of resistant alleles and/or phenotypes, but also to a large population size (e.g. more than 50% of the original population). In this paper we will by default adhere to this use but when necessary will specify the allele frequency and population size separately.

Is it possible to explain the difference in the resistance development of predators and prey without assuming genetic differences of prey and predator or different degrees of exposure to pesticides?

It is often argued that the lack of resistant predator species can be attributed to the retardation of their resistance development. This retardation might be caused by the different food conditions obtaining for prey and predators just after the first applications of pesticide. Surviving prey have an unlimited food supply for reproduction whereas surviving predators are faced with a reduced prey population. Predator species therefore cannot continue their resistance development until their prey species become resistant (food-limitation hypothesis) (Croft & Brown, 1975; Morse & Croft, 1981; Tabashnik & Croft, 1985).

Is the observed lack of resistant predators in the field only a temporary effect caused by retardation or is it a persistent phenomenon?

THE MODEL

We study the differential resistance development in the simplest possible model, i.e. a discrete-time Lotka-Volterra predator-prey model incorporating density dependence, which has been extended with immigration and allele frequencies (table 1). Thus without selection prey grows logistic and suffers predation, which is assumed to be a linear function of the number of predators. The number of predators in the next generation depends entirely on the predation ($e*X_t*Y_t$).

The phenotype resistance is assumed to depend on one locus with two alleles, R (resistant) and S (sensitive). On the assumption of Hardy-Weinberg equilibria (genotype distribution of RR, RS and SS can be regarded as p^2, 2pq and q^2 with p= R-allele frequency and q= S-allele frequency=1-p), selection can be easily incorporated in the model (Roughgarden, 1979). Survival of the three genotypes is assumed to be dose dependent. We assume that for the dose used all homozygote sensitives are killed (S=0), whereas all homozygote resistants do survive (R=1). As for the heterozygotes a certain fraction (H), which is directly dose dependent, will survive. If H is smaller than 0.5 we consider a high dose ,whereas a low dose means a H greater than 0.5. In our model it is assumed that there is selection once in every generation and that immigrants arrive after the pesticide application (which allows them to mate before being killed).

The model is initialised so as to make predator and prey as similar as possible, i.e. equal selection and immigration parameters, equal initial resistant allele frequency and same maximum growth rate (for prey when prey and predator density is low, and for predator when prey is at carrying capacity). All the resistant allele frequencies (Px, Py and Pim) are initialised as 10^{-2}. The resistance allele frequency in the field is reported to be between 10^{-2} and 10^{-13} (Georghiou & Taylor,1986; Roush & Croft, 1986). We used the highest value to ensure that there would be sufficient surviving prey and predators , and that the lack of resistant predators would not be caused simply by a lack of animals with a resistant phenotype after the first pesticide application. Prey and predator are assumed to be at equilibrium density before selection ($X=5*10^5$ and $Y=5*10^4$). The next generation can then be determined by multiplying the selection factor (the genotype distribution and their fitness parameters) with the population growth factor and adding the immigration. The resistant allele frequencies of the next generation can be defined as half of the number of the heterozygotes and the total number of the homozygote resistants in the next generation, divided by the total next generation. If there is no immigration, the resistant allele frequency is independent of the population size as the equations show.

ANALYSIS

The model and its extensions are investigated using the numerical package GRIND (De Boer, 1983), which involves dynamic and static analysis, i.e. numerical integration and zero-isoclines respectively.

A population is considered to be resistant if both the resistant allele frequency is greater than 0.5 and the population size is greater than 50 % of its original size.

Table 1 Main model

$$X_{t+1} = (a*X_t - b*X_t^2 - c*X_t*Y_t)*(Px_t^2*R + 2*Px_t*Qx_t*H + Qx_t^2*S)+IM$$

$$Y_{t+1} = (e*X_t*Y_t) * (Py_t^2*R + 2*Py_t*Qy_t*H + Qy_t^2*S)+IM$$

$$Px_{t+1} = \frac{(a*X_t - b*X_t^2 - c*X_t*Y_t) * (Px_t^2*R + Px_t*Qx_t*H) + Pim*IM}{(a*X_t - b*X_t^2 - c*X_t*Y_t)*(Px_t^2*R + 2*Px_t*Qx_t*H + Qx_t^2*S)+IM}$$

$$Py_{t+1} = \frac{(e*X_t*Y_t) * (Py_t^2*R + Py_t*Qy_t*H) + Pim*IM}{(e*X_t*Y_t) * (Py_t^2*R + 2*Py_t*Qy_t*H + Qy_t^2*S)+IM}$$

X_t	number of prey at time t (at t=0 : $X=5 * 10^5$)
Y_t	number of predators at time t (at t=0 : $Y=5 * 10^4$)
Px_t	resistant allele frequency of prey at time t (at t=0 : $Px= 10^{-2}$)
Py_t	resistant allele frequency of predator at time t (at t=0 : $Py= 10^{-2}$)
Qx_t	sensitive allele frequency of prey at time t, ($Qx = 1-Px$)
Qy_t	sensitive allele frequency of predator at time t, ($Qy = 1-Py$)
IM	number of immigrants per generation
$Pim = 10^{-2}$	resistant allele frequency in immigrants
$a = 3$	maximum rate of increase of prey
$b = 2 * 10^{-6}$	selflimiting factor in prey
$c = 2 * 10^{-5}$	decrease per prey per predator due to predation
$e = 2 * 10^{-6}$	increase per predator per prey due to predation ($e=0.1*c$)
$R = 1$	probability of survival of RR (homozygote resistants)
$S = 0$	probability of survival of SS (homozygote sensitives)
$H = 0$ to 1	probability of survival of RS (heterozygotes), depends on pesticide dose (high dose = $H < 0.5$; low dose = $H > 0.5$)

RETARDATION AS A PROPERTY OF PREDATOR-PREY INTERACTION

If there is no immigration a significant retardation of the resistance development of the predator population can be seen . Although already after two generations the resistant allele frequency of the predator (Py) becomes greater than 0.5, it takes about 50 generations to reattain a population size which is greater than 50 % of the original predator population size (figure 1). So, only after 50 generations the predator population is called resistant. This long period is caused by the very small predator population after the first pesticide applications. Thus, a priori we would expect a faster resistance development in the predators if there were some immigration.

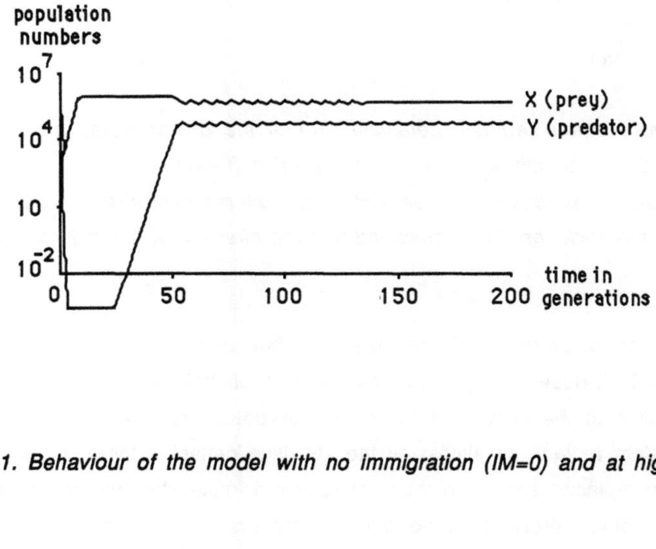

FIGURE 1. Behaviour of the model with no immigration (IM=0) and at high pesticide dose (H=0.2).

IMMIGRATION : STABILITY OF NON-RESISTANCE

However, immigration together with high dose of pesticide (H<0.5) gives rise to three stable equilibria in the system, as static analysis reveals (figure 2a).

There is an equilibrium at which both prey and predator are resistant (present in high numbers, high resistant allele frequency) (R-R) and a second one at which both predator and prey are sensitive (present in low numbers, low resistant allele frequency) (S-S). But there is also an equilibrium at which prey is resistant (large population, high resistant allele frequency), while predator is sensitive (small population, low resistant allele frequency) (R-S equilibrium).

Dynamical analysis shows that this last equilibrium (R-S) is reached under most conditions. The first equilibrium (R-R) is only reached at an unrealistic low immigration ($IM<5*10^{-9}$), whereas the second one (S-S) is only reached if there is very high immigration ($IM>5*10^{3}$).Thus with arbitrary small immigration, resistance in predators will never develop (figure 2b).

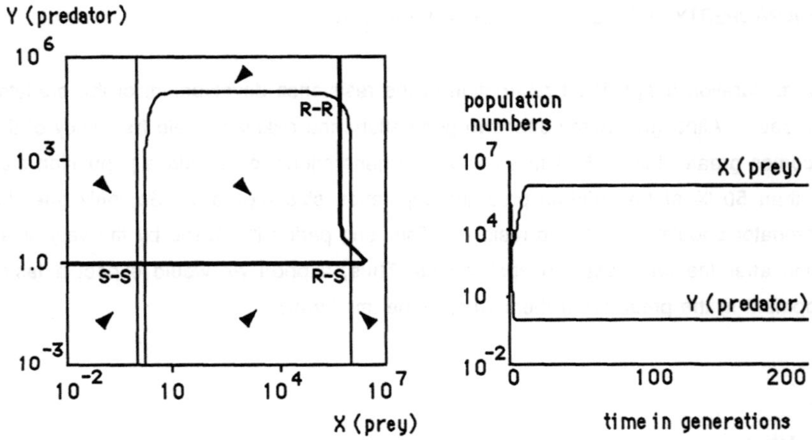

(a.) (b.)

FIGURE 2. 2D-zero-isoclines (a.) and behaviour (b.) of the predator-prey model at high pesticide dose (H=0.2) and with arbitrary low immigration (IM=1).
For the static analysis we assume quasi steady state for both resistant allele frequencies. The predator isocline is the thick line. The arrows indicate the direction of the trajectories.

The mechanism which is responsible for this result is often studied with one-population models and referred to as the 'high dose strategy' (Comins, 1977; Tabashnik & Croft, 1982, 1985; Taylor & Georghiou, 1982). Due to the high dose (most heterozygotes are killed) resistant genes are removed from the system as well as almost all the sensitive genes. If there is immigration, the sensitive genes will be replaced each time they are removed, meanwhile the resistant genes will slowly disappear. Of course there must be enough immigrants to produce this result. The immigration level in one-population models and the immigration level for the prey (in our model) are often too high ($IM>5*10^3$), but predator size decreases so dramatically after the first pesticide applications that a sufficient immigration level for predators is very low ($IM>5*10^{-9}$).

PARAMETER SENSITIVITY

Rate of increase

Simple models in the form of difference equations often show chaotic behaviour if the rate of increase becomes higher. So it is possible that overshoots of prey population size will result in a bypass of the R-S equilibrium. The results of analysis using higher rates of increase (up to 10) turn out to be the same. The equilibrium is still reached, no bypassing occurs.

Prey equilibrium size

In the basic model the initial prey equilibrium size is 50% of the carrying capacity. The R-S equilibrium disappears if this ratio becomes too small (less than 30%), as is illustrated in figure 3.

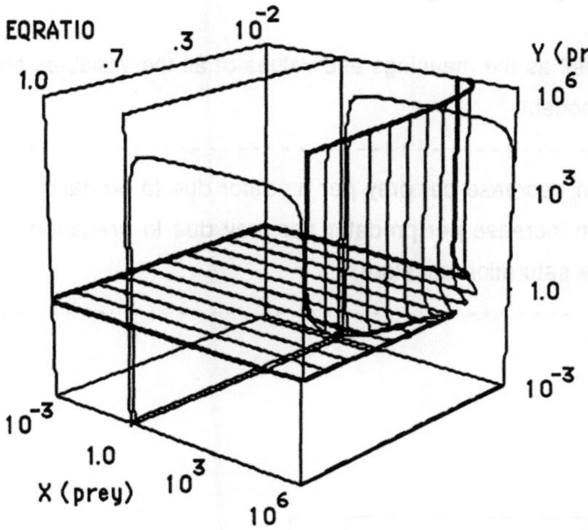

FIGURE 3. Presence of the R-S equilibrium in relation to the ratio of prey equilibrium size/prey carrying capacity (EQRATIO).
Predator and prey isoclines in 3D space of predator, prey and EQRATIO. The predator isocline plane is shaded and is thickly drawn at intersections with the cube. Its intersection with the X=10^6 plane shows at which values of EQRATIO there is an R-S equilibrium.

MODEL SENSITIVITY

Holling type 2 response

An obvious extension of an insect predator-prey model is to use the Holling type II functional and numerical response (May, 1975). The altered equations and parameters are presented in table 2. As "external equivalence" (Irvine & Savageau, 1985) between this model and the basic model, we used : equal prey carrying capacity, prey and predator equilibrium size and predator efficiency (g = 0.1 * f).

Static analysis shows that the previous results are even more robust in this model (figure 4). The R-S equilibrium cannot be bypassed, since the zone of decreasing predator population extends to an arbitrary large prey population. Furthermore, the presence of this equilibrium does not depend on the prey equilibrium size/carrying capacity ratio.

Table 2 Holling type II model

$$X_{t+1} = (a*X_t - b*X_t^2 - f*(X_t/(k+X_t)*Y_t)*(Px_t^2*R+2*Px_t*Qx_t*H+Qx_t^2*S)+IM$$

$$Y_{t+1} = (g*(X_t/(k+X_t)*Y_t) * (Py_t^2*R+2*Py_t*Qy_t*H+Qy_t^2*S)+IM$$

Equations for Px and Py as well as the meanings and values of all the variables and parameters are the same as in the main model.

f = 20	maximum decrease per prey per predator due to predation
g= 2	maximum increase per predator per prey due to predation
$k = 5 * 10^5$	response saturation constant

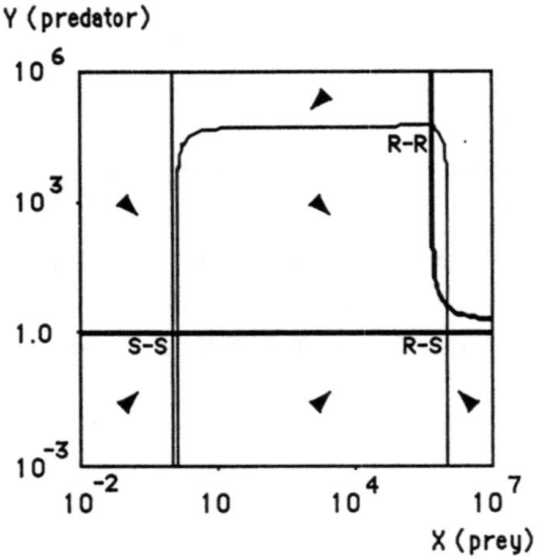

FIGURE 4. 2D-zero isoclines of the model including Holling type 2 response, at high pesticide dose (H=0.2) and with arbitrary low immigration (IM=1).
Compare with figure 2a.

Other extensions

We studied the Nicholson-Bailey model (Nicholson & Bailey, 1935; May, 1981) and a model with alternative prey, but our results were essentially the same as those obtained with the basic model.

CONCLUSIONS

In our basic model the lack of resistant predators is a stable phenomenon. Moreover it is a stable property of the model and remains valid for a large class of models and for a large class of parameter values. In the case of Holling type II response the results become even more robust. This suggests that the observed lack of resistant predators in the field is not a temporary phenomenon but will persist.

ACKNOWLEDGEMENT : We thank Miss S.M. McNab for linguistic advice.

REFERENCES

COMINS, H.N. (1977). The development of insecticide resistance in the presence of migration. J.Theor.Biol. 64 : 177-197.

CROFT, B.A. & A.W.A. BROWN (1975). Responses of arthropod natural enemies to insecticides. Ann.Rev.Entomol. 20 : 285-335.

CROFT, B.A. & K. STRICKLER (1982). Natural enemy resistance to pesticides : documentation, characterization, theory and application. In : Pest Resistance to Pesticides, G.P. Georghiou and T. Saito eds., New York - Plenum, p. 669-702.

DE BOER, R.J. (1983). GRIND : Great integrator differential equations. Bioinformatics Group, University of Utrecht, The Netherlands.

GEORGHIOU, G.P. (1986). The magnitude of the resistance problem. In : Pesticide Resistance : Strategies and tactics for management, National Academy Press, Washington D.C., p. 14-43.

GEORGHIOU, G.P. & C.E. TAYLOR (1986). Factors influencing the evolution of resistance. In : Pesticide Resistance : Strategies and tactics for management, National Academy Press, Washington D.C., p. 157-169.

IRVINE, D.H. & M.A. SAVAGEAU (1985). Network regulation of the immune response: alternative control points for suppressor modulation of effector lymphocytes. J.Immunol. 134 : 2100-2116.

MAY, R.M. (1975). Stability and complexity in model ecosystems. Princeton University Press, Princeton, New Jersey.

MAY, R.M. (1981). Theoretical Ecology - Principles and applications, Blackwell scientific publications, Oxford.

MORSE, J.G. & B.A. CROFT (1981). Developed resistance to azinphosmethyl in a predator-prey mite system in greenhouse experiments. Entomophaga 26 : 191-202.

NICHOLSON, A.J. & V.A. BAILEY (1935). The balance of animal populations. Part I., Proc.Zool.Soc.Lond.,551-598.

ROUGHGARDEN, J. (1979). Theory of population genetics and evolutionary ecology : an introduction. MacMillan Publishing Co., Inc., New York.

ROUSH, R.T. & B.A. CROFT (1986). Experimental population genetics and ecological studies of pesticide resistance in insects and mites. In : Pesticide Resistance : Strategies and tactics for management, National Academy Press, Washington D.C., p.257-270.

TABASHNIK, B.E. & B.A. CROFT (1982). Managing pesticide resistance in crop-arthropod complexes : Interactions between biological and operational factors. Environ.Entomol. 11 : 1137-1144.

TABASHNIK, B.E. & B.A. CROFT (1985). Evolution of pesticide resistance in apple pests and their natural enemies. Entomophaga 30 : 37-49.

TAYLOR, C.E. & G.P. GEORGHIOU (1982). Influence of pesticide persistence in evolution of resistance. Environ.Entomol. 11 : 746-750.

Collective Intelligence in Evolving Systems

H.-P. Schwefel

Fachbereich Informatik, Universität Dortmund,
Postfach 500500, D-4600 Dortmund 50, Fed. Rep. of Germany

1. Introduction

Ecosystems, comprising diverse living beings with complex, ever changing, compartments and interrelationships, cannot be handled sufficiently by the same kind of models as mechanical systems. Static or equilibrium models may describe short term adaptation phenomena adequately, but in the long term the openness of ecosystems allows them to reach ever new states and structures. Only evolutionary paradigmata can help in understanding ecodynamics and in developing adequate adaptive management strategies. This has been emphasized by biologists like Dobzhansky [1] as well as social scientists like Boulding [2]. Eigen and Winkler-Oswatitsch [3], moreover, have shown how to interprete natural phenomena in the framework of evolutionary chance- and-necessity games. Even if the model, presented here, arose from the inverse goal to use nature's learning strategy for technical meliorization, it may serve as well to learn about the learning process of ecosystems by comparing the effectiveness of variants of the evolutionary strategies. There have been several attempts to do so [4,5,6,7].

2. Evolution: Search for Improvement

A group or population consists of different individuals, the phenotype of each of which may be represented by a state vector x with n components x_1 to x_n . The group thus forms a 'cloud' in $\mathbb{R}^n$ which moves from one generation to the next according to mutations Δx with components Δx_i, and selection. The fitness function F(x) here will be taken as constant over time. All observations in the following do not rely, however, upon the specific functions used, but are more general in nature.

Mutations in a descriptive, not explicative, way are modelled as random samples from a normal distribution with zero mean and a standard deviation σ . In order to model the genetically programmed role of repair enzymes, σ is added to the set of individual characteristics x, and mutated as well, using a logarithmic normal distribution in order to avoid a drift over time which is not caused by selection.

Later on, individual σ_i for all Δx_i separately will be taken into consideration, and even covariances of the mutation vectors Δx will be allowed for. The latter are realized by means of a linear coordinate transformation of the Δx_i with respect to corresponding angles of rotation α_j - which again become genetically controlled individual attributes. Thus not only a proper scaling can be learned but even a positive definite metric. Another type of variability emerges as soon as one takes into account sexual propagation with recombination, may it be intermediate or discrete.

3. Selection schemes

Throughout this paper it will be assumed that the population size is constant, i.e. the numbers of parents (μ) as well as of descendants (λ) do not change from one generation to the next.

The simplest version of the evolution strategies (E.S.) was called two-membered or (1+1)- E.S. [6], since μ=1 parent is allowed to have λ=1 offspring within one generation, selection letting survive the fitter of both $\mu+\lambda=2$ individuals to the next time step according to a very strict imitation of the misleading slogan 'survival of the fittest'. More general is a ($\mu+\lambda$)- E.S. [7,8] with both μ and λ greater than one. Special cases are (μ+1)- and (1+λ)- versions.

As soon as one takes into account the limited life span of each individual, one comes to another class of strategies which will be termed (μ,λ)- E.S.. Now λ must be greater than μ, since only the μ best of the λ offspring become parents of the next generation. For $\mu=\lambda$ the evolutionary process would turn out to be a random walk, and for $\lambda<\mu$ the population would die off. Whereas with a ($\mu+\lambda$)- E.S. an improvement never will go lost, intermediate deterioration might happen in the (μ,λ)-case.

4. Numerical experiments

As far as possible, analytical results for the convergence rates of such evolutionary processes in typical situations, i.e. with respect to typical fitness functions F(x), were derived. Rechenberg [6] treated the (1+1)- E.S., Schwefel [7,8] the (1+λ)- and (1,λ)- versions. In case of $\mu>1$ so far only computer experiments are available. In the following, three functions F(x), to be minimized, are used

$$F_1(x) = \sum_{i=1}^{n} x_i^2 \;; \qquad F_2(x) = \sum_{i=1}^{n} i\, x_i^2 \;; \qquad F_3(x) = \sum_{i=1}^{n} \Big(\sum_{j=1}^{i} x_j \Big)^2 .$$

Instead of looking to the final results of the minimization process, attention was given to the convergence velocities under different mutability and selection schemes, and to the steadiness of that process. In all three cases both quantities heavily depend on the self-learning capability of the simulated populations. The following properties can/must be learned:

F_1 : one common standard deviation σ (mean step size of mutations),

F_2: a proper scaling, i.e. individual σ_i, i=1...n ,

F_3: a proper metric, i.e. individual σ_i as well as n (n−1)/2 covariances or angles α_j of inclination of the hyper-ellipsoids which form the constant density surfaces of the mutation probability.

For F_1 and F_2 n was taken to be n=30, for F_3 n=10. Using smaller problems , the results do not show the characteristics clearly enough, and they largely depend on the pseudo-random numbers used. But with n that large, the underlying randomness is averaged sufficiently.

5. Results

No effort was spent to include the simplest (1+1)- E.S., since in that case only exogenous control of the variance would be possible, i.e. by adjusting one common σ to a specified success/failure rate [6].

The first test with F_1 was performed to compare a (1+10)- with a (1,10)- strategy. At first glance the latter should be slower, since regress is possible if all 10 descendants turn out to be inferior to their ancestor. The opposite is true, however, when the variance σ^2 of the mutations is adapted permanently during the search process. If it happens that an offspring gains a superior position with a badly adapted step size and is allowed to become parent for further generations until one offspring surpasses his fitness, this hampers the combined search and adaption process. It is better to forget that superior position together with the unsuitable step size. With finite life span the evolution is more gradual as well as quicker overall, even if inbetween regress may occur.

In the second test F_2 was used as objective function. Now not only one common step size σ has to be learned but a proper combination of n=30 different ones. Only if the learning process of that scaling problem is tackled, then evolution speeds up sufficiently. A (1,10)- E.S. now gets stuck prematurely due to overadaptation of the scaling parameters. Since the progress rate generally decreases with the number n of variables involved, selecting the best of λ=10 offspring as the one and only parent of the next generation gives an opportunistic advantage to those descendants which search in a subspace $\mathbb{R}^m \subset \mathbb{R}^n$. But, reducing largely some of the step sizes σ_i, in the long term leads to a relative optimum in that subspace only, since one or even more degrees of freedom are omitted. Then a drastic rescaling would be necessary to make any further progress, and this turns out to be a rare event. Increasing λ alone does not help anyhow. Only if λ as well as μ are increased simultaneously, the situation can be improved. Not only the best, but others too, out of the set of offspring must become parents of the next gerneration. 'Survival of the fittest' once again turns out to be a bad advice, or a misinterpretation of evolution.

In order to enhance further the learning-by-doing of the proper scaling, recombination must be applied. Since none of the offspring has got a correct scaling information or 'internal model' of its topological environment, a genetic mix of that dispersed knowledge turns out to be of utmost importance. This gives a hint as to why sexual propagation has been introduced rather early during organic evolution, and has persisted thereafter. Evolution itself has learned to improve its own principles, one might say.

With a third test series it was studied whether there exists an optimum with respect to the selection pressure, and how far a learning population could approach an evolutionary performance with pre-known proper scaling. Using F_2 again and holding $\lambda = 100$ constant, only μ was varied. With prefixed scaling the mean progress rates monotonically decrease with increasing μ, both for optimum scaling and thus maximum convergence rates, and for some arbitrary scaling which yields largely slower progress rates.

Allowing to learn the proper scaling by simultaneous mutations of all σ_i as well as all x_i within the multimembered (μ,λ)- E.S. including recombination, one finds a maximum rate of evolutionary progress for μ between μ=12 and μ=20. That optimum being rather flat, its slopes bend downwards if μ>20 as well as, and even more drastically, if μ<12. Before approaching μ=1 the process stagnates or even recedes, i.e. it exhibits a negative convergence rate. Overall, most surprising is the fact that the maximum progress rate for a learning population (in the vicinity of μ=15 out of λ=100) is nearly equal to that for μ=1 and prefixed optimum scaling, whereas in the latter case with μ=15 it is less than half of the maximum value. The collective learning process reveals an important synergistic effect, mediated by recombination. This cannot be modelled appropriately on a higher aggregated level.

The last test with fitness function F_3 should reveal the ability of the collective learning process to learn an even more appropriate 'internal model' of its environment by allowing for covariances of the phenotypic mutation steps. Now any of the mutually independent genetic parameter changes may influence all phenotypic changes Δx_i at the same time (pleiotropy) and vice versa (polygeny), a phenomenon really observed in organic evolution, which may be interpreted as some kind of learning of a locally-temporally adapted metric. The search effort of mutations thus may be concentrated towards those regions of $\mathbb{R}^n$ which ly uphill. One might compare this with anticipation guided by past experiences. Again, it turns out that the selection pressure must not be too strong in order to enable learning. And, most importantly, the search process really can be accelerated beyond that margin which is achievable by scaling only [9].

6. Conclusions

Self-learning of an internal model of the environmental response surface is possible within an evolutionary mutation-and-selection search game. Requisite to an efficient learning-by-doing process are the following ingredients:

- diversity of the population at every instant of time,
- greater chances for the currently better individuals or 'ideas' (which is obvious), but likewise, too:
- fertility of quite a large portion of the new mutants, not only the best ones (which is not so obvious, at first glance even counter-intuitive),
- forgetting of even the best old positions ,
- recombination of the knowledge gathered in the gene pool.

If the ability to 'grope in the dark' efficiently, i.e. to ride a ridge by simultaneously exploring it and learning an internal model of its topology, is called intelligence, then 'natural intelligence' turns out to be a collective phenomenon. 'Survival of the fittest' was found to be a misleading description of evolutionary processes. Since nobody is perfect, even the currently fittest may rely upon an internal model which is not adequate for further explorations. Mixing up fuzzy knowledge is the best advice.

7. References

[1] Dobzhansky, Th. (1962) Mankind Evolving, Yale University Press, New Haven, Connecticut
[2] Boulding, K.E. (1978) Ecodynamics, Sage Publications, Beverly Hills, California

[3] Eigen, M. and Winkler, R. (1975) Das Spiel, Piper, München
[4] Bremermann, H.J. (1962) Optimization through Evolution and Recombination, pp. 93-106 in: Selforganizing Systems, Yovits, M.C. et al. (eds.), Pergamon, Oxford
[5] Holland, J.H. (1975) Adaptation in Natural and Artificial Sytems, University of Michigan Press, Ann Arbor, Michigan
[6] Rechenberg, I. (1973) Evolutionsstrategie, Frommann-Holzboog, Stuttgart
[7] Schwefel, H.P. (1975) Evolutionsstrategie und numerische Optimierung, Dr.-Ing. Diss., Technical University of Berlin
[8] Schwefel, H.P. (1981) Numerical Optimization of Computer Models, Wiley, Chichester
[9] Schwefel, H.P. (1987) Collective Phenomena in Evolutionary Systems, paper presented at the 31st Annual Meeting 'Problems of Constancy and Change' of the Int'l. Society for General Systems Research, Budapest, June 1-5, 1987

Part II

Marine and Lake Ecodynamics

Biological-Physical Interactions in the Sea: Marine Ecodynamics

A. Okubo

Marine Sciences Research Center, State University of New York, Stony Brook, NY 11794, USA

Abstract

In this article the emphasis is on the importance of organism-organism interactions in examining the energy flow in marine ecosystems. The ecodynamical study must include physical, chemical, and biological processes as they are coupled in a nonlinear fashion. In particular, the fluid dynamical understanding of the flow pattern around organisms is needed for providing insights into the interaction processes.

1. INTRODUCTION

The interaction between physical and biological processes has been one of the main subjects in the marine ecosystem, where the primary and secondary producers are mostly planktonic and thus their distributions are subject to the flow of the environmental water. The distribution of marine planktonic organisms is usually patchy. This spatial heterogeneity occurs in many scales and depends upon the nature of the response of organisms to their environment.

Haury et al. /1/ conveniently categorize marine organism patchiness into six scales of spatial pattern, ranging from 10^4 km down to 1 cm. The smallest scale of the categories is micro-scale, having dimensions of 1 cm to 1 m. There is a great deal of evidence indicating persistence of micro-scale patches in the turbulent sea. Even smaller micro-scale patchiness of the order of 1 mm or less may occur in marine bacterial aggregations /2/. Behavioral factors are suggested to play an important role in the formation and maintenance of these micropatches. These patches may be of great significance in ecosystem function.

Recently much attention has been paid to planktonic organism interactions on the individual level. The time is now ripe to study the details of organism-organism interactions for understanding the energy flow in planktonic ecosystems. If these interactions of microscales of 1 cm down to 100μ really exist, then our traditional concept of planktonic community structure in the sea based on more or less continuous (on these microscles) distributions should be modified. This view is shared by Jackson /3/, and he further emphasizes the importance of theoretical calculations on the interactions simply because of the apparent difficulty of performing experiments on these microscales. The theoretical study must include physical, chemical, and biological processes in a fashion of nonlinear coupling. In this context the fluid dynamical understanding of the flow pattern around marine organisms will provide better insights into the interaction processes.

2. FLUID DYNAMICS RELEVANT TO MARINE ORGANISMS

In almost all motions in marine ecology we may assume that the environmental flow is incompressible and Newtonian and that variations in the fluid density are neglected in so far as they influence the inertia. For the movement of marine organisms the effect of the Coriolis force can be ignored in comparison with either the inertial or viscous forces /4/. Thus the equations of motion for the environmental water read

$$\partial \underset{\sim}{u}/\partial t + \underset{\sim}{u}\cdot\underset{\sim}{\nabla}\underset{\sim}{u} = -\rho_0^{-1}\underset{\sim}{\nabla}p + \nu\nabla^2\underset{\sim}{u} + (\rho-\rho_0)\underset{\sim}{g}/\rho_0 \qquad (1)$$

$$\underset{\sim}{\nabla}\cdot\underset{\sim}{u} = 0 \qquad (2)$$

where $\underset{\sim}{u}$: fluid velocity, p: pressure, ρ : fluid density, ρ_0 : reference density, $\underset{\sim}{g}$: acceleration of gravity, ν : kinematic viscosity of sea water. The buoyancy term, i.e. the last term on the right-hand side of (1) plays no role in our present problem since the term could be absorbed into the pressure gradient therm /5/, /6/.

A marine organism is usually considered as an entity separated from the environmental water and can, for the most part, be taken as an inpenetrable region on whose boundary the fluid has no slip, i.e. no relative velocity at the organism surface /6/.

The relative importance of the inertial and viscous forces is measured by the Reynolds number expressed by $R_e = UL/\nu$, where U and L are certain representative scales of velocity and length in the problem. Reynolds numbers associated with movement of marine organisms are estimated for a size range from bacteria to whales given by /4/

$$R_e = 270\, L^{1.86} \quad (L \text{ in cm}) \qquad (3)$$

Thus the movement of bacteria ($0.1\mu \leq L \leq 1\mu$ in the order of magnitude) and phytoplankton ($1\mu \leq L \leq 100\mu$) is characterized by the low Reynolds number flow, where viscous forces dominate inertial forces. On the other hand, the movement of zooplankton (typically L~1 mm or larger) ranges from low to intermediate Reynolds numbers. Zooplankters whose length is about 1 mm experience Reynolds numbers of the order of unity in their swimming behavior.

The knowledge of the flow pattern around a marine organism is indispensable for calculating the concentration distribution of a chemical substance released from a plankton cell, movement of a planktonic prey particle in the feeding current of zooplankton, the orientation of bacteria toward a phytoplankton cell, among others. The problem of the concentration distribution of a chemical is mathematically treated by the advection, diffusion, and reaction equation such as

$$\partial C/\partial t = -\underset{\sim}{u}\underset{\sim}{\nabla}C + D\nabla^2 C + R_a \qquad (4)$$

where $\underset{\sim}{u}$: advection velocity, D: molecular diffusivity and R_a represents the reaction rate.

3. MICROSCALE INTERACTION OF ZOOPLANKTON AND PHYTOPLANKTON

Recent progress in microcinematographic techniques has allowed us, for the first time, direct observations of small zooplankton (copepods) feeding on even smaller algae /7/, /8/, /9/, /10/. These observations reveal that instead of the traditional view of passive filter feeding, the copepod is capable of perceiving the presence of an algal particle at a distance of 0.1-1 mm away, reorienting itself, and adjusting the feeding current with respect to the prey particle so as to draw it into the current more efficiently. Chemoreception and mechanoreception have been suggested for copepod's sensory mechanisms controlling particle detection and size selection. Both mechanisms involve fluid mechanical considerations of the animal's behavior in the associated flow characteristic of low Reynolds numbers.

A phytoplankton cell is surrounded by an "active space" /11/ within which the concentration of simple sugars, amino acids and other excudates exceed their background levels (see also Section 4). Chemoreception of zooplankton depends primarily upon the chemical leaked from an algal cell. The active space will be deformed by the presence of a shear flow associated with the feeding current, and initially spherical isoconcentration surfaces will be elongated in the direction of the flow to such an extent that the feeding animal may be able to detect the presence of prey by chemoreception at a distance.

This problem is mathematically treated with the advection and diffusion equation of the chemical concentration C

$$\frac{\partial C}{\partial t} = -\underset{\sim}{u}\,\underset{\sim}{\nabla} C + D\nabla^2 C \tag{5}$$

which can be solved under given feeding current fields. This model predicts the distance of detection as a function of the prey size given the concentration of detection threshold. An example of calculation shows the detection distance of about 1 mm, which agrees well to the observed distance of early detection. One of the weak points in this model is the assumed value of the threshold concentration, which seems rather small compared with concentrations of amino acids for behavioral responses of the animal /12/.

The other mechanism of prey detection, i.e. mechanoreception has been proposed by Legier-Visser et al. /12/. It is based on the pressure field induced by a prey particle. For a spherical particle embedded in a low Reynolds number flow, the Stokes solution of the pressure field gives

$$\Delta p = 3\eta\, a\, U \cos\theta / r^2 \tag{6}$$

where Δp: pressure difference between flow in the absence of particle and flow deformed by the presence of particle of which radius is a, η : viscosity of water, θ:angle between the particle and the mechanoreceptor of copepod, r: distance from the center of the particle to the receptor of copepod measured perpendicular to the antenna, U: flow velocity far away from the particle.

One interesting feature results from this model. Letting d be the distance from the forward edge of the wall of the approaching cell of phytoplankton to the receptor on the antenna, we can rewrite (6) and solve for d,

$$d = (k/\Delta p_m)^{1/2} a^{1/2} - a \qquad (7)$$

where

$$k = 3\eta U \cos\theta \qquad (7')$$

For a given minimum pressure sensitivity of copepod mechanoreceptor, Δp_m, and k, the distance d increases with the prey size a, reaches a maximum value, and decreases toward zero as the prey size increases. Thus we expect that copepods would have the greatest detection ability in a narrow range of particle size. This maximum detection distance and optimal size of prey are determined in part by the intensity of the feeding current which may, in fact, be altered by the animal /12/.

Gill /13/ examined by scanning and transmission electron microscopy the fine structure of the antennule, setae and pegs of a copepod Temora longicornis Muller and suggested the antennule to be the major site for mechanoreception in copepods, although mechanoreceptors might also exist on other parts of the body.

Another type of microscale interaction between zooplankton and phytoplankton has become one of the recent issues. McCarthy and Goldman /14/ suggested that microscale patches of nutrients excreted by zooplankton should be an important nutrient source for phytoplankton in the oligotrophic ocean, e.g. central oceanic regions, where nitrogeneous nutrients in the near-surface waters in a large scale are below the limit of detectability even though considerable photosynthetic activity has been noted. To resolve this dilemma of "missing mass" in the ocean, it is hypothesized /14/ that small-scale nutrient plumes or patches excreted locally by zooplankton should be continuously utilized by adjacent phytoplankton cells at rates that could keep the concentration averaged over large water volumes at the level below detection limits.

Mathematical modelling the nutrient plumes or patches depends upon the advection-diffusion-reaction equation for the concentrations of nutrient excreted from zooplankton

$$\frac{\partial C}{\partial t} = -u\frac{\partial C}{\partial x} + D\nabla^2 C - V_m P \frac{C}{C+K} \qquad (8)$$

where C: nutrient concentration, u: advection velocity in x-direction behind the zooplankter, D: molecular diffusivity, V_m: maximum uptake rate, P: concentration of phytoplankton cells, K: half-saturation constant for uptake. The assumption of molecular diffusion (rather than turbulent diffusion) might be appropriate for the low Reynolds number regime.

Calculations based on (8) can be performed for two extreme cases of excretion. One is a continuous release of constant rate, and the other is an instantaneous release (once a day) with the same total amount of excretion. By choosing the threshold nutrient concentration to be 10 times the background level, we can estimate the size of plume or patch effective to phytoplankton uptake.

The calculation shows that, for a continuous release, the effective volume of nutrient plume occupies only about 0.01% of total volume of sea water for a mean zooplankton density of one animal per liter. For an instantaneous release once a day, on the other hand, the calculation results in the effective volume of 4.5% of the total with one animal per liter.

The real situation must be an intermittent excretion by zooplankton. Therefore the value of the effective volume of nutrient patch should be somewhere between these two limits. Above all, turbulent diffusion is ignored in these calculations, so that the calculated values should be considered as optimistic.

According to /14/, phytoplankton need to be exposed to a high concentration of ammonia during only about 3% of a doubling period of approximately one day. The caluculations presented here, even though ignoring turbulent diffusion, do not seem to support McCarthy and Goldman's hypothesis. Jackson /15/ pointed out by a similar calculation that such micropatches of nutrient should dissipate by diffusion too quickly to be used by phytoplankton. On the other hand, Lehman and Scavia /16/ show both theoretically and experimentally that algae can exploit micropatches of dissolved phosphate produced by zooplankton. Currie /17/ argued that a nonhomogeneous nutrient supply regime should actually decrease phytoplankton growth rates, regardless of the physical characteristics of the patches. The controversy is by no means settled yet.

In addition to the problem of diffusion the patch encounter rate of algal cells must be studied in more detail by taking into account the flow field around a zooplankter. In this context Strickler /8/ shows using high speed microcinematography that feeding currents of copepods result in the concentration of algal cells near the animal and most of these cells appear to survive the encounter. This implies that phytoplankton cells are favorably entrained into the excreted micropatches by the feeding current /14/.

4. INTERACTION OF PHYTOPLANKTON AND BACTERIA AND BACTERIAL BEHAVIOR IN THE MICROZONES

Marine bacteria occur mostly free in the environmental water rather than attached to suspended particles such as plankton cells. Nonetheless many studies show that peaks of bacterial cell number in the depth profile often coincide with those of phytoplankton cell numbers /18/. Also bacteria productivity closely followed phytoplankton productivity /19/.

Azam and Ammerman /20/ noted that marine bacteria appear to grow rapidly in the environmental water, although the background concentration of dissolved organic matter is too low to support significant bacterial growth. They suggested that there should exist microzones around phytoplankton cells having utilizable dissolved organic carbon orders of magnitude above background concentrations and that bacteria should cluster in the microzones.

Phytoplankton cells release products of photosynthesis at a variable rate (1 - 70%), and culture experiments demonstrated that bacteria readily utilized a significant portion of the released organic matter as source of nutrients and energy for their growth and metabolic activities. The released organic matter can return to the food web through utilization and transformation by bacteria and thus an intimate algal- bacterial interaction may take place.

Mitchell et al. /21/ estimated the size of the microzones on the basis of advection and diffusion equation, Eq. (5). For a steady-state distribution of nutrient around a phytoplankton cell at rest, the radius of the microzone R* is estimated by

$$R^* = \frac{Q}{4\pi D C^*} \tag{9}$$

where Q : the total dissolved organic carbon (DOC) flux per cell and C* : the threshold concentration taken as 10% above the background concentration. Given Q = 10^{-17} mol/sec, D = 10^{-5} cm^2/sec and C* = 10^{-12} mol/cm^3, it is calculated /21/ as

$$R^* = O\ (1\ mm) \tag{10}$$

Note that (10) also gives the maximum extent of the microzone behind a sinking phytoplankton cell. Solving (5) with $\underset{\sim}{u} = w_s \underset{\sim}{k}$ (w_s : settling velocity, $\underset{\sim}{k}$: unit vertical vector), we obtain the concentration of DOC around a sinking cell

$$C(\underset{\sim}{x}) = (4\pi D R)^{-1} Q \exp\left\{-w_s (R - z)/2D\right\} \tag{11}$$

where $x^2 + y^2 + z^2 = R^2$ and the cell is sinking in z-direction.

Interestingly enough the value of the microzone range R* is nearly the same as the length scale of Kolmogorov micro-eddies, i.e. the smallest modes of turbulence found in the upper mixed layer of the ocean /21/. However, in the deep waters of the oceans and in the thermocline below the upper mixed layer, low-level turbulent conditions prevail and the characteristic length scale of the Kolmogorov eddies is of the order of 1 cm, and the deep water, in particular "sheets" and "layers" composing the thermocline are the most likely location for microzones /21/.

Yet the mechanism of clustering of bacteria around phyplankton cells is still unknown. Chemotaxis has been proposed as a possible mechanism for bacterial clustering in the vicinity of algal cells /20/. Jackson /3/ has recently developed a computer model which simulates the behavior of marine bacteria around an algal cell by means of chemotaxis. The simulation study concludes that bacteria initially uniformly distributed around a relatively large cell at rest can aggregate toward the cell. Thus, for a 10μ radius alga, increased leakage rates of DOC cause increased drift toward the alga, but for a 2.5μ radius alga, there is no drift toward the cell even for high leakage rates. This suggests that there is a minimum size of alga below which chemotactic behavior cannot be used for bacteria to sense and cluster. For large cells, on the other hand, their sinking velocities can exceed bacterial drift due to chemotaxis so that bacteria may not be able to catch up the falling cells /21/. Thus there is also a maximum size of alga beyond which bacteria will be left behind. In the oceanic mixed layer an average non-motile phytoplankter, such as a diatom, with a mean diameter of 30μ, sinks at a rate of about 10μ/sec /22/, which is about three times the speed of a bacterium moving chemotactically toward a nutrient source /21/. This suggests that the range of phytoplankton cells appropriate for bacterial clustering by chemotaxis would be from 3μ to 30μ in diameter. However, the hydrodynamics of sinking particles is complex and should be more explicitly considered in a study of chemotactic behavior /3/, /23/.

Another possible mechanism for bacterial clustering is gyrotaxis. This mechanism is independent of chemotaxis. Gyrotaxis is a directed locomotion resulting from the orientation of cell's axis by compensating gravitational and viscous torques /24/, /25/. Consider a bacterium of spherical shape. Relative to the center of buoyancy or drag, the center of gravity of the bacterium is displaced due to an asymmetric distribution of heavy organelles in the cell. The force of gravity then exerts a torque so as to orient the cell axis vertically, and the bacterium tends to swim vertically upward. However, when the environmental water has spatial variation in its velocity, the bacterium experiences an additional hydrodynamic torque. At the balance of these two torques the following equation is obtained /24/

$$4\pi\eta a^3 \underset{\sim}{\omega} + \tfrac{4}{3}\pi a^3 \Delta\rho\, \underset{\sim}{L} \times \underset{\sim}{g} = 0 \qquad (12)$$

where a: radius of bacterial cell, $\underset{\sim}{\omega}$: vorticity of the environmental water, $\Delta\rho$: density difference between bacterial cell and water, and $\underset{\sim}{L}$: displacement vector from the center of buoyancy to the center of gravity. This equation determines the orientation of the cell axis and thus the direction of swimming.

Consider a situation where a phytoplankton cell of radius R is sinking vertically downward with a settling velocity of u_0, and at a great distance from the cell, bacteria are swimming vertically upward at a speed of v. The system is equivalent to that of the phytoplankton cell being at rest and the environmental water being moving upward with a speed of $u_0 + v$, in which bacteria are embedded. Application of Stokes' solution to this equivalent system yields the vorticity of the flow as

$$\omega = -\tfrac{3}{2}\, u_0 \,(R/r^2) \sin\theta \qquad (13)$$

where r and θ are radial and tangential coordinates, respectively, and the coordinate origin is taken at the center of the phytoplankton cell (Fig. 1).

Substitution of (13) into (12) results

$$\sin\phi = \frac{4.5\,\eta\, u_0 R}{\Delta\rho\, L\, g\, r^2} \sin\theta \qquad (14)$$

where ϕ is the angle of deviation of the bacterium cell axis from the vertically upward orientation, and the angle is taken in such a fashion that the bacterium tends to orient toward the phytoplankton cell (Fig. 1). As L is proportional to the size of organism, a, the gyrotactic orientation is more effective in microorganisms rather than the larger organisms. In fact, calculations using an appropriate range of the parameters /24/ show that bacteria of size 1μ would be able to orient toward a phytoplankton cell of size 30μ at a great distance of a few hundred micrometers.

This suggests that as a bacterium approaches closer to a phytoplankton cell by gyrotaxis, the viscous torque will dominate gravity and the bacterium will make a tumbling motion, so that the real trajectories of bacteria in orienting swimming may not be so simple as illustrated in Fig. 1. Another complication in gyrotaxis will arise from the fact that bacteria are generally of nonspherical shape /26/.

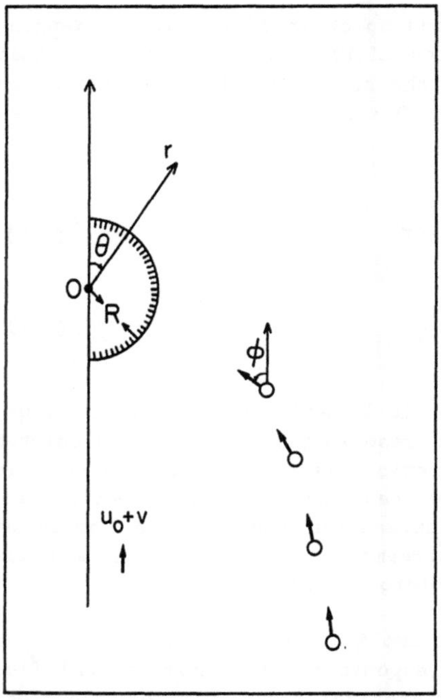

Fig. 1. Orientation of a bacterium toward a phytoplankton cell by gyrotaxis

A simple orienting effect can occur in any vorticity fields, provided the balance between gravity and viscous drag torques is established. For instance, self-focused bacterial streamers /24/ may spontaneously generate in intermittent fully developed turbulence in the sea /2/.

Local fluctuations in cell concentration produce locally higher concentrations of cell and the denser part sinks, thereby generating a fluid velocity field somewhat similar to that generated by a sinking cell of phytoplankton. By gyrotaxis more cells are drawn into this part of the fluid to accumulate. The process is characteristic of a positive feedback. Thus initially uniformly distributed cells tend to break up into downwelling columns (streamers) containing higher than average cell concentrations. The phenomenon of self-focused streamers is observed in a laboratory culture of marine bacteria (Dr. James Mitchell: personal communication). It may also occur in the sea, where small-scale turbulence is intermittent.

Turbulence consists of hierarchies of eddies of various sizes, where a small-scale eddy is generated from a larger eddy and distintegrated into smaller eddies (Cascade process). There is strong evidence that the small-scale structures of turbulence become less and less space filling as the scale size decreases. Let L be the size of the largest energy-continuing eddies. At each step of the cascade process any n-eddy of size $L_n = L_0 2^{-n}$ produces on the average N number of (n+1)-eddies. If the cascade process is space filling, $N = 2^3$. The β-model proposed by Frisch et al. /27/ assumes

$$N = 2^D \qquad (D < 3) \tag{15}$$

where D need not be an integer value; in fact, D is Mandelbrot's fractional dimension /28/.

At high Reynolds numbers there is a very broad spectrum of turbulence ranging from the energy-containing eddies to the Kolmogorov micro-eddies. This means that the spatial distribution of small-scale eddies in the sea is highly non-uniform and 'spotty' or 'intermittent'. Given the value of $D < 3$ we can estimate the mean distance of n-eddies (d_n) by

$$d_n = L_n 2^{(3-D)n/3} \quad \text{or} \tag{16}$$

$$d_n = L_o 2^{-Dn/3} \tag{17}$$

As a typical case of oceanic turbulence take L = 10 km and D = 2.66 /28/. Thus, for the Kolmogorov micro-eddies of size 1 cm, the mean eddy distance is calculated as 4.4 cm and for those of size 1 mm, the mean distance is calculated as 7 mm. In other words, the Kolmogorov micro-eddies in the sea are extremely spotty, and implications are that the space between the micro-eddies should be a domain of temporary quiescence, where the microorganism streamers may survive for a while before breaking up by the sporadic action of the micro-eddies.

In conclusion, gyrotaxis has very important implications for the ecology and evolution of aquatic bacteria and microalgae in response to the environmental flow field. Also it might have some relevancy to microalgal mass culture technology /25/, /29/.

5. CONCLUSION

Planktonic organisms in the sea are patchy on almost any temporal and spatial scales that we can imagine. At the smallest end of the scale (order of 1 mm or less), interactions and behavior of individual organisms play an important role in the ecosystem. Since the movement of these organisms is often controlled by the motion of the environmental fluid, we need to have biofluidmechanical understanding of the behavior of the organisms. Successful integration of the study of fluid motion in theoretical ecology will broaden our knowledge and gain insights into the dynamics of aquatic ecosystems.

REFERENCES

/1/ Haury, L.R., McCowan, J.A., and Wiebe, P.H.: Patterns and processes in the time-space scales of plankton distributions. In: Spatial pattern in plankton communities. Steele, J.H. (ed.). Plenum Press, New York, 1978, p. 277-327.

/2/ Mitchell, J.G.: A new mechanism for generating plankton heterogeneity on small scales. Ph.D. dissertation, State University of New York at Stony Brook, 1987.

/3/ Jackson, G.A.: Physical and chemical properties of aquatic environments. In: Ecology of microbial communities. Cambridge University Press, Cambridge, 1987, p. 213-233.

/4/ Okubo, A.: Fantastic voyage into the deep: marine biofluid mechanics. In: Mathematical topics in population biology, morphogenesis and neurosciences. Teramoto, E., and Yamaguti, M. (Ed.), Springer-Verlag, Heidelberg-New York, 1987, p. 32-47.

/5/ Batchelor, G.K.: An introduction to fluid dynamics. Cambridge University Press, Cambridge, 1981.

/6/ Childress, S.: Mechanics of swimming and flying. Cambridge University Press, Cambridge, 1981.

/7/ Paffenhöfer, A.G., Strickler, J.R. and Alcaraz, M.: Suspension-feeding by herbivorous calanoid copepods: a cinematographic study. Mar. Biol. 67 (1982) 193-199.

/8/ Strickler, J.R.: Calanoid copepods, feeding currents and the role of gravity. Science 218 (1982) 158-160.

/9/ Koehl, M.A.R. and Strickler, J.R.: Copepod feeding currents: food capture at low Reynolds number. Limnol. Oceanogr. 26 (1981) 1062-1073.

/10/ Price, H.J. and Paffenhöfer, G.A.: Effects of feeding experience in the copepod _Eucalanus_ _pileatus_: a cinematographic study. Mar. Biol. 84 (1984) 35-40.

/11/ Andrews, J.C.: Deformation of the active space in the low Reynolds number feeding current of calanoid copepods. Can. J. Fish. Aquat. Sci. 40 (1983) 1293-1302.

/12/ Legier-Visser, M., Mitchell, J.G., Okubo, A. and Fuhrman, J.A.: Mechanoreception in calanoid copepods: a mechanism for prey detection. Mar. Biol. 90 (1986) 529-535.

/13/ Gill, C.W.: Suspected mechano-and chemosensory structures of _Temora longicornis_ (Copepoda:Calanoida). Mar. Biol. 93 (1986) 449-457.

/14/ McCarthy, J.J. and Goldman, J.C.: Nitrogeneous nutrition of marine phytoplankton in nutrient-depleted waters. Science 203 (1979) 670-672.

/15/ Jackson, G.A.: Phytoplankton growth and zooplankton grazing in oligotrophic oceans. Nature 284 (1980) 439-441.

/16/ Lehman, J.T. and Scavia, D.: Microscale nutrient patches produced by zooplankton. Proc. Nat. Acad. Sci. USA 79 (1982) 5991-5005.

/17/ Currie, D.J.: Phytoplankton growth and the microscale nutrient patch hypothesis. J. Plankton Res. 6 (1984) 591-599.

/18/ Fuhrman, J.A., Ammerman, J.W. and Azam, F.: Bacterioplankton in the coastal euphotic zone: distribution, activity, and possible relationships with phytoplankton. Mar. Biol. 60 (1980) 201-207.

/19/ Albright, L.J.: Heterotrophic bacterial dynamics in the lower Fraser River, its estuary and Georgia Strait, British Columbia. Mar. Biol. 39 (1977) 203-211.

/20/ Azam, F. and Ammerman, J.W.: Cycling of organic matter by bacterioplankton in pelagic marine ecosystems: microenvironmental considerations. In: Flows of energy and material in marine ecosystems. Fasham, M.J.R. (ed.). Plenum Press, New York, 1984, p. 345-360.

/21/ Mitchell, J.G., Okubo, A. and Fuhrman, J.A.: Microzones surrounding phytoplankton form the basis for a stratified marine microbial ecosystem. Nature 316 (1985) 58-59.

/22/ Smayda, T.J.: The suspension and sinking of phytoplankton in the sea. Oceanogr. Mar. Biol. Ann. Rev. 8 (1970) 353-414.

/23/ Sanderson, B.G.: personal communication (1987).

/24/ Kessler, J.O.: Individual and collective fluid dynamics of swimming cells. J. Fluid Mech. 173 (1986) 191-205.

/25/ Kessler, J.O.: The external dynamics of swimming micro-organisms. Progress in Phycological Res. 4 (1986) 257-307.

/26/ Pedley, T.J. and Kessler, J.O.: The orientation of spheroidal microorganisms swimming in a flow field. Proc. Roy. Soc. Land. B231 (1987) 47-70.

/27/ Frisch, U., Sulem, P.-L. and Nelkin, M.: A simple dynamical model of intermittent fully developed turbulence. J. Fluid Mech. 87 (1978) 719-736.

/28/ Mori, H. and Fujisaka, H.: A simplified theory of intermittent fully-developed turbulence. In: Pattern formation and pattern recognition. Haken, H. (Ed.). Springer-Verlag, Heidelberg-New York, 1979, p. 82-89.

/29/ Soeder, C.J.: An historical outline of applied algology. In: CRC handbook of microalgal mass culture. Richmond, A. (Ed.). CRC Press Inc., Boca Raton, Florida, (1986) p. 25-41.

Understanding the Baltic Sea: Systems Ecology in Theory and Practice

F. Wulff

The Askö Laboratory, Institute of Marine Ecology,
University of Stockholm, S-10691 Stockholm, Sweden

INTRODUCTION

Studies on whole ecosystems have usually been motivated by man's growing ability to perturb larger and larger systems. Experience have shown that we cannot explain these changes based on studies on organisms and populations alone (Mann 1982). However, the theoretical basis for explaining natural processes becomes weaker and weaker as one moves up the ecological hierarchy, from individuals to populations and finally to whole ecosystems. Studies of whole ecosystems are usually both difficult and expensive and necessitate interdisciplinary cooperation of a nature that most ecologists, usually with their roots in "biological natural history", are quite untrained for. Ecosystem ecologists have often been criticised for a lack of good scientific foundations in their work (Lehman 1986). In the 1960's, many ecosystem projects were started where ecologists and systems engineers were working together hoping to develop predictions of ecosystem changes using powerful computers and large complex mathematical models. These reductionistic models had little practical success (Patten and Finn 1979) and it has become more and more apparent that a holistic approach, where the properties of the whole system is considered, is necessary (Platt *et al.* 1981)

I will here discuss some of the difficulties and progress encountered in marine ecosystem research by describing how the understanding of one particular system, the Baltic Sea, has evolved during the last decades. An exploited ecosystem such as the Baltic may offer unique opportunities for testing ecosystem level hypotheses (Legget *et al.* 1985).

THE BALTIC SEA

The Baltic Sea (Fig 1) is a large non-tidal estuary (373,000 km^2) with narrow and shallow connections to the Kattegat and the North Sea. The drainage area, inhabited by over 70 million people in seven countries, is about four times as large as the surface area. The system is composed of three major basins: the Baltic proper, the Bothnian Sea and the Bothnian Bay. The positive freshwater balance and the limited inflows of saltwater through the narrow and shallow connections to the ocean keep the system brackish. The salinity of the surface water is about 7 - 8 ^{o}/oo in the Baltic proper and decrease to less than 1 ^{o}/oo in the northern parts of the Bothnian Bay. The vertical salinity stratification is most pronounced in the Baltic proper where the salinity is between 10 - 15 ‰ below a permanent halocline, situated at about 65 meters depth. Ice usually covers the Bothnian Bay and Sea in winter but usually only the coastal zones of the Baltic proper. The Baltic communities are composed by a low diversity mixture of euryhaline marine and brackish estuarine species, freshwater species and glacial marine or freshwater relicts (Segerstråle 1957).

Fig 1. Map of the Baltic Sea. Dotted contour = 40 m, solid contour = 100 m depth. The insert shows the Danish sounds with the shallow and narrow entrances to the Baltic limiting the water exchange. The Baltic proper is the area between the Bothnian Sea and the Belt Sea, including the Gulf of Finland.

In the late 1960's the decrease in oxygen concentrations in the deep basins of the Baltic proper (Fonselius 1969) started a debate whether man, though eutrophication, or natural changes in the intermittent supply of oxygenated deepwater was the cause. During the 1970's, ecosystem oriented marine research projects were started around the Baltic with the aim to identify and quantify the major subsystems (Jansson 1972, 1984; Rumohr *et al.* 1987). These projects were stimulated to a great extent by the ecosystem approach advocated by the Odum brothers (e.g. Odum 1971a). Compartmental flow diagrams using the energy circuit language (Odum 1971b) were used as planning tools, prior to field and experimental studies aiming at quantifying biomasses, production and respiration rates etc. Early mathematical models (Sjöberg *et al.* 1972), based on existing knowledge, focused on the phosphorus and oxygen dynamics of the system. Although coarse, these models were valuable in the process of identifying critical areas for research, emphasising the need for a better understanding of biological as well as chemical and hydrodynamical processes, controlling this system. Although the Baltic has been considered as one of the best known marine areas in the world, we quickly became aware of the almost complete lack of quantitative information in many critical areas (Wulff 1976).

Time consuming methodological work, testing for example the methods to quantitatively estimate biomasses and production had to be made, before "good" data could be collected. It was obvious that the "Baltic problem" could not be solved by a single laboratory or country and the international scientific cooperation increased (Dybern 1980) and a joint monitoring programme was established (Hägerhäll 1980).

COMMUNITY STRUCTURE AND ENERGY FLOWS

Gradually, our knowledge of the community has reached a stage where it is possible to describe overall structures and trophic interactions in terms of energy flow diagrams.

Fig 2 (from Elmgren 1984) illustrates this. For the Baltic proper (Fig 2B) the overall net primary production is estimated to 160 g C $m^{-2}yr^{-1}$, of which phytoplankton production constitutes more than 95 %. Allochtonous inputs add another 5 g C $m^{-2}yr^{-1}$. About 15 % of the phytoplankton production is released as dissolved organic matter (DOM) which is rapidly utilized by pelagic bacteria grazed by flagellates, which are consumed by ciliates which in turn are eaten by larger zooplankton. Thanks to improved microbiological techniques, the classical notation of a phytoplankton-zooplankton-fish food web, has been replaced by a new more complex paradigm where bacterial production plays a central role (Azam *et al.* 1983). The Baltic food web shows strong couplings between the pelagic and benthic subsystem, through heavy sedimentation of organic matter after the plankton spring bloom and by filter feeding of mussels. The blue mussel (*Mytilus edulis*) totally dominates the hard bottom communities at all depths due to the lack of competitors and predators (excluded by the low salinity) that normally limits its distribution in more marine habitats (Kautsky 1981). The soft bottom benthic communities are dominated by a filter feeding bivalve (*Macoma balthica*) and by two deposit feeding amphipods (*Pontoporeia affinis* and *P. femorata*). Invertebrate benthic carnivores are scarce and echinoderms and decapod crustaceans are totally absent except in the southernmost part. Invertebrate plankton carnivores, primarily medusae (*Aurelia aurita*) in the southern parts and mysid shrimps in the coastal zones and in the northern areas might be responsible for the decline of the zooplankton populations in the

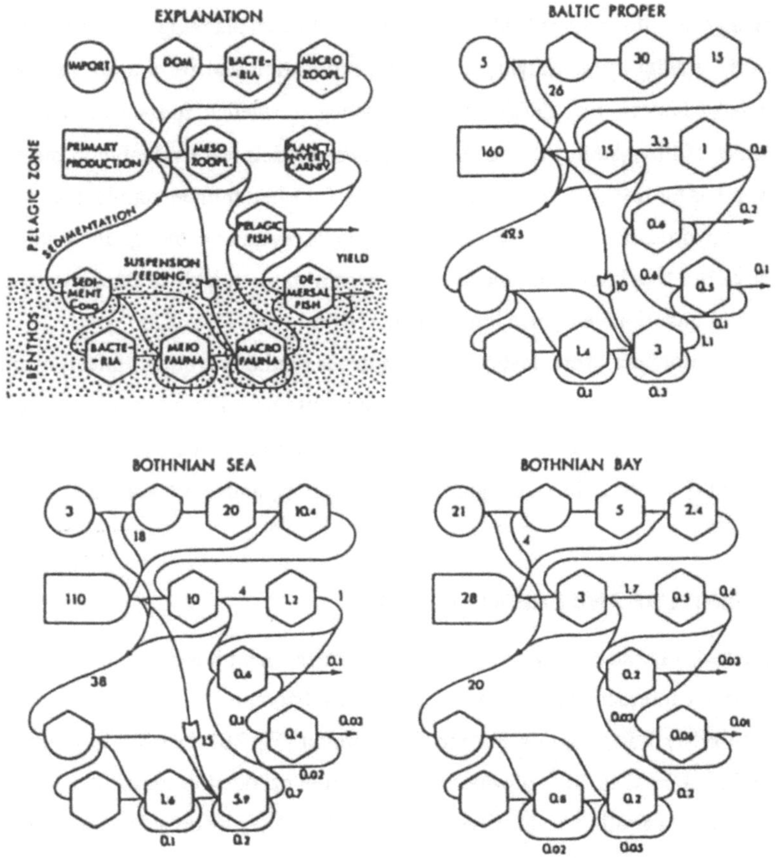

Fig 2. Energy flow diagrams of the Baltic Sea subareas, compiled by Elmgren (1984). Numbers (g C m^{-2} yr^{-1}) inside symbols are annual production estimates.

late summer and fall. The diagram for the Bothnian Sea (Fig 2B) is very similar to the one for the Baltic proper, with most figures scaled down by a factor of two-thirds. Since there are no anoxic areas devoid of macrofauna as in the Baltic proper, the benthic compartment is relatively more important here. Besides having a much lower overall productivity, the Bothnian Bay (Fig 2C) food web is very different because the lack of filter feeding benthos (excluded by the low salinity) eliminates an important efficient energy pathway. In a comparison of these Baltic flow diagrams with similar estimates from other marine areas, Elmgren (1984) concludes that the networks are strikingly similar. Only when a major component has been eliminated (the marine bivalves in the Bothian Bay) do distinct differences appear. Nixon *et al.* (1986) showed that the yield of fish per unit of primary production for all the Baltic subareas is very much in line with estimates from most other marine systems.

NUTRIENT BUDGETS

Knowing the major flows and stores makes it possible to ask questions about how these particular networks have evolved. What are the key processes controlling these flows? In order to understand what factors are controlling the overall productivity, a more holistic approach has been necessary where also the physical and chemical properties of the system are included. One essential cornerstone in such an approach is an overall nutrient budget of the system. The recent attempt, by Larsson *et al.* (1985) shows some critical features , including the impact of human activities (Table 1) on the Baltic Sea.

10^3 tons/year	Phosphorus	Nitrogen
External sources		
Municipal	+ 18.5	+89
Industrial	+ 2.5	+14
Rivers	+ 50.3	+635
Atmosphere	+ 5.5	+322
Internal sources and sinks		
Nitrogen fixation	-	+134
Denitrification	-	-470
Losses		
Sedimentation (net)	- 24	-60
Fishing	- 3	-30
Through the Danish sounds	- 1.5	-110
Net results		
Sources - sinks	+ 48.4	+523
Accumulation in the free water mass	- 15	-100
Unexplained net	33.4	423

Table 1. Nitrogen and phosphorus budgets for the Baltic Sea (from Larsson *et al.* 1985).

For nitrogen, almost 30% of a total input of 1.2 10^6 tons yr^{-1}, comes from direct precipitation. The atmospheric input is roughly equally divided between nitrate, originating from combustion of fossil fuels, and ammonia, originating from agriculture (intensive livestock farming). Thus, the classical sources of municipal, industrial and riverine inputs in the drainage area, have been supplemented by a new diffuse source adding pollutants from a much larger area. The internal source, through nitrogen fixation by bluegreen algae, is about equal to the direct municipal input. For phosphorus, inputs via rivers and municipal sources are the most important whereas industry and atmospheric fallouts are of minor importance.

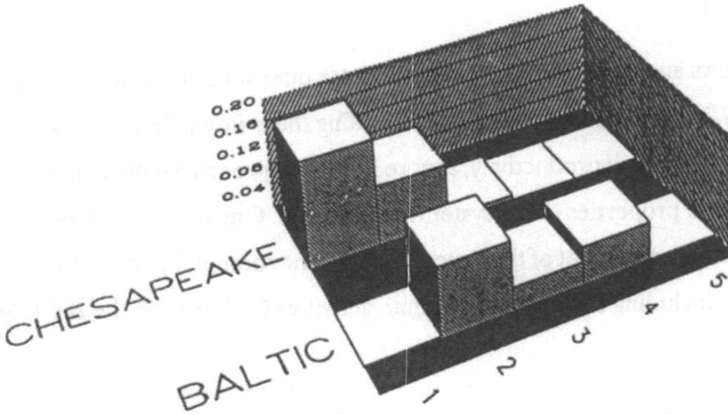

Fig 3. A comparison of recycling patterns in the Baltic proper and Chesapeake Bay (from Wulff and Ulanowicz 1988). The total amount of carbon flow (g C m^{-2} yr^{-1}) in cycles of different length (1-5), weighted for total system throughput (TST) in the two ecosystems.

Are there really no differences between the different marine areas or are our methods to measure and analyze them still to coarse? Techniques to analyze ecosystem flow data have been recognized as some of the most valuable tools recently developed within theoretical systems ecology (Ulanowicz & Platt, 1985). We have (Wulff and Ulanowicz 1988) tried some of these measures in a comparison of the Chesapeake Bay and the Baltic proper. We have used the same aggregation into 13 compartments in both systems. These analyses revealed several characteristic features of the networks which were not apparent by just looking at the data. For instance, an aggregation of the two networks into food chains with recycling (Ulanowicz 1987) showed higher trophic efficiencies and longer food chains and a predominance of herbivory over detrivory in the Baltic, compared to Chesapeake. A higher proportion of total activity is recycled in the Chesapeake network which would indicate a more "mature" system according to the now classical concept summarized by Odum (1969). But this index must be qualified by measures of the actual structure of the cycles. An analysis of cycles (Ulanowicz 1983) showed that most of the material recycling in the Chesapeake involved only a few compartments whereas the Baltic cycles were longer and thus distributed among more trophic levels (Fig 3). In summary, these analyses indicate that the Baltic proper is a more developed ecosystem with less signs of environmental perturbations in spite of its lower species diversity.

These very simplified budgets show considerable imbalances both for phosphorus and nitrogen. For phosphorus, estimated losses and an accumulation of about 15 000 tons per year averaged over the last 15 years, leave half of the load of almost 78 000 tons to be accounted for. The only reasonable explanation seems to be an accumulation of phosphorus in sediments in the still oxygenated parts of the Baltic. Shaffer and Rönner (1984) calculated the annual losses of nitrogen due to denitrification below the halocline to be about 470 000 tons. Since most of this denitrification occurs at the redoxcline in the sediments, it is likely that most of the nitrogen unaccounted for disappears through denitrification in the even larger sediment areas above the halocline. These budget calculations indicate that a considerable phosphorus accumulation must be taking place in the sediments. For nitrogen, the budget suggests that the nitrogen inputs are almost balanced by denitrification, resulting in a relatively small net accumulation.

Many studies have now shown that primary production in the Baltic proper and the Bothian Sea is limited by nitrogen, like in many other marine systems. The Bothnian Bay resembles most freshwater systems in being limited by phosphate. For the Bothnian Bay it appears that the shallow well oxidised sediments rich in iron, provides an effective trap for phosphate. However, the reasons for the different roles of the two nutrients in fresh- and salt-water systems are not yet understood (Smith 1984; Howarth & Cole 1985).

The Baltic Sea is not in steady state in terms of nutrients and all subareas show increasing concentrations of both phosphate and inorganic nitrogen (Nehring *et al.* 1987). Larsson *et al.* (1985) showed that the nitrogen load has increased by a factor of four and phosphorus load by a factor of eight during the last century. How has the biological communities reacted to this increase in load? Despite a large research effort very few convincing evidence for a large scale response in the pelagic community has emerged (Wulff *et al.* 1986) although many coastal areas show severe signs of eutrophication with toxic bloom and anoxic bottoms in the late 1970s and early 1980s. The most convincing evidence of eutrophication was shown by Cederwall & Elmgren (1980) in their comparison of benthic biomasses collected in 1920-23 with similar data collected in 1976-77. Near permanent anoxic bottom water has wiped out the macrofauna in all the deeper basins of the Baltic proper but above the halocline benthos biomass has increased considerably, between 3-5 times. Increased food supply due to increased sedimentation of organic matter to the bottoms seems to be the most reasonable explanation.

An increased primary production is difficult to document, not only because of methodological difficulties. The ratio of "new" to total production is often very small and most of the measured production is due to recycled nutrients both in the deep ocean (Eppley & Peterson 1979) as well as in costal areas (Wassmann 1986). An overall nitrogen budget for the Baltic proper (Wulff *et al.* 1986) shows that of an annual primary production of 1490 mmol N $m^{-2}yr^{-1}$, only 13 % is due to external inputs, less than 3 % to internal nitrogen fixation and about 11 % to benthic nutrient regeneration, the rest (73 %) is maintained by recycling within the trophogenic layer. Thus, a small shift in the processes controlling nutrient recycling may have a profound effect on net production within various parts of the community. Although empirical evidence show an increase in primary production in response to increased nutrient load for most marine (Nixon *et al.* 1986) as well as freshwater system (Vollenweider 1982) it is not evident that this will result in a linear response in all parts of the community. Although zooplankton production tends to increase, the efficiency decreases and a smaller proportion of the increased primary production is channeled to the higher trophic levels in the pelagic community. Empirical data from the Baltic (Ackefors *et al.* 1978), as well from freshwater (Hillbricht-Ikowska 1977) and marine areas (Cushing 1971), support this conclusion. A higher proportion of the new production is channeled via sedimentation to the benthos, a hypothesis suggested by the rapid response in benthic biomasses and production, usually encountered (Pearson and Rosenberg 1978).

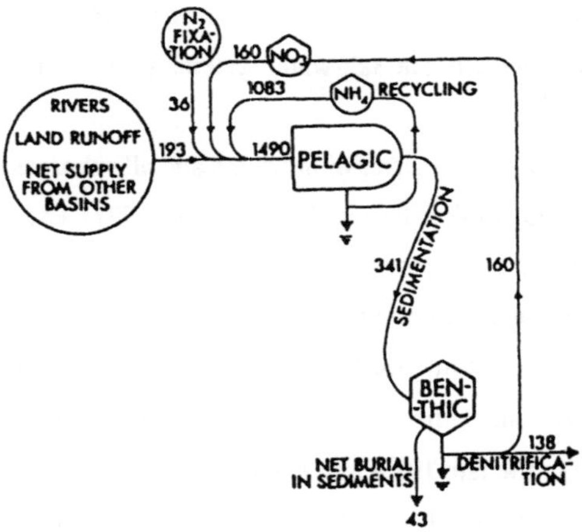

Fig 4. A nitrogen budget of the Baltic proper, expressed in mmol N m^{-2} yr^{-1} (from Wulff *et al.* 1986).

Obviously, the explanations for these responses cannot only be sought in changes in the composition of the biological communities. The overwhelming empirical evidence from limnological studies show that the responses of the ecosystem to changing nutrient load must be related to hydrological and topographical differences between the systems. Detailed models describing the biological changes that occur in lakes in relation to nutrient enrichment have had little success, compared to empirical regression that relates the trophic state to the phosphorus load and the mean depth/water retention time (Vollenweider 1982). It is tempting to apply the same approach to the Baltic and the Chesapeake Bay. Such calculations (Fig 5) show that the Baltic has moved from an oligotrophic to a moderately eutrophic state whereas the Chesapeake is far into the range that is considered excessively eutrophic. This simple dose-effect approach supports our earlier conclusions based on comparisons of the flow networks, but should be looked upon with caution since both these system are nitrogen and not phosphorus limited.

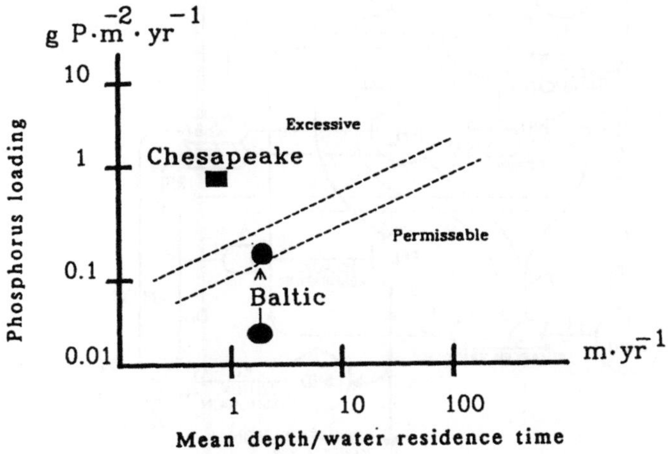

Fig 5. Vollenveider diagram relating the trophic status of the Chesapeake Bay and the Baltic Sea today and before 1900 to phosphorus loading and mean depth/water residence time (modified after Wulff and Ulanowicz 1988; Larsson *et al.* 1985)

PROCESS MODELS

C,N,P cycles are intimately coupled, but with different control processes and differ in importance in various part of the system. Obviously, N, P sources and sinks in the system occurring primarily in sediments, are controlled by redox conditions. Oxygen conditions are thus not only directly critical for the survival of the benthic fauna but also for the overall nutrient conditions and productivity of the system.

Almost all oxygen brought into the deep basins comes with new salt water through the Danish straights and Öresund. Recent estimates of the oxygen consumption (Rahm 1987) show that the supply of oxygen is equivalent to the total oxygen consumption below the halocline. In spite of this, most of the deep basins are anoxic since most of the new water is not dense (salt) enough to penetrate into the deeper parts. Hydrodynamical models describing mixed layer dynamics (Stigebrandt 1985), the exchange of water between the Baltic and the Skagerrak (Stigebrandt 1983), the vertical circulation of the deep water (Stigebrandt 1987a) and the inflows of dense water (Stigebrandt 1987b) have now been developed that describe the salt and temperature distributions in the Basins.

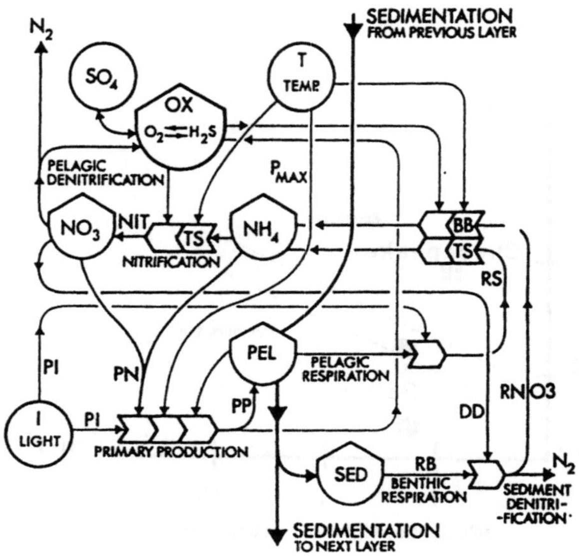

Fig 6. Biogeochemical processes included in the model of the Baltic proper (from Stigebrandt and Wulff 1987).

With these models describing physical transports, we have now reached a stage were it again is fruitful to use mathematical modelling and simulations as a tool to understand the nutrient and oxygen dynamics of the Baltic. Such a model, where the physical processes has been coupled to descriptions of biogeochemical processes controlling the nitrogen cycle, has now been developed for the Baltic proper (Stigebrandt & Wulff 1987). The model is horizontally integrated but with a high vertical resolution taking into account the distributions of volumes of water and sediments in 250, 1-meter thick layers. Primary production is controlled by light, temperature, inorganic nitrogen and density stratification (critical depth). Nutrient recycling, nitrification and sedimentation are described with the most simple general formulations possible (Fig 6). Denitrification in the water and sediment is controlled by the concentrations of oxygen, nitrate and organic substrate and by temperature. Sulphate reduction to hydrogen sulphide occurs during anaerobic conditions once nitrate has been depleted.

When compared to field data, the computed dynamics of primary production, nutrients and oxygen appear quite realistic, both in the mixed layer and in the deep basins (Fig 7). This suggests that we are now beginning to understand the coupling of hydrodynamic, chemical and biological processes that control the overall productivity of the system.

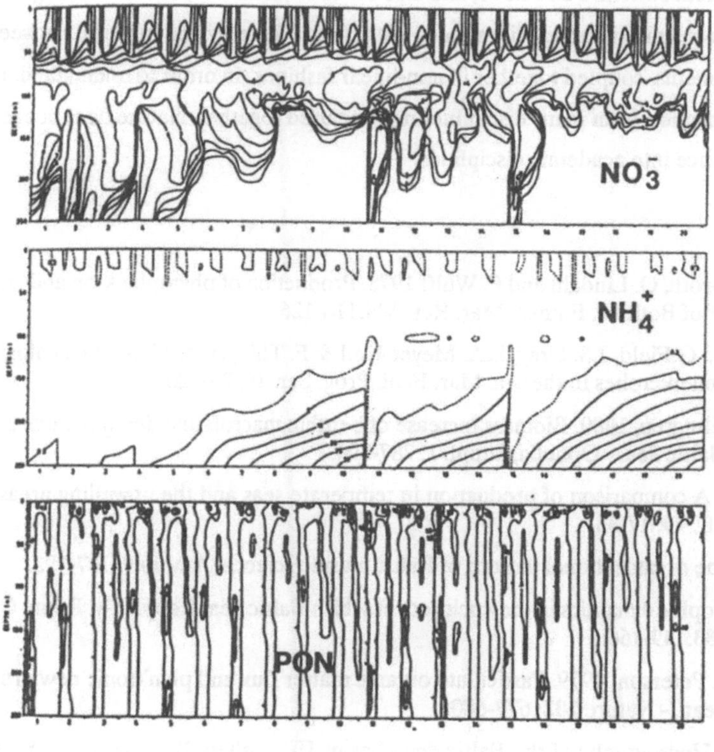

Fig 7. Model predictions of the vertical distributions (0-250 m) in the Baltic proper of nitrate, ammonia and particulate organic matter during a 20-year period (from Stigebrandt and Wulff 1987).

In limnology, the success in predicting the effect of perturbations on lakes, comes primarily from studies on a few variables describing overall biological, hydrodynamical, chemical and morphometric properties (Peters 1986). Although empirical regression models have prevailed so far, mathematical models describing the functional relationships between these variables are now being developed and successfully used in lake restoration management (Imboden 1985). These approaches ignore the fundamental genetic properties of organisms which hitherto has been the basis for the development of ecological theory. The challenge that lies ahead seems to be in finding a way to merge these schools of thought, both in limnology and oceanography (Lehman 1986).

CONCLUSIONS

Where do we stand now? The Baltic experience shows that we are just beginning to have the information of overall ecosystem structure and flows, and also the tools, necessary for developing ecosystem theory. An ecosystem is a multi-commodity system, with flows of energy and material between living and non-living components, coupled together in non-linear fashions. In order to understand how ecosystem operate, information from many disciplines must be used together. Nature does not recognize our separation of science into academic disciplines!

REFERENCES

Ackefors, H., L. Hernroth, O. Lindahl and F. Wulff 1978. Production of phytoplankton and zooplankton in the Gulf of Bothnia. Finnish Mar. Res. **244**:116-126.

Azam,F., T. Fenchel, J.G.Field, J.S. Gray, L.A. Meyer-Reil & F. Thingstad, 1983. The ecological role of water-column microbes in the sea. Mar. Ecol. Prog. Ser. **10**, 257-263.

Cederwall, H. & R. Elmgren, 1980. Biomass increase of benthic macrofauna demonstrates eutrophication of the Baltic Sea. - Ophelia, Suppl. **1**: 287-304.

Cushing, D. H. 1971. A comparison of production in temperate seas and the upwelling areas. Trans. Roy. Soc. S. Afr. **40**: 17-33.

Dybern, B. I. 1980. The organizational pattern of Baltic marine science. Ambio **9**: 187-193.

Elmgren, R., 1984. Trophic dynamics in the enclosed, brackish Baltic Sea. Rapp. P.-v. Réun. Cons. int. Explor. Mer **183**:149-166.

Eppley, R. W. & B. J. Peterson, 1979. Particulate organic matter flux and planktonic new production in the deep ocean. - Nature **282**: 677-680.

Fonselius,S.H., 1969. Hydrography of the Baltic deep basins III. Fishery Bd. Sweden, Ser. Hydrogr. **23**.

Hillbricht-Ilkowska, A. 1977. Trophic relations and energy flow in pelagic plankton. Pol. ecol. Stud., **3**: 3-98.

Howarth, R. W. & J. J. Cole 1985. Molybdenum availability, nitrogen limitation, and phytoplankton growth in natural waters. Science **229**, 653-655.

Hägerhäll, B. 1980. International cooperation to protect the Baltic. Ambio **9**: 183-186.

Imboden, D. M. 1985. Restoration of a Swiss lake by internal measures: can models explain reality. EWPCA: International Congress: Lakes Pollution and recovery. Proceedings-preprints, 29-40.

Jansson, B.-O. 1972. Ecosystem approach to the Baltic problem. Bull. Ecol. Res. Comm./NFR 16:1-82.

Jansson, B.-O. 1984. Baltic Sea ecosystem analysis: critical areas for future research. Limnologica (Berlin) 15(2):237-252.

Kautsky, N., 1981. On the trophic role of the blue mussel (*Mytilus edulis* L.) in a Baltic coastal ecosystem and the fate of the organic matter produced by the mussels. Kieler Meeresforsch., Sonderh. 5:454-461.

Larsson, U., R. Elmgren and F. Wulff 1985. Eutrophication and the Baltic Sea: causes and consequences. Ambio, 14, 9-14.

Legget, W.C., W.R. Boynton, V.F. Gallucci, L.Johnson, R.Margalef, J.McGlade, D.Mikelecky, R.Rosen, R.O.Ulloa & J.Watson,1985. I.Hypothesis testing and sampling design in exploited ecosystems. *In:* Ulanowicz, R.E. & T. Platt (eds.) Ecosystem theory for biological oceanography. Can. Bull. Fish. Aquat. Sci. 213: 237-240.

Lehman, J.T., 1986. The goal of understanding in limnology. Limnol. Oceanogr., 31(5),1160-1166.

Mann, K. H. 1982. Ecology of Coastal Waters. A systems approach. Studies in Ecology, Volume 8. Blackwell, 322 pp.

Nehring, D., G. Aertebjerg, P. Alenius, V. Astok, S. Fonselius, M. Hannus, V. Tervo, A. Troszinska, P. Tulkki & A.K. Yurkovskis, 1987. Nutrients - *In:* Lassig, J. (ed). First periodic assessment of the state of the marine environment of the Baltic Sea area, 1980-1985; Background document. Baltic Sea Environm. Proc. 17B: 35-81.

Nixon, S. W., C.A. Oviatt, J. Frithsen & B. Sullivan, 1986. Nutrients and the productivity of estuarine and coastal marine ecosystems. J. Limnol. Soc. sth. Afr. 12(1/2),43-71.

Odum, E. P. 1969. The strategy of ecosystem development. Science 164:262-270.

Odum, E.P. 1971a. Fundamentals of ecology. Saunders, 574 pp.

Odum, H.T. 1971b. Environment, power and society. Wiley-Interscience, 331 pp.

Pattern, B.C. and J.T. Finn 1979. Systems approach to continental shelf ecosystems. In: Theoretical systems ecology. (E. Halfon, ed.) pp. 183-212. Academic Press, New York and London.

Pearson, T. H. and R. Rosenberg 1978. Macrobenthic succession in relation to organic enrichment and pollution of the marine environment. Oceanogr. Mar. Biol. Ann. Rev. 16:229-311.

Peters, R.H., 1986. The role of prediction in limnology. Limnol. Oceanogr., 31(5),1143-1159.

Platt, T., K.H. Mann and R.E. Ulanowicz 1981. Mathematical models in biological oceanography. UNESCO Monographs in Oceanographic Methodology, UNESCO Paris, 156 pp.

Rahm, L. A. 1987. Oxygen consumption in the Baltic proper. Limnol. Oceanogr., 32(4):973-978.

Rumohr, J., E. Walger & B. Zeitzschel, 1987. Seawater-sediment interactions in coastal waters. An interdisciplinary approach. Lecture notes on Coastal and Estuarine Studies, 13. Springer Verlag, 338pp.

Segerstråle, S.G., 1957. Baltic Sea. Mem. Geol. Soc. America 67: 751-800.

Shaffer, G. & U. Rönner 1984. Denitrification in the Baltic proper deep water. Deep-sea Res., 31,197-200.

Sjöberg, S., F. Wulff & P. Wåhlström 1972. Computer simulation of hydrochemical and biological processes in the Baltic. - Contrib. Askö Lab. Univ. Stockholm 20:1-99.

Smith, S. V., 1984. Phosphorus versus nitrogen limitation in the marine environment. Limnol. Oceanogr. 19, 1149-1160.

Stigebrandt, A. 1983. A model for the exchange of water and salt between the Baltic and the Skagerrak. J. Phys. Oceanogr., 13, 411-427

Stigebrandt, A. 1985. A model for the seasonal pycnocline in rotating systems with application to the Baltic proper. J. Phys. Oceanogr., 15, 1392-1404.

Stigebrandt, A. 1987a. A model for the vertical circulation of the Baltic deep water. J. Phys. Oceanogr., in press.

Stigebrandt, A. 1987b. Computations of the flow of dense water into the Baltic from hydrographical measurements in the Arkona basin. Tellus, 39A, 170-177.

Stigebrandt, A. & F. Wulff 1987. A model for the dynamics of nutrients and oxygen in the Baltic proper. J. Mar. Res., 45, 729-759.

Ulanowicz, R. E. 1983. Identifying the structure of cycling in ecosystems. Mathematical Biosciences 65: 219-237.

Ulanowicz, R. E. 1987. Ecosystem trophic foundations: Lindeman exonerata. In: Pattern, B.C. and S. Jorgensen (eds.) Progress in systems ecology. Elsevier, Amsterdam, in press.

Ulanowicz, R. E. and T. Platt (ed.) 1985. Ecosystem theory for biological oceanography. Can. Bull. Fish Aquat. Sci. 213: 260p.

Vollenweider, R. A. 1969. Possibilities and limits of elementary models concerning the budget substances in lakes. Arch. Hydrobiol. 66(1): 1-36.

Vollenwieder, R. A. (ed) 1982. Eutrophication of waters. Monitoring, assessment and control. OECD, Paris, France. 154 pp.

Wassmann, P., 1986. Benthic nutrient regeneration as related to primary production in the west-Norwegian coastal zone. - Ophelia 26: 443-456.

Wulff, F. 1976. Ecological data and the model. Ambio Special Rep., 4: 171-177.

Wulff, F., G. Aertebjerg, G. Nicolaus, Å. Niemi, P. Ciszevski, S. Schultz & W. Kaiser, 1986. The changing pelagic ecosystem of the Baltic Sea. Ophelia, Suppl. 4: 299-319.

Wulff, F. and R. E. Ulanowicz 1988. A comparative anatomy of the Baltic Sea and Chesapeake Bay ecosystems. *In:* Wulff, F., J.G. Field and K.H. Mann (eds.). Flow analysis of marine ecosystems. Theory and practice. Lecture Notes in Coastal and Estuarine Studies. Springer Verlag, in press.

Ultimate Causes of Vertical Migration in Zooplankton: An Evaluation by Evolutionary Game Theory

W. Gabriel and B. Thomas[1]

Max-Planck-Institute for Limnology,
Department of Physiological Ecology,
Postfach 165, D-2320 Plön, Fed. Rep. of Germany and
[1]Institut für Entwicklungsphysiologie der Universität Köln,
Gyrhofstr. 17, D-5000 Köln, Fed. Rep. of Germany

INTRODUCTION

Diel vertical migration is a widespread but not well understood behaviour in freshwater and marine zooplankton (Pearre 1979a and 1979b). Many taxa avoid the warm and food rich upper waters during day, but at dusk they swim long distances upward and stay in the upper water layers during night. Around sunrise they descend again and stay in colder and food scarce waters during day. At least cladocerans can not compensate for the food shortage during day by increasing feeding rates and by storing food during night (Lampert and Taylor 1985). The extra swimming for migration seems not to be very costly in terms of energy consumption. Most zooplankters carry their eggs with them. As egg development time is inversely proportional to temperature, a lowered temperature increases generation time drastically but may reduce metabolic costs.

To find ultimate causes for the migratory behaviour, the mentioned disadvantages have to be offset by fitness components, which increase because of vertical migration. Several hypotheses have been proposed: metabolic advantages or better utilization of resources (McLaren 1963 and 1974, Kerfoot 1970, Enright 1977, Enright and Honegger 1977), avoidance of starvation (Geller 1986), and avoidance of visual predators (Zaret and Suffern 1976, Wright et al. 1980). The validity and relative importance of these arguments can be tested only by quantifying the relative strength of the various selective forces. There are good reasons to assume migratory behaviour has a strong genetic component (Weider

1984, Gliwicz 1986). Therefore, a most promising framework for such a problem is the concept of evolutionarily stable strategies (=ESS) (Maynard Smith and Price 1973, Maynard Smith 1982, Thomas 1984).

Gabriel and Thomas (1988a) developed an ESS-model on vertical migration of zooplankton which is able to explain the observed coexistence of two similar *Daphnia* species, one migrates while the other does not migrate (Stich and Lampert 1981). After a short description of the model parameters we will derive equations for a discussion of the ultimate causes of vertical migration of zooplankton.

MODEL PARAMETERS

A detailed discussion and description of the model is given in Gabriel and Thomas (1988a). Migrating and non-migrating behaviour are treated as two distinct strategies. The model calculates the food uptake and its conversion into successful reproductive output for both strategies. Fitness is calculated in terms of intrinsic growth rate on a time scale of 24 hours. The interaction of zooplankton with its algae food is considered in detail according to the well studied feeding physiology of zooplankton. The payoffs depend on the following parameters which all are known from laboratory experiments and field studies:

- t : egg development time,
- A : algal density,
- N : population density of zooplankton,
- p : predation risk for zooplankton by optically orientated predators like fish,
- r : partial intrinsic growth rate of algae,
- T_n: relative length of night (in parts of 24 hours),
- ß : conversion efficiency of food uptake to successful reproductive output (mortalities other than considered under p are taken into account),
- g : maximal grazing rate of zooplankton.

Some of these parameters, especially t and ß, vary strongly for the different strategies.

Due to self-interaction through food competition and frequency dependent predation risk, the payoff to the two strategies is dependent on their relative frequencies. In fact, the payoff difference is a non-linear function.

BASIC EQUATIONS TO STUDY ULTIMATE CAUSES

To study the ultimate causes of vertical migration let us assume that a population consists only of non-migrating ethotypes and let us ask for conditions allowing a successful invasion of migrating ethotypes. This is only possible when the payoff for migrating ethotypes is larger than for non-migrating ones. Therefore, we can derive a minimal condition for vertical migration by setting the payoff difference equal to zero (at relative frequency of non-migrating zooplankton arbitrarily close to $x_s=1$). From the payoff equations given by Gabriel and Thomas (1988b), we have

(1) $$0 = (1 - p)\beta_s a_d - 1.5p\beta_s(a_n + a_d)t_s + \beta_s a_n - \beta_v a_n w .$$

Subscript s is used for the (stationary) strategy of non-migrating zooplankton and subscript v for the strategy of vertical migration. The food-uptake during night (=n) and day (=d) in the upper water layers is denoted as a_n and a_d, respectively, and is a function of zooplankton density, daylength, algal growth rate, maximal filtration rate of zooplankton, and availability of algal food. A slow-breeding correction factor w has to be applied to measure the fitness disadvantage of prolonged generation time (for detailed discussion see Gabriel and Thomas 1988):

(2) $$w = (1 + \beta_v a_n t_v) / (1 + \beta_v a_n t_s)^{t_v/t_s} .$$

The food uptake is calculated from the interaction of zooplankton and algae. At low food levels the ingested food is proportional to the actual food concentration. This results in a food uptake during day of

(3) $$a_d = gA[\exp\{(r_p - gx_sN)(1 - T_n)\} - 1]/(r_p - gx_sN) ,$$

and during night (for non-migrating zooplankton) of

(4) $$a_n = A\exp\{(r_p - gx_sN)(1 - T_n)\}[1 - \exp\{-gNT_n\}]/N .$$

At algal concentrations above a certain limit (A_{lim}) the food uptake during day (= a_d) and night (= a_n) is independent of the actual algal density but only a function of this limiting concentration level. For high food concentrations a_d and a_n are then simply given by

$$(5) \qquad \begin{aligned} a_d &= gA_{lim}(1 - T_n) \qquad &(\text{for } A > A_{lim}) \\ a_n &= gA_{lim}T_n \qquad &(\text{for } A > A_{lim}). \end{aligned}$$

For any situation, viz. for any given parameter set, we can now solve equation (1) in order to calculate threshold values. Whether these threshold values are maximal or minimal values depends trivially on the derivative of the payoff difference with respect to the parameter in question. To discuss ultimate causes we solve equation (1) for the ratio between the strategy dependent conversion efficiencies ß. From the model equations, it follows that $ß_s$ is always positively correlated with the non-migration strategy and any increase in $ß_v$ is in favour of vertical migration. Therefore, the threshold ratio $ß_v/ß_s$ is a minimum value: vertical migration can never be an ESS for values below this threshold. In the following we will discuss the consequences of this threshold ratio which can be calculated from equation (1) :

$$(6) \qquad (ß_v/ß_s)_{thres} = [\ 1 + a_d/a_n - p\{a_d/a_n + 1.5\ t_s(\ 1+ a_d/a_n)\}]/w\ .$$

RESULTS AND DISCUSSION

Lets first consider the simplest case of very low mortality caused by visually orientated predators like fish; i.e. p=0. The threshold ratio (6) then gives the value that compensates for the lower food uptake and the prolonged development time of vertically migrating zooplankton. The value for this threshold ratio is then

$$(ß_v/ß_s)_{thres} = [\ 1 + a_d/a_n\ ]\ /\ w \qquad (\text{for } p=0)$$

and is reduced further in the case of high food concentration to the simple expression

$$(ß_v/ß_s)_{thres} = (\ T_n w\ )^{-1} \qquad (\text{for } p=0 \text{ and } A>A_{lim})$$

because of $a_d/a_n=T_n^{-1}-1$. At lower food concentrations ($A<A_{lim}$) this

is also a valid approximation under the condition that $gNT_n<<1$ and $(r-gN)(1-T_n)<<1$; this means biologically that the interaction of algae and zooplankton is not too far from a steady-state condition and that the mortality imposed on the algae by zooplankton is not too large.

The necessary advantage for vertically migrating zooplankton in β_v, the efficiency of converting food uptake into successful reproduction, can be lowered drastically for p>0. For high food concentrations we get from equations (6) and (5)

$$(\beta_v/\beta_s)_{thres} = [\ 1 - p\{1 - T_n + 1.5\ t_s\}]/wT_n.$$

This is again approximately true also at low food concentrations under the above mentioned conditions.

If predation pressure is high enough, vertical migration is favoured even for $\beta_v<\beta_s$. For further increasing p, the necessary β_v-value to establish vertical migration as an ESS becomes smaller and smaller. From equation (6) it can be seen that there is an upper limit for p

$$p_{tol}=(a_d + a_n(1 - \beta_v w/\beta_s))/(a_d(1 + 1.5t_s) + 1.5a_n t_s)$$

above which the threshold ratio would become negative. This means that at predation pressure above this value the strategy of non-migrating zooplankton can never be an ESS irrespective of the value for β_v, the conversion efficiency for migrating ethotypes. (This limit on p is discussed by Gabriel and Thomas (1988b) as the tolerable predation pressure for the non-migratory strategy.) In Figures 1 and 2 the threshold ratios are shown for values of p varying from p=0 to $p=p_{tol}$. They are plotted as functions of t_v/t_s, and reflect various mean temperature differences between upper and lower water strata which are experienced by migrating and non-migrating zooplankton. From the figures it is immediately clear that it would be very difficult to compensate for the disadvantage of vertical migration without the added advantage of predation avoidance. It seems to be impossible to overcome this disadvantage simply by invoking a more favourable metabolic state. But there might be a chance to compensate for the disadvantage of vertical migration even at low predation pressure (small p) if juvenile mortality for non-migrating zooplankton is very high compared to migrating ones.

Another interesting aspect arises from the attempt to quantify the

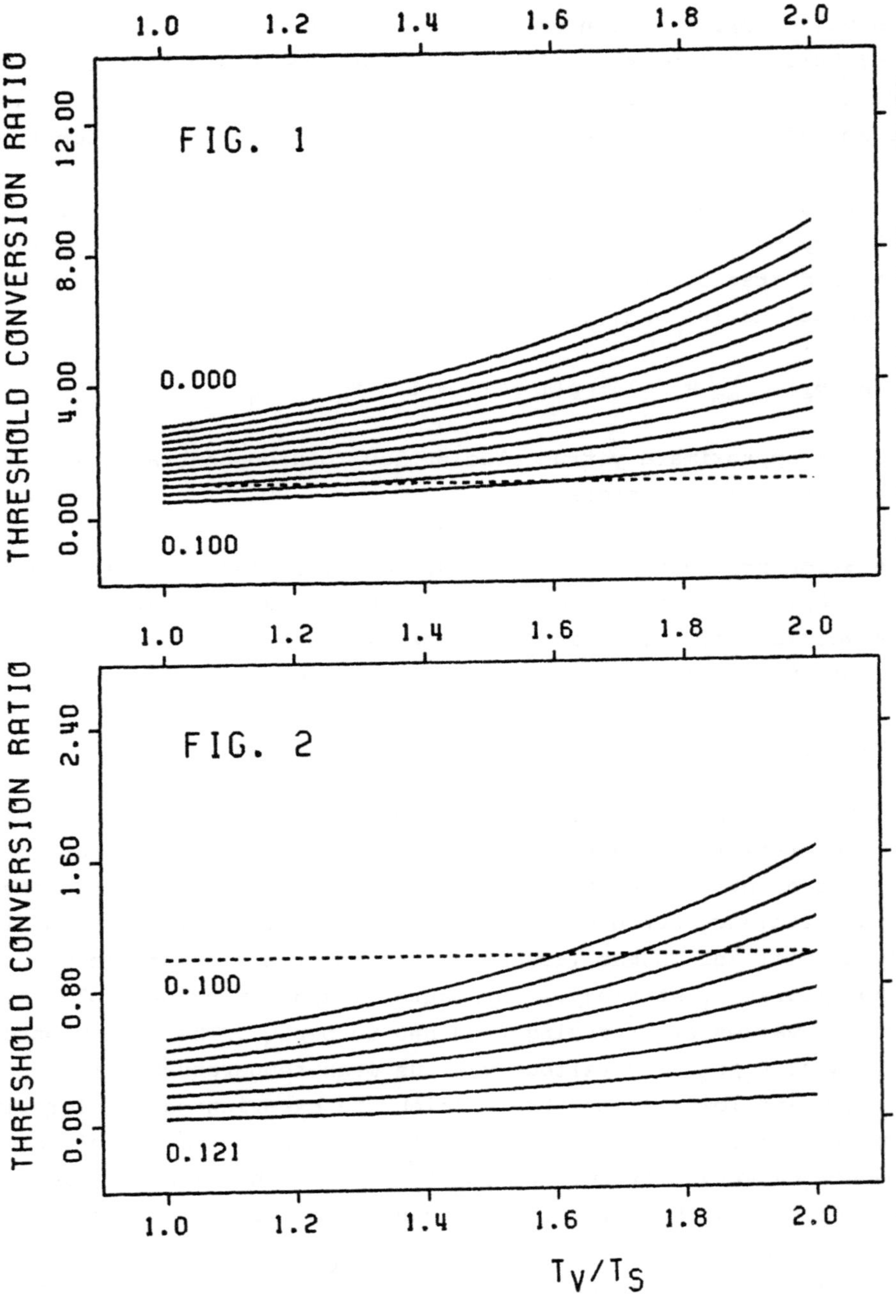

Fig. 1 and Fig. 2 :
Threshold ratio of the conversion efficiency β_v/β_s of food into succesful reproduction depending on the ratio t_v/t_s of egg development time of both strategies. The predation pressure is varied from 0 to 0.1 in Fig.1 and from 0.1 to 0.121 in Fig.2. The broken lines indicate equal conversion efficiencies $\beta_v=\beta_s$. (The following parameter values are used : g = 0.55, r = 0.35, N = 1, T_n = 0.4, t_s = 5, A = 0.5 A_{lim}.)

necessary compensation just for the reduced food uptake during vertical migration. Let us assume equal conversion efficiencies for this purpose. Therefore, we put $(\beta_v/\beta_s)_{thres} = 1$. Furthermore let us assume equal developmental time for both strategies which is equivalent to a negligible temperature difference between upper and lower water strata. Therefore, let us put $t_s = t_v$ which implies w=1. With equation (6) we can now calculate a predation pressure p_{com} that compensates only for the lower food uptake of migrating ethotypes:

$$p_{com} = (1 + 1.5(a_n/a_d + 1)t_s)^{-1} .$$

For high food concentrations (and also approximately for low food concentrations) we get

$$p_{com} = (1 + 1.5t_s/T_n)^{-1} \quad \text{(for A>Alim)}.$$

It may be counterintuitive that this value is still dependent on t_s, the egg development time for non-migrating zooplankton. But the energy stored in the eggs and ovaries is lost in the case of predation and this energy is a function of the moulting period which is roughly equivalent to t_s. Therefore, at lower temperatures in the upper waters, fish predation can compensate more easily for the reduced food supply due to vertical migration. This is completely independent of the temperature in the bottom waters.

SUMMARY

The model equations demonstrate that it is extremely unlikely that metabolic advantages by themselves are the ultimate causes of diel vertical migration in zooplankton. Various selective forces interact in a complex way to select this behaviour, which only seems to be disadvantageous at the first glance. Relative strength and interaction of single components of these selective forces can be quantified by applying evolutionary game theory. This concept allows one to calculate boundary conditions for the invasion of vertically migrating ethotypes into a population consisting only of non-migrating zooplankton. A most useful quantity to study ultimate causes of this behavioural phenomenon is the threshold ratio of conversion efficiency of food into reproduction. From this analysis, we suggest that the reduction of predator-induced mortality is one of the most important selective forces influencing vertical migration.

Acknowledgement

We thank L.J. Weider for improving the manuscript.

LITERATURE CITED

Enright, J. T. 1977. Diurnal vertical migration: adaptive significance and timing. Part 1. Selective advantage: A metabolic model. Limnol. Oceanogr. 22:856-872.
Enright, J. T., and H. W. Honegger. 1977. Diurnal vertical migration: adaptive significance and timing. Part 2. Test of the model: Details of timing. Limnol.Oceanogr. 22:873-886.
Gabriel, W. and B. Thomas. 1988a. Vertical migration of zooplankton as an evolutionarily stable strategy. Am.Nat. (in press).
___________________________. 1988b. The influence of food availability, predation risk, and metabolic costs on the evolutionary stability of diel vertical migration in zooplankton. Verh.Internat.Verein. Limnol.23 (in press).
Geller, W. 1986. Diurnal vertical migration of zooplankton in a temperate great lake (L. Constance): A starvation avoidance mechanism? Arch.Hydrobiol./Suppl. 74:1-60.
Gliwicz, M. Z. 1986. Predation and the evolution of vertical migration in zooplankton. Nature 320:746-748.
Kerfoot, W. B. 1970. Bioenergetics of vertical migration. Am.Nat. 104: 529-546.
Lampert, W., and B. Taylor. 1985. Zooplankton grazing in a eutrophic lake: implications of diel vertical migration. Ecology 66:68-82.
Maynard Smith, J. 1982. Evolution and the theory of games. Cambridge University Press, Cambridge.
Maynard Smith, J., and G. R. Price. 1973. The logic of animal conflicts. Nature 246:15-18.
McLaren, I. A. 1963. Effects of temperature on the growth of zooplankton and the adaptive value of vertical migration. J.Fish.Res.Bd.Can. 20:685-727.
_____________. 1974. Demographic strategy of vertical migration by a marine copepod. Am.Nat. 108:91-102.
Pearre, S. Jr. 1979a. On the adaptive significance of vertical migration. Limnol.Oceanogr. 24:781-782.
_____________. 1979b. Problems of detection and interpretation of vertical migration. Journal of Plankton Research 1:29-44.
Stich, H.-B., and W. Lampert. 1981. Predator evasion as an explanation of diurnal vertical migration by zooplankton. Nature 293:396-398.
Thomas, B. 1984. Evolutionary stability: states and strategies. Theor. Pop.Biol. 26:49-67.
Weider, L. J. 1984. Spatial heterogeneity of genotypes: Vertical migration and habitat partioning. Limnol.Oceanogr. 29:225-235.
Wright, D., W. J. O'Brien, and G. L. Vingard. 1980. Adaptive value of vertical migration: A simulation model argument for the predation hypothesis. Pages 138-147 in W.C. Kerfoot, ed. Evolution and ecology of zooplankton communities. Univ. Press of New England, Hanover, New Hampshire.
Zaret, T. M., and J. S. Suffern. 1976. Vertical migration in zooplankton as a predator avoidance mechanism. Limnol.Oceanogr. 21:804-813.

Trophy-Balanced Turnover Velocities in a Two-Species System of Competing Daphnids: A Test of the "Energy Residence Time" Concept*

W. Geller

Limnological Inst. der Universität, D-7750 Konstanz, Fed. Rep. of Germany

The primary and secondary productivity of the pelagic zone of Lake Constance has been investigated since 1980. During the growing seasons (May to autumn) two Daphnia species were the dominant herbivores. The productivity of the two Daphnia populations was estimated from the dynamics of their population densities, their individual body weights, and their development times. The seasonal course of the population densities can be generalized as a logistic growth with a log-phase in spring and a subsequent food-limited equilibrium density throughout the summer. Body weights varied according to nutritional state (as measured by the average clutch size) /1/. Development times were estimated as function of body size and environmental temperature /2/, the latter of which was monitored in a detailed investigation of the specific vertical migration of the zooplankton populations and simultaneous recording of the lake temperatures in 10-min-intervals

Year	Biomass (g DW/m²) D.g.	D.h.	D.	Production D.g.	D.h.	D.
1979	1.26	1.44	2.70	32.6	13.9	46.5
1980	0.51	2.02	2.52	13.4	24.5	37.9
1981	1.13	1.33	2.46	29.9	14.9	44.8
1982	0.73	1.57	2.30	20.2	18.7	38.9
Average 1979-82	0.91	1.59	2.50	24.0	18.0	42.0
± % SD	38 %	19 %	7 %	37 %	27 %	10 %

Tab. 3: Seasonal mean of Daphnia population density and annual production from 1979 to 1982 in Lake Constance: 'D.g.' - Daphnia galeata, 'D.h.' - D. hyalina, and 'D' - sum of both populations.

*) Abbreviations in the text: DW: dry weight; B: standing crop biomass (g DW/m²); P: cumulative annual production (g DW/m².year); P/B: turnover rate and B/P: turnover time (days), both refer to the growing season of 200 days; B/R: energy residence time = relation between pool size (B) and flow of energy through the pool (R: cumulative annual respiration).

within the upper 40 meters of the water column /3/. Daphnia galeata showed no migratory behaviour and, living in the warm epilimnic layers, had short development times. D. hyalina began to migrate in June when the population density reached equilibrium. Its development was prolonged, therefore, by the low hypolimnetic temperatures as experienced during the days.

Long-term averages (1979-1982, Tab.1) of mean biomass (B) were 0.9 g DW/m² & 1.6 g DW/m² in D. galeata & D. hyalina, respectively. The cumulative annual production (P) was 24 g & 18 g DW/m² per year. This productivity resulted in P/B turnover rates of 27x & 11x, corresponding to a season length of the growing period of 200 days and in turnover times (B/P) of 7.5 & 18 days in D. galeata & D. hyalina, respectively. On a seasonal average the costs to maintain a biomass unit of standing stock, as measured by the amount of necessary replacement production, were 2.5 times higher in D. galeata than in D. hyalina. In mid-summer this relation reached a maximum of 4x, as shown by the respective generation times of 11 & 43 days during late July in D. galeata & D. hyalina. Within the three years the cumulative annual primary productivity (PPR) /6/ varied between 260 and 330 g C/m².year (Fig. 1). The common standing stock of both daphnids was nearly constant at 2.5 g DW/m², but their common production followed the PPR trend. This was achieved by an abundance-shifting between the fast-growing D. galeata and the slow-growing D. hyalina: at a PPR of 260 g C/m².y, D. hyalina was 4 times more abundant, and at 330 g C/m².y both species were about equally abundant.

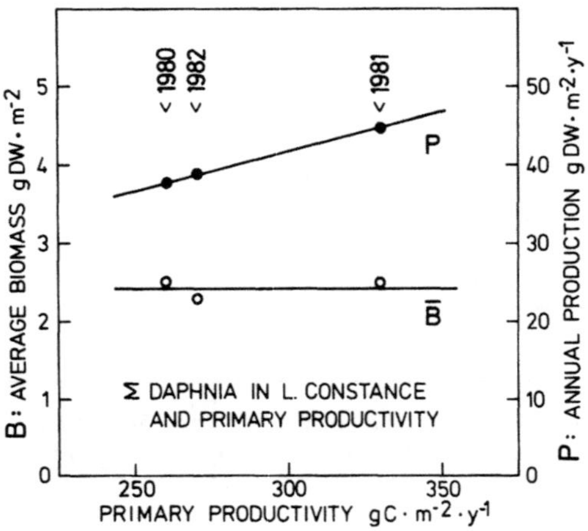

Fig.1: Relationship among primary production of phytoplankton /6/, production and standing crop of Daphnia (sum of two species) in Lake Constance 1980 - 1982.

The observed constant standing stock can be interpreted as an adjustment through long-term adaptation following the eutrophication of the lake through decades. The shift between the two **Daphnia** species can be interpreted as a competitive adjustment, occurring in a temporal scale of weeks to months, between the two pools with different reaction times. Such mechanism would be in accordance with the hypothesis on system thermodynamics and ecological organisation level as established by /4/ and /5/. They expected the 'energy residence time' (B/R) (a measure of an ecosystem's distance from thermodynamic equilibrium) to increase with increasing nutrient limitation and increasing strength of competitive interrelationships (a measure of the degree of ecological organization). Because of the strong dependence of respiration on production /7/, the biomass turnover times (B/P) of **Daphnia** may be taken as indicative of the appropriateness of this hypothesis in the three-compartment system of primary producers and two competing **Daphnia** species: with high food supply the flow of energy through the system component '**Daphnia**' is predominantly directed through the faster energy processor sub-component '**D. galeata**'. With stronger food limitation, the energy increasingly flows through '**D. hyalina**' which is the more economically working processor of energy. The shift between the two species can be understood as the internal mechanism to regulate the velocity of energy flow through the functional component '**Daphnia**'.

References

/1/ Geller, W. and Müller, H.: Seasonal variability in the relationship between body length and individual dry weight as related to food abundance and clutch size in two coexisting **Daphnia** species, J. Plankt. Res. 7 (1985) 1-18.

/2/ Geller, W.: On estimating the age and the development time of **Daphnia** as function of body size and temperature, J. Plankt. Res. 9 (1987) 1225-1230.

/3/ Geller, W.: Diurnal vertical migration of zooplankton in a temperate great lake (L. Constance): A starvation avoidance mechanism?, Arch. Hydrobiol. Suppl. 74 (1986) 1-60.

/4/ Morowitz, H.J.: Energy flow in biology: Biological organization as a problem in thermal physics. Acad. Press, New York 1968.

/5/ Cheslak, E.F. and Lamarra, V.A.: The residence time of energy as a measure of ecological organization. In: Energy and ecological modelling. Mitsch, W.J. and Bossermann, R.W. (Eds.). Elsevier, Amsterdam 1981.

/6/ Tilzer, M.M.: The importance of fractional light absorption by photosynthetic pigments for phytoplankton productivity in Lake Constance, Limnol. Oceanogr. 28 (1983) 833-846.

/7/ McNeill, S. and Lawton, J.H.: Annual production and respiration in animal populations, Nature 225 (1970) 472-474.

Ecodynamic Changes in Suburban Lakes in Berlin (FRG) During the Restoration Process After Phosphate Removal

G. Klein

Institut für Wasser-, Boden- und Lufthygiene, Corrensplatz 1,
D-1000 Berlin 33, Germany

1. INTRODUCTION

Ecodynamic processes display two major regulatory aspects:

- the level of process activity
- the time constants of process kinetics.

Many ecosystems have been described extensively as far as the level of activity is concerned. In most cases however there is not sufficient time available to study time-response relationships going beyond population dynamics and species interaction. Thus our understanding of time dependence of ecological changes in polluted systems is quite meagre and the same holds for the kinetics of restoration of disturbed or destroyed ecosystems.

Of all ecosystems lakes are known best with regard to the abovementioned aspects because of their more or less definite borders. This applies especially to studies concerning the eutrophication of lakes in consequence of excessive phosphorus input.

2. THE CASE : "LAKE RESTORATION IN BERLIN"

After many decades of severe eutrophication of all aquatic systems in the surroundings of the crowded area of "Greater Berlin" a strategy for solving this problem had been developed during the last twenty years. Based on the scienitific knowledge of phosphorus related eutrophication processes a technological solution was found for keeping away the waste-water borne nutrients from the sensitive lakes (KLEIN 1988).

The procedures to follow for phosphate elimination from eutrophied surface waters were being discussed up to a few years ago, especially with the respect to more comprehensive sewage treatment, however, with considerably fewer requirements. The requirements usually drew to a close at a level which could be obtained fairly easily without seriously interfering with the existing biological treatment plants. However in restoring lakes, we have to contend with a concentration range which can only be attained by technology suited to produce drinking water. Due to the fact that we have to reckon with many sources of nutrients (e.g. recycling of phosphate from the sediments, nutrient input from precipitation) the phosphate elimination plants have been constructed in such a manner that there is a residue of less than 10 µg/l per litre of total phosphate after treatment. This can only be achieved by a high performance chemical flocculation followed by a rapid sand filtration (HÄSSELBARTH 1979).

Two Phosphate Elimination Plants (PEA) have been constructed in Berlin. One has been in operation since 1985, treating the inflow to Lake Tegel in the north of Berlin. This plant has to be optimized to reach the requiered level of phosphate elimination, which is already the case for the PEA at the Grunewald Lake Chain , which has now been in operation for more than five years. During that time, Schlachtensee has shown definite positive results which prove that the restoration concept was correct and the technology which was used is in fact effective. With a bit of delay the Krumme Lanke has started to stabilize on the level of a reduced trophic condition.

3. RESULTS UP TO PRESENT

Two years prior to the planned starting up of PEA in Beelitzhof, a research programme was started to determine the status quo ante. This programme was conducted over a period of more than eight years and it provided for a comprehensive answer to the open questions.

3.1. REACTION OF THE PHOSPHATE BUDGET

The phosphate content in the lakes has been reduced very quickly, with a reduction of the total phosphorus influx flowing from Wannsee into Schlachtensee from originally about 1 500 kg per year (1980/1981) to about 30 kg per year in 1986.

The annual mean value of the total-phosphorus-concentration lies at about 30 µg/l P only a bit over the goal of about 20 µg/l P , based upon more than 600 µg/l P before the restoration (fig. 1).

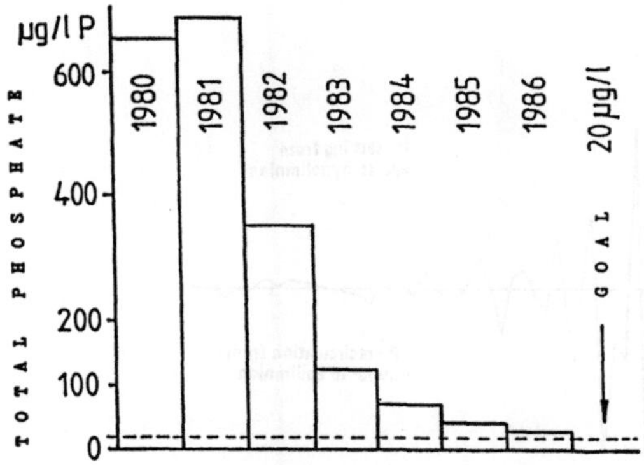

Fig. 1: Reduction of the average total phosphate concentration in Lake Schlachtensee after installation of the phosphate elimination plant in 1981

3.2. MODELLING PHOSPHATE DYNAMICS BY A COMPARTMENT MODEL

Using a specially developed compartment model, the phosphate dynamics in the restored lakes have been calculated based on morphological and monitoring data. The vectors as indicated in fig. 2 show the transfer activities for phosphate between three epilimnetic and three hypolimnetic sections of Schlachtensee.

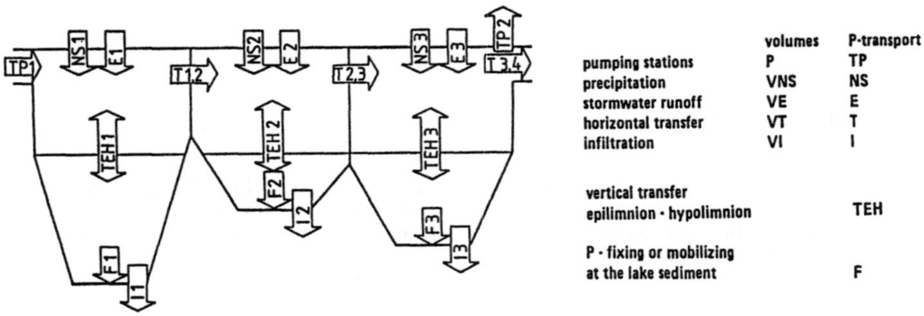

Fig. 2: P-balance model for Lake Schlachtensee

Each of these vectors represents a group of reactions. E.g. the transfer between epi- and hypolimnetic compartements is determined by the settling velocities of plankton and detritus, chemical precipitation and coagulation of organismes (P-settling from epi- to hypolimnion) on one hand, and convection, reactivation of chemically bound phosphate by red-ox-processes in the water column and at the sediment-water interface, mineralization of organically bound phosphate by decomposition of organic matter on the other hand (P-recirculation from hypo-to epilimnion).

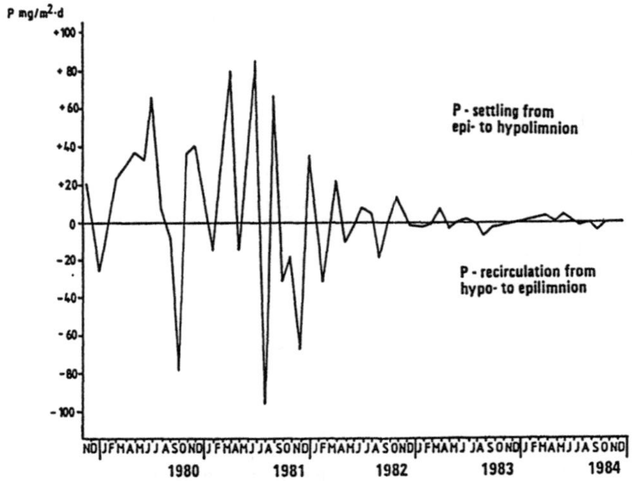

Fig. 3: P-exchange at the thermocline of Lake Schlachtensee

As fig. 3 demonstrates, the exchange activity at the thermocline of Lake Schlachtensee has been reduced from about 100 mg / $m^2 \cdot d$ to only a few mg/$m^2 \cdot d$ within only two years from the beginning of the phosphate removal. These calculations have recently been confirmed by mesurements of phosphate settling rates in sediment traps: During summer 1987 settling rates between 0.34 and 0.93 mg / $m^2 \cdot d$ have been determined.

3.3. RELATIONS BETWEEN OXYGEN DEFICIENCY AND P - ACCUMULATION IN THE HYPOLIMNION

According to the reduced sedimentation of organic matter and nutrients the oxygen consuming processes fell to a considerably lower activity. Before the treatment Lake Schlachtensee had an anaerobic hypolimnion from mai to september which spread quite rapidly over the total sediment surface below 5 m depth. In some years the autumn circulation was not strong enough to saturate the whole water body with oxygen, and anoxic conditions occured under ice cover. The time and area related extension of anaerobic conditions has been combined to an oxygen deficiency index:

$$I_{od} = \frac{F_{od}}{F_c} \times \frac{t_{od}}{t_{str}}$$

with

I_{od} = oxygen deficiency index

F_{od} = area of hypolimnetic sediment surface below the level of the isopleth with an oxygen content of 0 mg/l (in m^2)

F_c = total sediment surface area of the lake compartment (in m^2)

t_{od} = time during which a certain area (F_{od}) is to be found under anoxic conditions (d)

t_{str} = duration of thermal stratification of the lake compartment (d)

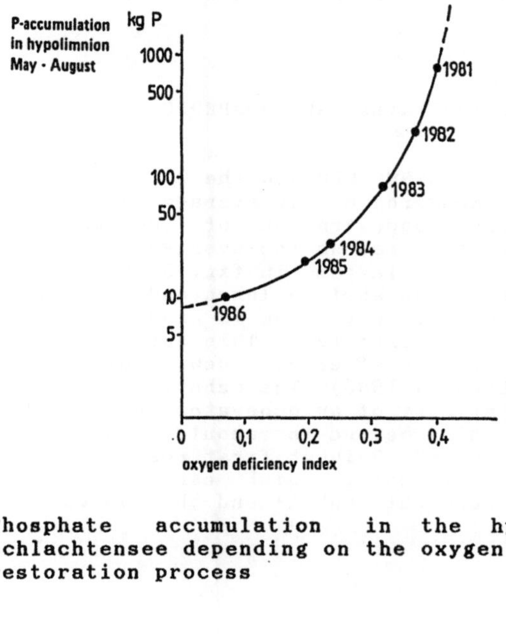

Fig. 4: Phosphate accumulation in the hypolimnion of Lake Schlachtensee depending on the oxygen deficiency during the restoration process

This index covers all oxygen consuming activities in the hypolimnion and at the sediment-water interface and gives an integrated result of the transport of oxydizable organic and inorganic material to the deep layers of the lake. One of the most important effects of oxygen deficiency in eutrophic lakes is the remobilization of phosphate which provides nutrients for the next algal blooms after the circulation. During the restoration we could report the dependence of phophate accumulation of the oxygen deficiency index (see fig. 4). This demonstrates two interesting effects:

- Even though we still have a temporary anoxic hypolimnion, the reactivation of phosphate from the sediments is continuously reduced during the years after beginning of the restoration.
- Even in the case of oxic conditions at the sediment surface throughout the year, there will be a considerable accumulation of phosphate in the hypolimnion due to equilibria which stabilize between mineralization activities and chemical exchange processes

4. RESILIENCE PHENOMENA AND BIOLOGICAL RESPONSE

All reactions within the ecosystem of the restored lake show a special time dependence. While dilution of the nutrients in the water column and distribution of the inflowing water with low nutrient content just follow physical influences like wind, temperature stratification and morphology of the lake, the complex interaction between nutrient concentration, nutrient availability, transfer in the food web and sediment-water interaction show different reaction velocities. All these processes are connected in a manner we do not understand sufficiently. But the development in Lake Schlachtensee gives some deeper insight into the interdependences.

4.1. RELATIONS BETWEEN CHLOROPHYLL AND PHOSPHORUS

One of the approaches of VOLLENWEIDER and the OECD study group connects the ecosystem reaction, measured as the average annual chlorophyll-a-concentration, with the concentration of the main stressor of eutrophying lake systems, measured as the average annual in-lake total-phosphorus concentration. The diagram in fig. 5 shows the distribution curve along which eutrophying shallow lakes and reservoirs seem to group. If the phosphorus concentration is reduced, a proportional ecosystem reaction was to be expected. This has been the case on a gradual scale in several systems after e.g. sewage diversion (Veluvemer, Alderfen Broad, SAS & VEERMEIJ 1986). The behaviour of Lake Schlachtensee demonstrates the elasticity of an ecosystem when short term changes take place without reaching beyond threshould values of the growth limiting factor. The reduced P-input first results in a shift of phosphorus utilization efficiency, which can be calculated as the quotient between the average chlorophyll and phosphorus concentrations from year to year:

	(Chl.a) µg/l	(Pt) µg/l	(Chl.a)/(Pt)
1980 (before treatment)	78	600	0.13
1982	57	200	0.35
1984	56	50	1.1
1986	8	25	0.35

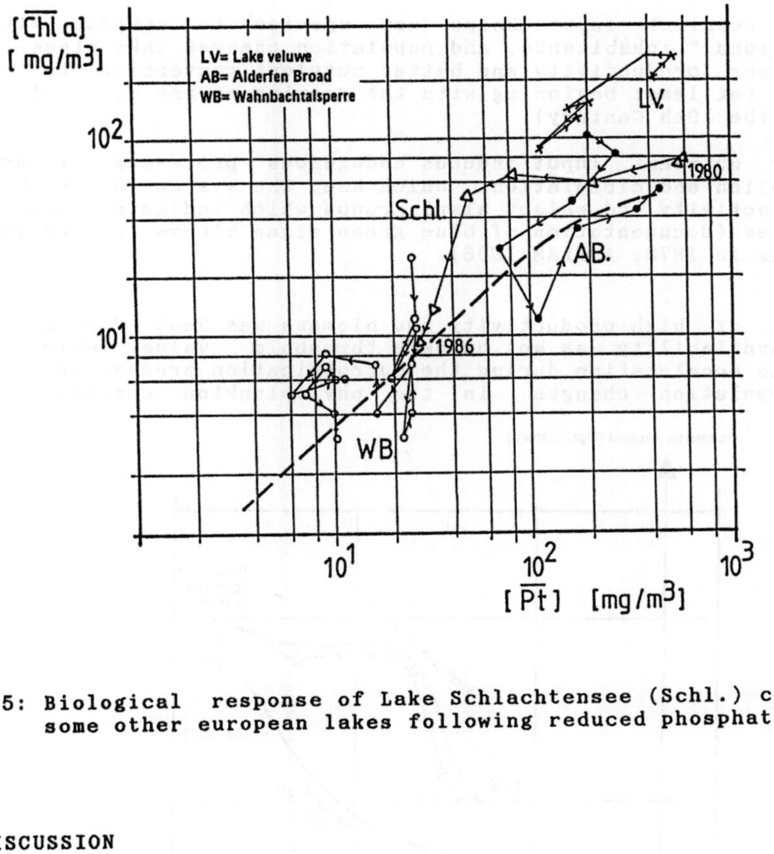

Fig. 5: Biological response of Lake Schlachtensee (Schl.) compared to some other european lakes following reduced phosphate load

5. DISCUSSION

All chemical, physical and biological processes show strong interference during the process of oligotrophication of lakes. This can be demonstrated with strong evidence at the restoration process of Lake Schlachtensee, which is by far the most demanding in phosphate elimination of any restoration measures taken up to now in the whole world. And after five years of phosphate removal from the inflowing water it is also the most advanced as regards the effects achieved by restoration as expressed by the phosphate concentration and the development of algae (qualitatively by changing the population to more favourable species and quantitatively by reducing the biomass of algae).

Nevertheless there were some findings of great importance for future planning of other phosphate elimination plants and for our understanding of lakes during environmental stress.

As we could imagine from data that had been collected in former decades and centuries, eutrophication is a long term process, which may have followed several steps during the development of lakes in the Berlin area:

1. Nutrient input from the catchment area increases their availability for organisms, which grow in the lake, but which have in the case of phosphate to compete with very efficient chemical processes (since the forest clearing from about 1200 p.Chr.).

2. Excess nutrient input helps new species to compete with the "traditional" inhabitants, and population changes take place leading to higher productivity and better nutrient conversion to organic matter (at least beginning with the growing of the city of Berlin during the 19th Century).

3. Further nutrient input induces endogenous processes of nutrient utilization and circulation, which keep the system on a high level of bioactivity and select algal groups which indicate rapid ageing processes (documentation of blue green algae blooms in Lake Tegel as early as in 1878, ANKLAM 1906).

This level of high productivity and biomass was kept for until the nutrient availability was set below a threshold value, which either triggers the acceleration during the eutrophication process or which may induce population changes in the phytoplankton community during

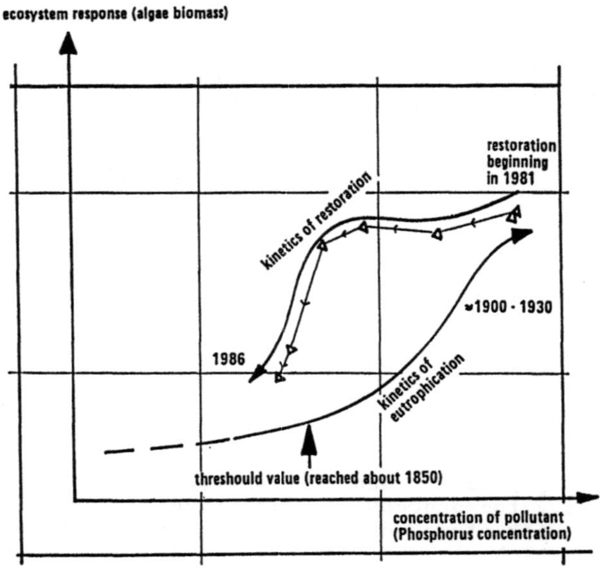

Fig. 6: Kinetics of eutrophication (in a historical context) and restoration of lakes in the Berlin area (according to the data of Lake Schlachtensee restoration)

oligotrophication. In the latter case a strong biological response as shown by Lake Schlachtensee indicates the success in our efforts to change the limiting factor, which had been light, nitrogen or the carbondioxide after the primarily phosphate limited system had been destabilized. This destabilization can take centuries in great lakes, but it can happen within a few years, as it occured in many man-made lakes and reservoirs throughout the world.

The efficiency of restoration is depending on the knowledge of the stress factor to be changed, the threshold level where this factor becomes a regulator and on interdependences between different characteristics of the complex ecosystem. It seems to be quite simple in the case of lake eutrophication and restoration - nevertheless only few cases of efficient lake restoration have been successfully carried out. In the case of the Baltic or the North Sea eutrophication and pollution we have to be aware of very complicated conditions as well as in the case of the "Waldsterben".

In any case it will be important to reduce ecosystem stress factors to a very low level. Our present view of destroyed ecosystems represents the effects of a stress factor or even a combination of stress factors, which had triggered ecodynamic changes a long time ago. Thus we could follow two ways of understanding the changes:

- The ecological strategy, which looks upon data we have about the normal and the stress function of the ecosystem.
- The historical strategy, which helps us to define an undestroyed situation of the ecosystem according to our knowledge about historical development of stressors.

In the case of Lake Restoration in Berlin we were successful in following an ecological strategy, which could be confirmed by a historical analysis of the development and ecological studies of the changes that took place.

6. REFERENCES

ANKLAM, G.: Die Wasserversorgung Berlins bisher und in der Zukunft. Ber.Dtsch.Bot.Ges. 26: 505 - 519 (1906)

HÄSSELBARTH, U.: Die Phosphoreliminierung aus dem Zufluß von Seen zur Sanierung stehender Gewässer. Z.Wasser Abwasser Forsch. 10: 120 - 125 (1977)

KLEIN, G.: Seensanierung in Berlin. Wasserkalender 1988: 46 - 74 Erich Schmidt-Verlag Berlin (1987)

SAS, H. & VERMEIJ, S.: Second Interim Report of the project "Eutrophication Management in International Perspective", Inst.v. Milieu- en Systeemanalyse, Amsterdam (1986)

A Comparison of Fish Size-Composition in the North Sea and on Georges Bank

J.G. Pope[1], *T.K. Stokes*[1], *S.A. Murawski*[2], *and S.I. Idoine*[2]

[1] Ministry of Agriculture, Fisheries and Food,
Fisheries Laboratory, Lowestoft, Suffolk NR33 0HT, UK

[2] National Marine Fisheries Service,
Northeast Fisheries Center, Woods Hole, MA 02543, USA

ABSTRACT

Aggregated length distributions of all fish species combined may be obtained from groundfish surveys conducted by the UK and the USA for the North Sea and Georges Bank respectively. In both areas, the aggregate logarithmic numbers at length decline linearly over the fully selected size range. Comparison of these distributions through time and between areas indicates that slope varies considerably between areas but much less through time within an area. Such observed changes are consistent with the slope of the line being a broad indicator of the exploitation regime, the more heavily exploited North Sea having a steeper decline with size than that of Georges Bank. The slopes of distributions on Georges Bank show more gradual declines in years after management regimes were imposed. It seems possible, therefore, that aggregate size distribution may be a characteristic property of exploited fisheries systems - perhaps an emergent property.

INTRODUCTION

Biomass at length distributions have been shown to be conservative features of unperturbed ecosystems [1,2]. It is interesting to ask if this is also the case for exploited fisheries systems. Pope and Knights [3] showed that aggregated numbers at length distributions seem to be a consistent feature of results from North Sea groundfish surveys. They also showed that systematic differences existed between such results for surveys of Faroe Bank and the northern North Sea. Those differences were arguably consistent with the different exploitation histories of the two areas, with the heavily exploited northern North Sea having less large piscivores and more small fish prey than the less exploited Faroe Bank. They suggested, however, that a more useful comparison might be made by considering an area for which groundfish survey data spanned periods of different exploitation levels. In particular, they identified the US surveys of Georges Bank [4] and the Scottish surveys of the northern North Sea [5] as being most suitable for such an investigation.

In this paper, the Woods Hole groundfish survey data for Georges Bank and the Lowestoft groundfish survey data for the North Sea are presented for the years 1963-86 and 1977-86 respectively. During that time, Georges Bank has been subjected to various exploitation regimes. Specifically, from the late 1950s to 1970, management was by mesh regulation. During the period 1963-70 (for which we have data), however, there was a rapidly increasing international exploitation of the fishery, and in 1970 the International Commission for the Northwest Atlantic Fisheries (ICNAF) instigated management by catch quotas. Starting in 1977, the USA initiated the US coastal state management regime which exists to the present day.

By contrast, the North Sea has not seen such large shifts in exploitation over the period for which we have groundfish survey data. The herring fishery was closed between 1977 and 1983 and is now recovering, and in 1983 management by single-species total allowable catches (TACs) was started. It is doubtful, however, that TACs were effective until 1986. Therefore, although the North Sea has had officially sanctioned changes in management strategy, exploitation has not seen major shifts. There have, though, been large fluctuations in recruitment of commercially important species. Cod, for instance, has recently displayed a record low recruitment followed by very high recruitment years [6].

It is worthwhile investigating whether these highly variable single-species effects are noticeable in the species-aggregated system or if the features of the aggregate system dominate, to the extent of being characteristic (emergent) properties.

RESULTS

Aggregate fish numbers at length distributions (on a logarithmic scale) for the North Sea from 1977-86 and Georges Bank from 1963-86 are shown in Figure 1. The points to note are the regularity of numbers at length from year to year for each system and that the distributions decline linearly over the fully selected size ranges. With regard to the first point, although the distributions cover a range of three to four orders of magnitude, there is rarely a difference of more than a factor of four inter-annually at any length group. This consistency occurs despite the highly variable year-class strengths and the polymodal length distributions of individual species. In the case of Georges Bank, it also spans three different management regimes. The linear declines of the length distributions are seemingly consistent from year to year within systems but different between systems; this is possibly a reflection of different exploitation levels.

The above consistencies in the length distributions are more notable when the species compositions for the two areas are observed. Percentage composition for the two areas, by broad taxonomic grouping for earlier and later years, is shown in Figure 2. In both areas, there are marked changes in the proportions of prime commercial fish of large size. In particular, the gadoids (cod, haddock, whiting, saithe) are reduced in the high length groups in the more recent years, their place substantially being filled by the commercially less exploited rays and dogfish. In the North Sea, the remaining change is in the number of 'other' fish which are (at least currently) commercially uninteresting. We can also note the recovery of the herring stocks in the North Sea.

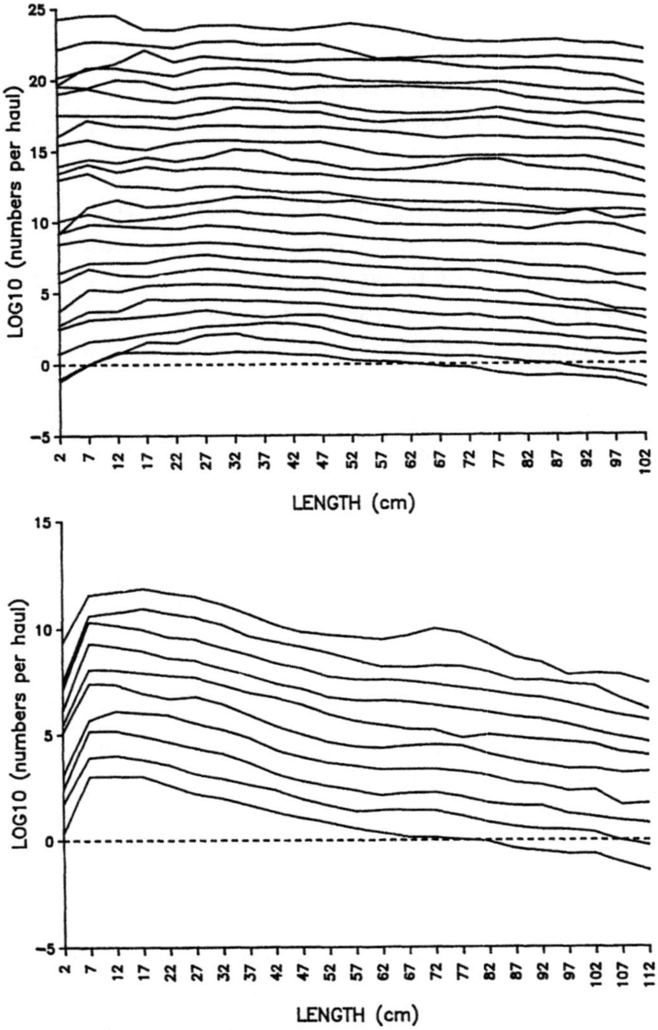

Figure 1: Annual, aggregated, fish length distributions (normalised to numbers per haul) on a log10 scale. The bottom lines refer to the earliest years and each successive, higher line is the successive year with 1 added to the ordinate scale: top - Georges Bank; bottom - North Sea.

If we perform analyses of covariance (ANCOVA) on the log numbers at length data with length as covariate, year as a factor and a length.year interaction, we can study more closely the effects of exploitation on the log length distributions. For our purposes, we have performed ANCOVAs for lengths between 20 and 110 cm (the commercially selected size range), which removes old fish for which there are relatively few data and 0-group fish (fish in their first year), for which sampling is less satisfactory and the numbers of which are highly variable. Plots of the length.year interactions (slopes) are shown in Figure 3.

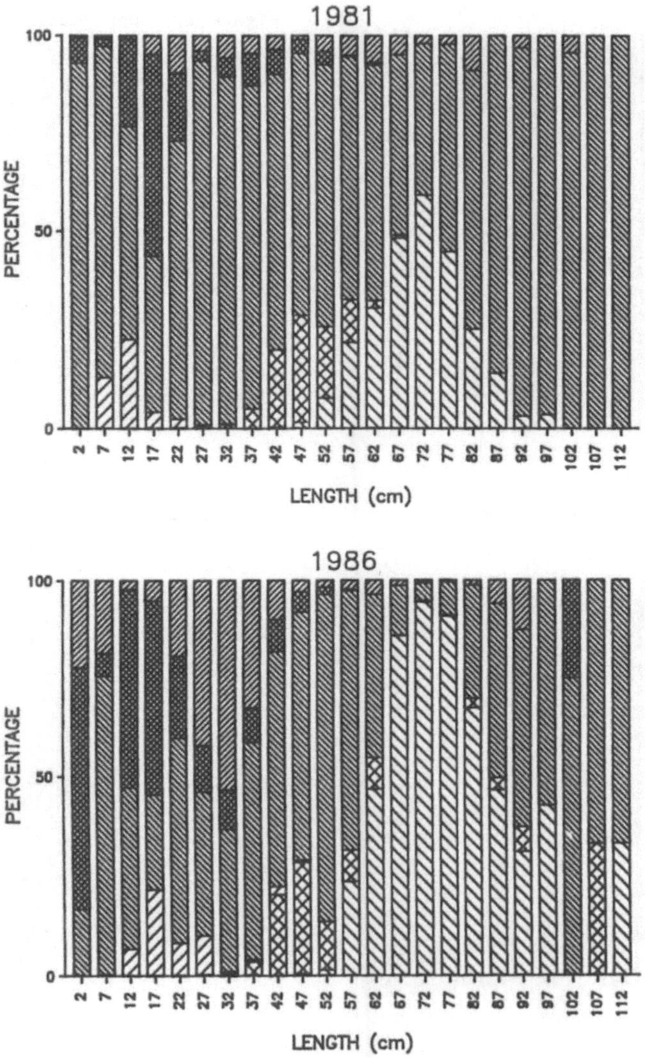

Figure 2: Percentage size compositions by broad taxonomic grouping for: North Sea (1981 and 1986); Georges Bank (1969 and 1982).
Key: ◪ dogfish; ⊠ rays; ▨ clupeids; ▧ gadoids; ▩ flatfish; ▨ others.

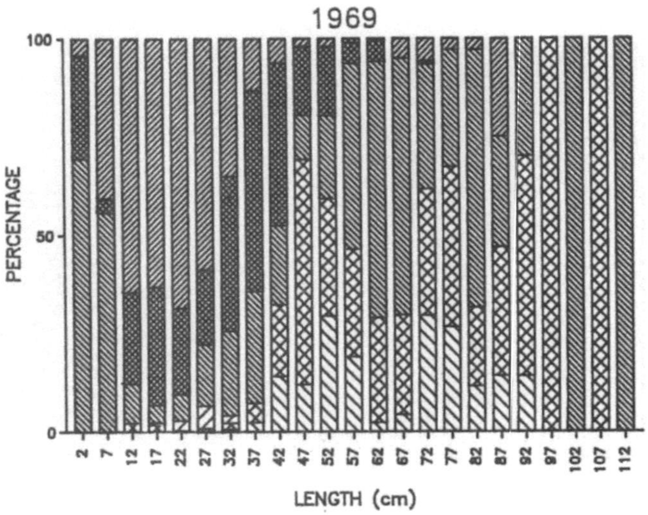

Figure 2: Continued.

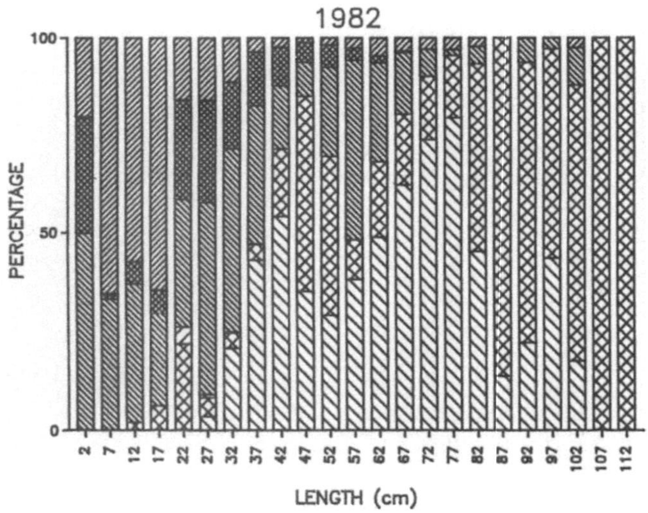

For the North Sea data, there is no apparent annual pattern and no interaction term is significantly different at the 5% level. The log-linear decline in numbers against length is approximately 0.1 per length group in all years. The observed fluctuations may be noise or possibly due to the variable numbers of small fish that are recruited to the fishery.

The picture for Georges Bank is more striking and needs more explanation. Firstly, the observed slopes are not as steep as those for the North Sea analysis. This may be because the North Sea fisheries tend to exploit smaller fish, which will lead to a cumulative biomass at length curve which is more skewed to the left [3]. Secondly, there are apparent changes in slope which tie in with the changes in exploitation des-

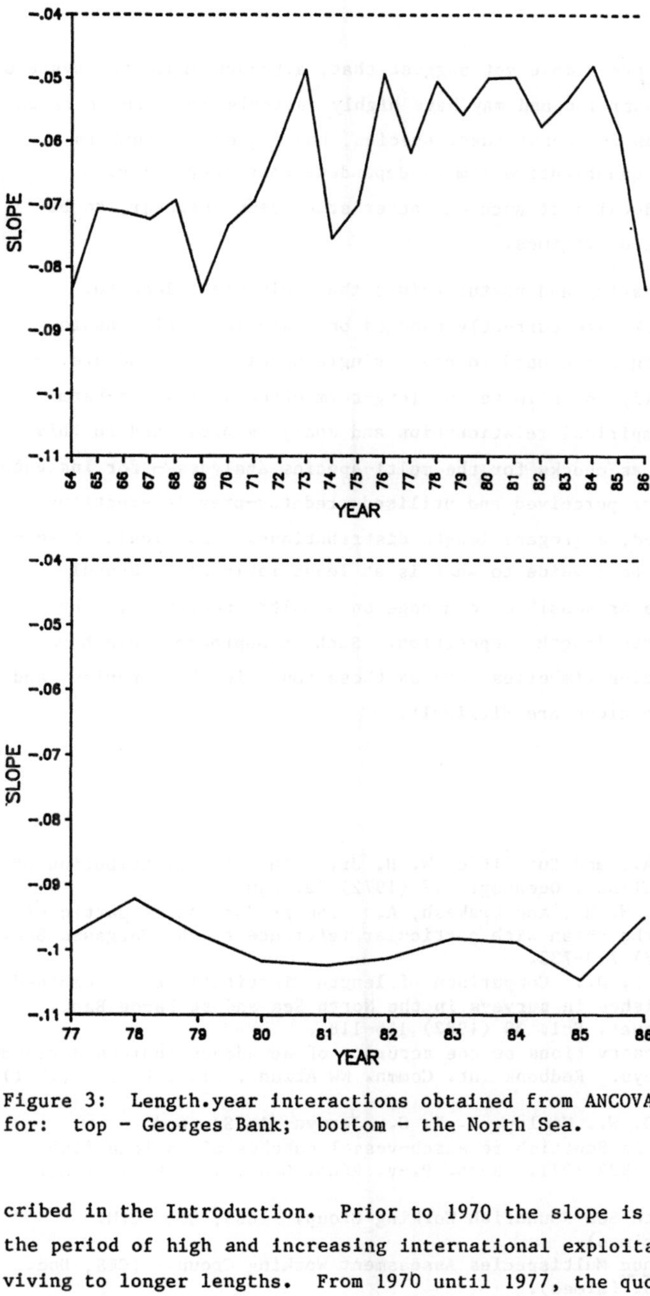

Figure 3: Length.year interactions obtained from ANCOVAs as explained in the text for: top - Georges Bank; bottom - the North Sea.

cribed in the Introduction. Prior to 1970 the slope is steep - this corresponds to the period of high and increasing international exploitation when fewer fish were surviving to longer lengths. From 1970 until 1977, the quota management appears to have softened the over-exploitation and more fish survived to larger sizes, the differences, however, are not significant at the 5% level and the pattern is erratic. From 1977 onwards, under the coastal state management scheme, the slope has remained at a fairly steady, low value (which is significantly different to previous years at the 5% level), except for the last year which shows a return to previous low levels. Whether this is a real effect or an artefact is not yet known.

DISCUSSION

The data that are presented are simple yet suggest that, although exploited systems display changes in species composition and may have highly variable recruitment rates and polymodal size distributions for constituent species, the aggregated numbers at length display qualitative and quantitative time-independent constancy - i.e. a characteristic property. The details of such characteristics seem, however, to be alterable by changing exploitation regimes.

There are a number of interesting and useful things that this might lead to. Firstly, for example, fish stocks are currently managed on a single-species basis although multi-species assessments are used to check single-species work and are starting to be used strategically to estimate the long-term effects of short-term management options [7]. The empirical relationships and analyses presented in this paper might prove to be useful as checks for the multi-species analyses - for instance in questions such as whether our perceived and utilised predator-prey interactions make sense in terms of predicted, aggregate length distributions. Empirically observed distributions should serve as a guide to what is at least rational. Secondly, whether it might prove possible or sensible to manage on a multi-species level by aiming for a particular aggregate length composition. Such an approach could have uses in very complex multi-species fisheries such as those found in the 'Tropics' and where single-species data collections are difficult.

REFERENCES

[1] Sheldon, R. W., Prakash, A., and Sutcliffe, W. H. Jr.: The size distribution of particles in the ocean. Limnol. Oceanogr. 17 (1972) 327-340.

[2] Sheldon, R. W., Sutcliffe, W. H., and Prakash, A.: The production of particles in the surface waters of the ocean with particular reference to the Sargasso Sea. Limnol. Oceanogr. 18 (1973) 719-733.

[3] Pope, J. G., and Knights, B. J.: Comparison of length distributions of combined catches of all demersal fishes in surveys in the North Sea and at Faroe Bank. Can. Spec. Publ. Fish. Aquat. Sci. 59 (1982) 116-118.

[4] Grosslein, M. D.: Some observations on the accuracy of abundance indices derived from research vessel surveys. Redbook int. Commn. NW Atlant. Fish. Part 3 (1971) 249-266.

[5] Richards, J., Armstrong, D. W., Hislop, J. R. G., Jermyn, A. S., and Nicholson, M. D.: Trends in Scottish research-vessel catches of various fish species in the North Sea, 1922-1971. Rapp. P.-v. Réun. Cons. int. Explor. Mer 172 (1975) 211-224.

[6] Anon.: Report of the North Sea Roundfish Working Group. ICES, Doc. C.M. 1986/Assess:16, 163 pp. (mimeo).

[7] Anon.: Report of the ad hoc Multispecies Assessment Working Group. ICES, Doc. C.M. 1987/Assess:9, 171 pp. (mimeo).

Generic Models of Continental Shelf Ecosystems

W. Silvert

Modelling and Statistics Group, Biological Sciences Branch, Bedford Institute of Oceanography, P.O. Box 1006, Dartmouth, Nova Scotia, B2Y 4A2, Canada

Ecosystem modelling work at the Bedford Institute of Oceanography within what was formerly called the Marine Ecology Laboratory has been based on comparative studies of similar types of ecosystems. Through cooperative work with Dutch, German, and British scientists a generic approach to estuarine modelling was developed in which the first step in creating a new site-specific simulation model involves reparametrization and modification of an existing model. This methodology was successfully tested by constructing an ecosystem model of the Cumberland Basin in the Bay of Fundy (Gordon *et al.* 1986)

The same approach is now being used for the development of continental shelf models. This presentation outlines the development of a simulation model of the Grand Banks ecosystem designed to simulate the potential effects of oil spills associated with exploitation of the Hibernia oil field on the northeast part of the shelf, an area of high biological productivity and strong currents. It also covers the subsequent creation of a 'Theoretical Macrocosm', which is a generic model of a single spatial compartment of a continental shelf ecosystem which can be used both to investigate the biological processes occurring on the shelf and as a building block for the assembly of site-specific models.

The generic modelling approach requires a highly aggregated functional treatment of biological variables and is not appropriate for detailed population modelling of individual species. We have used a size-structured system description based on the flows of energy and nutrients. This works well for marine ecosystems because size is a good predictor of both trophic status and turnover time.

Present plans are to apply the 'Theoretical Macrocosm' approach to analysis of field data from the Scotian Shelf and Georges Bank, as well as to extend its use to coastal waters by investigation of several well-studied bays along the coast of Nova Scotia. Future applications are expected to lie in the rapid development of site-specific models for multi-species fishery analysis and environmental impact assessment.

Background

The decision to work on development of generic ecosystems models grew out of several years' modelling work, first with estuarine models and later with models of continental shelf ecosystems. It became evident that site-specific models had major similarities even though the sites appeared to be fundamentally very different, and that exploitation of these similarities would serve several valuable functions:

- Greatly reduce the amount of work that goes into developing each new model.

- Discourage "sins of omission", since justification is required to omit features that have been found relevant to other modelling programs and have therefore been incorporated in the standard modelling formulation.

- Foster a comparative approach, in which each new system is analysed in terms of its similarities and differences *vis a vis* other systems.

The decision to formalize this generic approach involved three projects:

1. In 1983, as a result of a comparative study (Gordon and Baretta 1982) of the Ems-Dollard estuary on the Dutch-German border and the Cumberland Basin in the upper reaches of the Bay of Fundy, an ecosystem model of the Cumberland Basin was developed. The approach taken was that a team of Dutch, German, and Canadian researchers first translated the Ems-Dollard model into a representation of the Cumberland Basin by simply reparametrizing it, and only after the team had agreed that the results looked promising was it decided to rewrite the Ems-Dollard model to fit the Cumberland Basin (Gordon *et al.* 1986, Keizer *et al.* 1987).

2. At the same time as the Cumberland Basin model was being developed, a team consisting largely of the same personnel undertook development of a model of the Grand Banks of Newfoundland to assess the possible impacts of oil spills on the marine ecosystem. The model was made as general as possible, and it was agreed at the outset that the structure of the model should be such that it could be applied to Flemish Cap and other important fishing grounds off the Newfoundland coast.

3. Upon completion of the Grand Banks and Cumberland Basin models in 1986, it was concluded that these experiences provided good justification for developing a more formalized generic approach to the creation of future models. Since it appeared that we were more likely to be involved in continental shelf models than estuarine models at that time, we began work on a generic model of continental shelf ecosystems, which is referred to as the "Theoretical Macrocosm".

This paper describes the structure and properties of the Theoretical Macrocosm and outlines present plans for further development and application of the model. Although the Theoretical Macrocosm is based on the Grand Banks model, it has developed beyond it in several respects, and present development plans are focussed on areas which were poorly covered in the Grand Banks work because of lack of adequate data.

Time and Space Scales

One of the greatest problems in ecosystem modelling is that the biological and physical processes involved cover an enormous range of time and space scales, from hours and centimeters for single-celled organisms to years and thousands of kilometers for the larger marine mammals (Steele 1978). The dominant scales for continental shelf models are determined by the size of these systems, typically several hundred kilometers, and we normally carry a simulation through one or more annual cycles with a resolution of one day.

At these scales the processes governing primary production and microbial decomposition can only be described in an aggregate way. Although the integration algorithm used in the BSIM simulation software package, forward differencing with a variable time step, permits arbitrarily small time intervals (Silvert 1986), the resolution of the model does not incorporate small-scale effects such as diurnal vertical migration and tidal cycles. Processes which depend on these effects, such as primary production, are modelled by routines which are based on analytical integration of light and other variables over the full day used as a time step (Brock 1981, Evans and Parslow 1985).

Similarly, large-scale processes such as the life cycle of fish and the migrations of whales are not intended to be represented by the dynamics of the model. These could be incorporated if the model were routinely run for intervals of several years, but processes at these time scales also tend to cover larger space scales than the region being modelled, and therefore are strongly dependent on boundary conditions.

Transport

Most continental shelf ecosystems are strongly affected by physical transport. Several different types of transport have to be accounted for in every marine model, and these have to be handled differently in the context of a generic model.

Advective exchange with adjoining regions is usually the dominant contribution to water mass property transport, although the magnitude and temporal variability of the advection are generally site-specific. Advection may be purely horizontal, but it generally has a vertical component (*e. g.*, upwelling and downwelling). On the Grand Bank, for example, the relatively steady horizontal flow of the Labrador Current is the major influence, particularly over the sides of the Bank, but time-varying flow events associated with the Labrador Current and with Gulf Stream rings also contribute to exchange. One the Scotian Shelf there is a seasonally varying along-shelf transport of outflow water from the Gulf of St. Lawrence as well as episodic exchange across the shelf break due to Gulf Stream rings and wind-induced upwelling.

In constructing the Theoretical Macrocosm we have dealt with different types of physical processes in the following way:

- Horizontal advective processes as described above are important and have to be taken into account in any marine ecosystem modelling study, but they are not built into the Theoretical Macrocosm. We have taken the view that it is essential to distinguish effects arising from internal dynamics from those caused by exchange with surrounding water masses. In fact, if one imagines the extreme case of a system totally dominated by strong advection, it is evident that the internal dynamics would become unimportant and that the internal structure of the system would closely reflect that of the boundary regions. The Theoretical Macrocosm provides a way to understand the *internal* dynamics of an ecosystem, but it cannot by itself resolve the relative importance of internal and external factors in the system dynamics.

- Vertical advection, such as upwelling and downwelling, give rise to net vertical transport within the Theoretical Macrocosm itself and therefore play a major role in the internal dynamics. Upwelling in particular accounts for a large part of the productivity of coastal waters, and therefore must be considered.

- Diffusion by small-scale turbulence is a mechanism of particular significance to vertical exchange over the water column. It has been a basic feature of the Grand Banks model and the Theoretical Macrocosm from the beginning. Unlike advective transport, it does not involve exchange with adjoining water masses, and therefore it can be viewed as affecting only the internal dynamics of the model.

- Frontal processes represent perhaps the most difficult physical effects to incorporate in continental shelf models, in part because of the strong spatial physical and biological inhomogeneities associated with them. There are many different types of fronts, and although their importance is well recognized, the mechanisms through which they affect biological productivity are as yet poorly understood. We hope that combined field and theoretical research planned for the coming years on frontal systems in the Gulf of Maine will help resolve this issue and improve the representation of fronts within the Theoretical Macrocosm.

Physical Structure

The Theoretical Macrocosm can be viewed as a large 'aquarium' positioned on the continental shelf which is uniform in the horizontal direction. The interior of the 'aquarium' is considered to be horizontally uniform, which can be interpreted as meaning that internal horizontal exchange is sufficiently rapid to homogenize the system on time scales of biological interest. It has not yet been resolved whether at a later stage it may prove necessary to incorporate horizontal structure within the Theoretical Macrocosm. The absence of exchange with surrounding water masses (*i. e.*, water transfer through the sides of the 'aquarium'), means that some processes of biological interest are not adequately represented and have to be incorporated separately when the Theoretical Macrocosm is developed into a site-specific model.

The internal vertical structure incorporates two spatial compartments representing surface and bottom mixed layers. Vertical exchange is characterized by a vertical mixing rate, which generates a symmetric coupling of the two layers, and an upwelling rate which corresponds to a net transfer from one layer to the other. This is adequate for describing such stratification-dependent processes as the termination of the spring bloom by depletion of the nutrients in the mixed layer (Evans and Parslow 1985).

Size Structure

One of the unusual features of the Grand Banks model and of the Theoretical Macrocosm is that they are highly aggregated on the basis of size, which is a reasonable indicator of trophic status in marine ecosystems (Silvert 1981). This avoids obvious problems in including species-specific structure in a generic model, and therefore the biological compartments are defined in terms of organism size rather than on taxonomic considerations. In the Grand Banks model all animals were grouped into three size classes of zooplankton and three of fish, as shown in Fig. 1. In the Theoretical Macrocosm this aggregation has been carried even further, and both categories are condensed into five size classes of pelagic grazers.

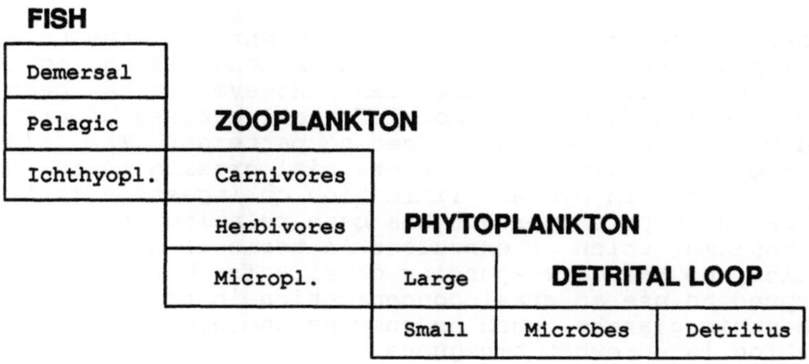

Figure 1: Size classes used in the Grand Banks model.

In both models the basic physiology of animals is expressed in terms of allometric relationships, and it was found that an allometric exponent of 0.824 proposed by Conover and Huntley (1980) on the basis of copepod feeding studies gave results consistent with physiological data on fish (Kerr 1982).

It proved difficult to parametrize a size-structured benthic submodel adequately, in large part because benthic communities exhibit characteristic size spectra which depend on sediment grain size and other site-specific attributes (Schwinghamer 1981). The present version of the Theoretical Macrocosm therefore uses three functional categories of benthic organisms: benthic detrital feeders, including most meiobenthos; benthic suspension feeders; and epibenthic carnivores, such as large crustacea.

Biological Submodels

Since the organisms in the Theoretical Macrocosm are described in very general terms, a uniform set of biological subroutines was developed to facilitate description of the living organisms within a wide range of different marine ecosystems. The basic categories are as follows:

- Animals
  - Fish
  - Zooplankton
  - Benthos
- Plants
  - Phytoplankton
- Microbes
  - Pelagic bacteria
  - Benthic bacteria

In addition, detrital matter is included as an important component of the system.

The phytoplankton subroutine is the mathematically most sophisticated, since algal growth depends on a complex interplay of light, temperature, and nutrient factors, and the submodel used is based on the work of Evans and Parslow (1985). The zooplankton and fish submodels developed for the Grand Banks model proved very similar to each other and were combined in a single submodel for the Theoretical Macrocosm. It incorporates physiological data for fish (Kerr 1982) and is based on a zooplankton grazing model of Conover and Huntley (1980).

The benthic subroutine is quite different from the combined zooplankton and fish routines, since the benthos is aggregated on functional rather than size categories. However it has been possible to develop a common feeding submodel for all classes of organisms, based on a matrix formulation of feeding patterns. The feeding model combines size selectivity with preferential grazing on the more abundant prey, with an overall limitation on ingestion rate. The major problem with this approach has been relating the density of pelagic organisms, which is expressed in terms of a volume concentration, to the corresponding density for benthic feeders. We have attempted to use an areal concentration in this case, although for some benthic grazers, such as suspension feeders, the interpretation is somewhat ambiguous.

Contaminant Effects

The Grand Banks model was developed to address the problem of oil spills on the Grand Banks, and it was recognized that contaminants can act at several hierarchical levels within an ecosystem:

- Ecosystem-level interactions between different species groups.
- Elevated mortality for some species.
- Physiological changes in individual organisms.

Mechanisms for addressing at least some of the processes acting at these different levels were developed for the Grand Banks model, and these mechanisms have been carried over to the Theoretical Macrocosm.

Mortality is relatively easy to incorporate, since it is simply added to the natural mortality terms already present in the model. Sublethal effects are more of a problem, and have been treated in two ways:

- Basic physiological quantities, such as basal respiration and growth efficiency, depend on degree of exposure. There is a fair amount of laboratory data on which to base models of the stress caused by various common contaminants.

- Grazing efficiency is the only ecosystem-level effect which is currently modelled as a function of exposure, and this is based on very limited data. Palatability is known to be affected by exposure to some substances, but this tends to be characteristic of particular predator-prey pairs and has not been incorporated separately in the model.

Subroutine structure

The subroutines in the Theoretical Macrocosm are organized around the scientific interests of the researchers involved in the project, and therefore reflect the disciplinary interests characteristic of oceanography. The major exception to this is in the subroutine SRGRAZ, which covers both fish and zooplankton.

The submodels are written in Fortran-77, since this is the language most familiar to the scientific community, and they operate within the framework of a special modelling package called BSIM, which was developed expressly for simulating marine ecosystems (Silvert 1986). Variables of interdisciplinary importance, such as nutrient levels, have been carefully identified in plenary meetings of all scientists in the project and are communicated between subroutines by the use of COMMON blocks.

The major subroutines are as follows:

PHYS Calculation of solar radiation, water temperature, and other important physical variables at each time step.

VERTMX Vertical mixing is dealt with separately from other physical processes. This subroutine also includes a mechanism for incorporating storm events, which are characterized at present as periods of increased mixing. In future versions of the models the relationship between storm events and horizontal transport, particularly upwelling, will be treated more fully.

SPILL Developed originally for the Grand Banks model, this subroutine calculates distributions of oil or other pollutants, based on different input scenarios. These distributions are used within the biological subroutines to calculate lethal and sublethal toxic effects.

SRBACT Biological subroutine for microbes and detrital carbon, describing the "detrital loop".

SRPHYT Biological subroutine for phytoplankton.

SRGRAZ Biological subroutine for zooplankton, fish and resident marine mammals.

SRBENT Biological subroutine for benthos.

EXPORT Losses to system due to fishing and migratory predators (birds and whales).

Several lower level subroutines contain code which is common to SRGRAZ and SRBENT, describing the feeding behaviour and physiological rates of animals and their interaction with oil spills. Another subroutine, RECYCL, is called by the other biological subroutines and handles the recycling of waste products into different forms of nutrients. A portion of the waste products is transferred directly to bacterial production as well.

Sensitivity Analysis

Extensive sensitivity analysis has been carried out on both the Grand Banks and Theoretical Macrocosm models (Hickman and Silvert, 1986). As has been found in other studies (e.g., Kremer and Nixon 1978), parameters which affect grazing and metabolic rates are among the most important. Small changes in growth efficiency lead to significant differences in the biomasses of the higher trophic levels and can also affect system stability.

In early versions of the Grand Banks model it proved virtually impossible to maintain zooplankton and fish stocks without careful analysis of the feeding rates and reevaluation of the laboratory data on which they were based. We concluded that the role of food patchiness had not been adequately incorporated in the model, and we had to add a factor to represent the degree of enhancement of feeding efficiency due to the ability of predators to find and exploit localized concentrations of prey. Sensitivity analysis on this factor has enabled us to estimate its value for the Grand Banks to within a fairly narrow range, but this will remain a focus of research attention as more detailed versions of the Theoretical Macrocosm are developed.

The results of the sensitivity analysis have played an important role in the development of the Theoretical Macrocosm and have also provided a valuable guide to data needs and experimental planning in the organisation of further modelling projects. We cannot overemphasize the importance of sensitivity analysis in the planning and evaluation of ecosystem modelling projects.

Summary

The development and use of the Grand Banks and Theoretical Macrocosm models has shown that a generic approach to modelling based on comparative studies is a productive and valuable way to analyze ecosystems. The Theoretical Macrocosm in particular is providing a useful framework for synthesizing multidisciplinary studies on continental shelf systems, and previous work with estuarine models has already established the value of the modelling approach for identifying research needs and for providing a quantitative framework for evaluating the significance of experimental results.

Research currently under way will establish how effective the Theoretical Macrocosm approach is to providing a basic building block for more detailed site-specific models. It clearly has a role in the research process, but the relative merits of alternate modelling approaches in answering specific questions about changes in the marine environment will require extensive application and evaluation to resolve.

Acknowledgments

I am indebted to many of my scientific colleagues for their help in developing the models described here. Special thanks are due to Don Gordon, Paul Keizer, Wolfgang Ebenhoeh, John Loder, Jake Rice, and Charles Hickman for the exceptional efforts which they all made to bring these projects to successful completion. I also thank John Loder, George White, and Paul Keizer for their helpful comments and suggestions about this paper.

References

Brock, Thomas D. 1981. Calculating solar radiation for ecological studies. Ecological Modelling 14:1-19.

Conover, R. J., and M. E. Huntley. 1980. General rules of grazing in pelagic ecosystems. In *Primary Productivity in the Sea*, P. G. Falkowski, Ed., Plenum Press, pp. 461-485.

Evans, Geoffrey T., and John S. Parslow. 1985. A model of annual plankton cycles. Biol. Oceanography 3:327-347.

Gordon, D. C. Jr., and J. W. Baretta. 1982. A preliminary comparison of two turbid coastal ecosystems: the Dollard (Netherlands-FRG) and the Cumberland Basin (Canada). Hydrobiol. Bull. 16: 255-267.

Gordon, D. C. Jr., P. D. Keizer, G. R. Daborn, P. Schwinghamer and W. L. Silvert. 1986. Adventures in holistic ecosystem modelling: the Cumberland Basin ecosystem model. Neth. J. Sea Res. 20:325-335.

Hickman, Charles, and William Silvert. 1986. Sensitivity Analysis of the Grand Banks Ecosystem and One-Box Models. Set of 5 MEL Internal Reports. Pp. 64.

Keizer, P. D., D. C. Gordon, Jr., P. Schwinghamer, G. R. Daborn and W. Ebenhoeh. 1987. *Cumberland Basin Ecosystem Model: Structure, Performance and Evaluation*. Can. Tech. Report Fish. Aquat. Sci. No. 1547.

Kerr, S. R. 1982. Estimating the energy budgets of actively predatory fishes. Can. J. Fish. Aquat. Sci. 39:371-379.

Kremer, J. N., and S. W. Nixon. 1978. *A Coastal Marine Ecosystem: Simulation and Analysis*. Springer-Verlag. Pp. 217.

Schwinghamer, P. 1981. Characteristic Size Distributions of Integral Benthic Communities. Can. J. Fish. Aquat. Sci. 38:1255-1263.

Silvert, William. 1981. Principles of ecosystem modelling. In *Analysis of Marine Ecosystems*, A. R. Longhurst, Ed., Academic Press, London., pp. 651-676.

Silvert, William. 1986. ***BSIM*** *Users' Guide and Reference Manual*. Pp. 36. Available from the author.

Steele, J. H. 1978. Some comments on plankton patches. In *Spatial Pattern in Plankton Communities*, J. H. Steele, Ed., Plenum Press, pp. 1-20.

Part III

Ecosystems Analysis

On Quantifying the Effects of Formal and Final Causes in Ecosystem Development

R.E. Ulanowicz[1] *and A.J. Goldman*[2]

[1]University of Maryland, Chesapeake Biological Laboratory, Solomons, MD 20688, USA
[2]Department of Mathematical Sciences, The Johns Hopkins University, Baltimore, MD 21218, USA

Abstract

In physics and in traditional biology it has sufficed until now to describe phenomena as the results of purely material or efficient causes. However, a growing number of biologists and philosophers think that a satisfactory description of biological development must also include reference to what Aristotle labeled formal and final (or teleonomic, sensu Mayr) causation. A measure called the network ascendency has been defined to track the changes in the system that result from positive feedback acting as an endogenous formal cause of system development. In turn, positive feedback appears to exert a selection pressure reminiscent of teleonomic final cause upon each of its constituent elements.

Associating development with the increase of network ascendency permits the modelling of final cause using the powerful modern tools of mathematical optimization. Cheung and Goldman have developed an algorithm to find a reconfiguration of any given starting network that locally optimizes its network ascendency. Typically, the optimal configuration, as demonstrated by two simple examples, is a "one-tree", that is, a single, directed cycle adjoined to a tree. Studying the differences between the observed and the optimal networks yields insights into the particular constraints acting on the system and reveals the most efficient pathways through the network.

Introduction

Ecodynamics, as conceived by Kenneth Boulding (1978), is a highly complex subject involving numerous new and unique features. But the name itself is not evocative of just how different ecological phenomena are from those included in classical dynamics. However, such inadequacy of name is quite understandable. As the late media analyst Marshall McLuhan (1973) was wont to point out, when faced with the radically new, we are often numbed into seeing the new in the guise of the familiar. For example, the name "quantum mechanics" stresses the formal, almost superficial connections of that field to its precursor, classical mechanics. The title tends to obscure the fact that the actual phenomena treated in both fields radically differ in qualitative and fundamental ways.

Exactly how does ecodynamics differ from the more familiar realm of classical dynamics? After all, is not significant effort now being expended in creating models of ecosystem behavior consisting of sets of coupled first-order differential equations--much as one would simulate chemical kinetics? Why is there any reason not to believe that such "mechanical" portrayals of ecodynamics will eventually predict ecosystem changes?

The answers to these questions are largely matters of experience. The recent observation is that such mechanical treatments of ecodynamics simply don't work very well, if at all. While there have been moderate successes in treating single ecological processes or populations in this manner, most projects seeking to simulate numerous coupled processes (such as comprise full ecosystems) have left much to be desired (Platt et al., 1981). It is tempting to ascribe the deficiencies of mechanical models to inadequate precision in the mathematical specifications of component processes. However, recent advances in classical mechanics (see Schaffer, this volume) indicate that ever-greater precision may not be the panacea for conventional modelling.

The reason why mechanical models remain inadequate for portraying ecodynamics may be that ecological succession and evolution are the results of causes that are not wholly mechanical in nature. It behooves us, therefore, to re-examine the nature of causality in scientific description.

The strictly mechanical, or newtonian view of the world identifies only material and proximate, or efficient causes. (Newton himself maintained a world-view that was quite unorthodox by the standards of his intellectual heirs, but that is another story.) During the Enlightenment, when the idea of the universe as a "cosmological clockwork" was at its zenith, to call someone a mechanic was one of the highest accolades one could bestow. Society, organisms and governments were perceived as machines, and one finds a substantial influence of this metaphor upon the leaders of the American Revolution (Wills, 1978).

Erosion of the clockwork image began early in the nineteenth century with the formulation of the second law of thermodynamics, i.e., the quantification of irreversibility in nature. The development of quantum mechanics later served to encourage a stochastic picture of the microscopic world. One is tempted to add the accidental to the list of natural causes, were it not for the fact that any particular manifestation of an accident can be described in terms of its material and efficient agents. Early attempts by the philosophers of biology to amend the clockwork image as it pertained to living organisms met with disaster. Vitalism and teleology, two patently non-mechanical views of the living world, were soundly rejected--in part for their lack of supporting evidence but also because of the distaste that contemporary biologists then bore for the theological extrapolations many drew from these theories. Darwin's exposition of evolutionary change survived for numerous reasons, not the least among them his determination to remain within the bounds of proximate causality. The physicist's awakening to stochastic influence passed without much resistance into biology, and the final piece of the neo-Darwinian synthesis appeared less than 40 years ago with the identification of the material locus (cause) of heredity at the biomolecular level. Thus, the entire framework of the accepted view of evolution rests within the confines of mechanical causality, and reverberations from the earlier rejections of non-mechanical hypotheses still pose substantial professional risks to anyone who would venture beyond those limits.

Such considerations notwithstanding, the modern synthesis of evolution presumes that the dynamics of heredity and change occur against a backdrop called the environment, which for explanatory purposes is taken as given and, once specified, remains outside the scope of discussion. Several decades ago, when the neo-Darwinian synthesis precipitated, insufficient mathematical and descriptive tools were available to address adequately the context within which the evolutionary game was played. There was then little reason to question the adumbrated role assigned to the environment. But today, with recent advances in the theories of cybernetics, hierarchies, information and networks, little excuse remains _not_ to consider the interplay between organism and environment, and the study of that interrelationship is called "ecology". However, in the process of fully incorporating the environment into discussion, it becomes difficult to describe reality in strictly mechanical terms.

Aristotlean Causalities

Rather than engage in radical speculation concerning the nature of non-mechanical causality, it is useful first to reconsider older models of how events occur. In particular, Aristotle's organismic view of nature seems appropriate to the subject of evolutionary change (Rosen, 1985). Aristotle recognized that the causes of a phenomenon were not always simple, and he proposed four types of cause that potentially could contribute to a single event: (1) material, (2) efficient, (3) formal and (4) final. The textbook example of an event is the building of a house. The material cause is obviously the bricks, mortar, lumber, etc. that go into the structure. The laborers who assemble the materials act as the efficient cause. The formal cause is usually taken to be the blueprints, or bauplan used by the laborers, and the final cause resides in the need for shelter by its eventual occupants.

With the overwhelming emphasis that newtonian science has placed upon the first two categories of causation, the perceived significance of the latter pair has atrophied. However, with the narrative powers afforded by modern cybernetics and hierarchy theory, it is now possible to entertain formal cause in a natural and rational manner. Even more importantly, recent progress in network analysis and information theory now encourages attempts to quantify and measure at least the effects of formal and final causality in evolving systems.

Formal Cause

A significant (but not exclusive) example of formal cause is the positive feedback loop. In its ideal form one can imagine positive feedback as a ring of processes or entities wherein the activity of each member positively catalyzes the activity of the next element in one direction around the loop until the last member catalyzes the first. Thus, each member in the loop engages in some degree of autocatalysis. It is not widely appreciated that positive feedback possesses at least six important characteristics:

In the ideal case, when the only causal links are those which form the chain, the system appears as a wholly (1) autonomous structure. Of course, no real causal structure is wholly autonomous, and any actual feedback loop always appears as a substructure embedded within a larger network of causalities, at least some of which must relate to the external world (cf, Goedel, 1931). Although this contingent nature of feedback serves to diminish its autonomy, it cannot be said to eliminate it. As any student of non-linear systems will recognize, systems with internal feedback are often less responsive to changes in their boundary conditions. Some particular nonlinear systems are even capable of generating output with no given input. In order to better identify the automonous elements in a causal network, it is helpful if the linkages can be quantified in some way (see below). Then algorithms exist that can separate the feedback loops from their associated tree of unidirectional causalities (Ulanowicz, 1983).

It is not difficult to explain how one can perceive the autonomy of positive feedback as an (2) emergent property of a system. If one defines the system so that only some of the elements of a particular loop are included, then that subset of elements will appear to function in strictly non-autonomous fashion at the behest of the boundary conditions driving them. However, once the system boundary is enlarged to include all the members of the feedback cycle, autonomy then "emerges" as a new attribute of the expanded system.

In response to stochastic influences on a system, positive feedback can serve to generate (3) selection pressure upon its composite members. This is because any accidental change in the properties of a loop element that abets its action upon the next member will be reflected positively upon itself (rewarded). Conversely, stochastic changes which diminish catalytic activity are self-inhibiting. Such selection pressure can act not only to change the attributes of a component, but it can also replace members by others that contribute more effectively to the positive feedback. For example, if a more efficient catalytic element should accidentally appear connecting any two nonsequential components, it will reinforce its own activity to the eventual exclusion of those less effective loop members it bypasses. This potential ability to replace components creates an extremely intriguing possibility. By induction, one can imagine a situation where all the original members of a loop have been replaced in turn by a new suite of elements. That is, the functional loop is seen to persist beyond the durations of all its individual components. One often observes, for example, that the cellular and material constituents of a complex organism are totally replaced over the lifetime of the organism, which nonetheless maintains its same basic structure over the longer interval. It is important to note that the context provided by the feedback cycle at any instant plays an active role in influencing what any next replacement parts might be. Not every arbitrary change is incorporated into the developing structure.

In particular, one of the attributes of a loop element subject to positive selection is its ability to bring external resources into the cycle. Hence, one effect of positive feedback can be to draw progressively more medium into the loop. In a system with fixed or diminishing resources available, this self-aggrandizing tendency will have the unavoidable consequence of (4) inducing competition among the various loops present in the system. The net effect of such competition is the tendency during the latter stages of system development towards a more highly articulated topology of network flows, wherein those pathways and loops with more efficient overall transfers would grow at the expense of less effective routes. The ultimate outcome of progressive articulation, if it were allowed to persist without mitigation, would be for a single loop to win out; that is, the resulting network configuration would be a "one-tree",or a causal tree with but a single cycle.

A typical extensive effect of feedback self-aggrandizement, and of positive feedback more generally, is to increase the level of total system activity, i.e., (5) growth enhancement.

Finally, it should be noted that positive feedback is inherently (6) formal in nature in that it is defined as a relational structure of the processes comprising it.

System Ascendency

Properties 1, 3, 4 and 5 enumerated above strongly suggest that positive feedback possesses agency. More specifically, it is a formal agent working at the level of the whole system to augment system activity and to streamline and make more efficient the associated network topology. With the qualitative nature of positive feedback as a formal agent now evident, it becomes fruitful to shift attention to quantifying its effects on the process of the "growth and development" of an evolving system.

In order to keep the discussion focused upon palpable entities, it is useful at this point to confine further discussion of species interactions to material causalities, i.e., the exchanges of materials or energy that transpire within an ecosystem. One must decide upon a particular medium to follow (e.g., energy, carbon, nitrogen, etc.);this restriction to a single medium is an artifact of the present stage of the research being reported. After numbering the elements of the system in some fashion, the flow of medium from species i to population j (if possible) can be denoted as T_{ij}, with i and j taking on values from 1 to n, the number of nodes in the network. Exogenous inputs will be presumed to flow from a fictitious compartment 0 (zero), so that T_{0i} will represent the external input to compartment i. Likewise, the fictitious compartments n+1 and n+2 will be assumed to receive the usable exports and the products of dissipation, respectively.

The level of activity in each compartment can be measured either by the total amount flowing out of it,

$$T_i = \sum_{j=0}^{n} T_{ij} , \qquad (1)$$

or by the aggregate entering it,

$$T_i' = \sum_{j=1}^{n+2} T_{ji} . \qquad (2)$$

These two measures are equivalent only when the system is at steady-state. Even if the system is not in balance, its unique measure of total activity is

$$T = \sum_{i=0}^{n} T_i = \sum_{i=1}^{n+2} T_i' , \qquad (3)$$

and is referred to as the total systems throughput. Growth in its purely extensive manifestation would translate as an increase in T.

"Development" refers predominantly to intensive characteristics of the network and will be taken here to mean an increase in articulation, or organization of the flow structure. The more articulated a network becomes, the less uncertain one is about the destination of a quantum leaving a particular compartment. Information theory provides the appropriate calculus to quantify uncertainty. The apriori uncertainty about the direction of a quantum leaving a particular compartment, when averaged over all compartments in the system, is captured by the Shannon-Wiener index of uncertainty

$$H = -K \sum_{i=0}^{n} p_i \log p_i, \tag{4}$$

where p_i is the probability that the quantum is leaving compartment i.

Knowing the (conditional) probabilities of which compartment will receive the flow from i reduces one's initial uncertainty by an amount known in information theory as the "average mutual information" (McElice, 1977),

$$AMI = K \sum_{i=0}^{n} \sum_{j=1}^{n+2} p_{ij} \log [p_{ij}/p_i p_j'], \tag{5}$$

where p_{ij} represents the joint probability that a quantum both leaves i and enters j, and p_j' is the probability that a quantum enters compartment j.

From the observed flows and their aggregates one can construct convenient estimators for the various probabilities in equations 4 and 5. The estimate for the probability that a quantum leaves node i (p_i) is T_i/T; that it enters j (p_j'), T_j'/T; that it both leaves i and enters j (p_{ij}), T_{ij}. As for the unspecified scalar, K, Tribus and McIrvine (1971) suggest that it be used to impart physical dimensions to the information measures, i.e., to scale it in the literal sense of the word. The appropriate size of a flow network is its total activity, T. Making these various substitutions into equation (5) yields a scaled quantity called the network ascendency (Ulanowicz, 1980),

$$A = T \sum_{i=0}^{n} \sum_{j=1}^{n+2} (T_{ij}/T) \log [T_{ij} T/T_i T_j'] \tag{6}$$

The ascendency has the dimensions of the constituent flows multiplied by the informational units specified by the base used to calculate the logarithms. For example, if the medium in question is energy and the base of the logarithms is 2, the ascendency has the units "power-bits". Ascendency is the product of an extensive factor (the "size", T) times an intensive factor that gauges the articulation of the network (the double summand). Figure 1 shows how the latter factor expresses the network articulation. All three hypothetical networks shown have the same total throughput (96 units) and differ only in their visible degrees of articulation, which are reflected in their respective values of ascendency.

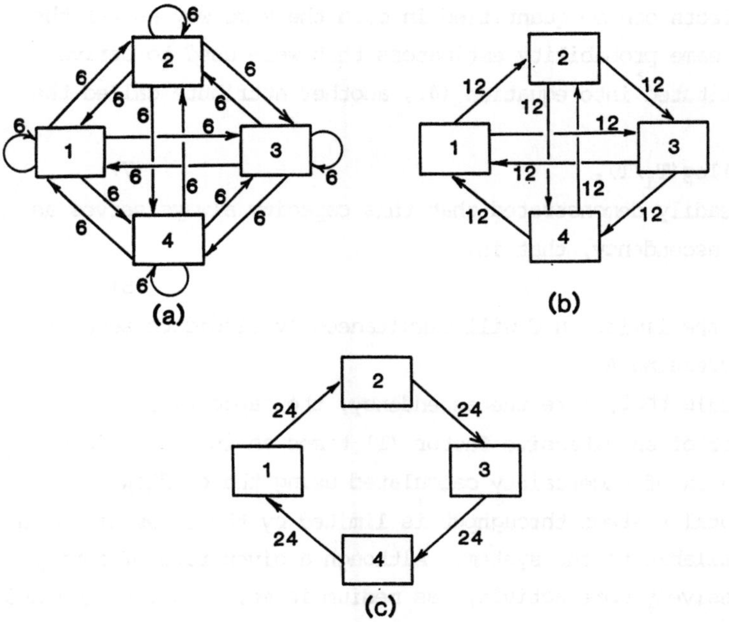

Figure 1: Three hypothetical configurations of 96 arbitrary units of flow among 4 compartments. (a) No articulation--ascendency is identically zero. (b) Moderate articulation--A=96 bits. (c) Maximal articulation--A=192 bits.

Because positive feedback tends to increase both factors in the ascendency, a rise in that measure can be said to quantify the effects of feedback as a formal agent imparting structure to evolving systems. Of course, other agents or accidental influences might also contribute to a rise in the observed ascendency, which serves as a phenomenological indicator in its own right. In fact, ascendency first appeared as a purely phenomenological quantity that unified most of Eugene Odum's (1969) 24 characteristics of "mature" ecosystems (Ulanowicz, 1980). The unifying power of this variable suggests the following ecological principle: In the absence of major disturbances ecosystems naturally tend in the direction of increasing network ascendency. Because this principle has been formulated from only the most general properties of ecosystems, it is likely to be applicable also to other evolving systems, such as one encounters in economics, sociology, and developmental biology.

Limits to Development

The proclivity of living systems to increase their ascendencies is only one half of the development story. Opposing this drive towards organization and efficiency is a host of countervailing entropic, or disorganizing effects.

Fortunately, these effects can be quantified in much the same way as was the ascendency. When the same probability estimators that were used to derive equation (6) are substituted into equation (4), another attribute called the development capacity,

$$C = -T \sum_{i=0}^{n} (T_i/T)\log(T_i/T), \qquad (7)$$

results. It can be readily demonstrated that this capacity always serves as an upper bound on the ascendency, that is,

$$C \geq A \geq 0. \qquad (8)$$

Therefore, specifying the limits on C will simultaneously elucidate some of the constraints on increasing A.

Equation (7) reveals that, like the ascendency, the capacity also consists of the product of an extensive factor (T) times an intensive index (the Shannon-Wiener index of uncertainty calculated using the of flow outputs). Now, the total systems throughput is limited by the rates at which resources are made available to the system. Although a given rate of supply can underwrite progressively more activity, as medium is more tightly retained and recycled within the system, the second law of thermodynamics still extracts a loss from each successive transfer, thus guaranteeing that the increase in T will have a finite limit. Despite the limit on T, the system capacity might still rise by dividing into ever-finer and more numerous compartments, which process drives up the second factor in the capacity (the Shannon-Wiener index). But the partition of a finite activity into sufficiently many components insures that eventually some of those components will become vanishingly small and subject to chance extinction by prevailing perturbations.

Even after the rise in C has somehow been arrested, the ascendency may still continue to increase by minimizing the difference between itself and its limit, C. This amount, C-A, is called the system overhead, and may be algebraically decomposed into 4 separate terms. Three of the overhead elements are generated by the exogenous transfers, while the remaining term represents the average redundancy in the number of pathways connecting any two arbitrary nodes. Not all of the overhead can be expended to create ascendency, however. Much of it stands as the system's "strength-in-reserve" with which to maintain structure in the face of unexpected perturbations. The hierarchical, thermodynamic and environmental interplay between ascendency and overhead is a long and interesting story that is developed more fully in Ulanowicz (1986).

Ascendency and Final Cause

Little mention has been made thus far of final cause. But the distinction between final, formal and efficient causes is sometimes mostly a matter of scale. For example, it has been argued earlier how a feedback loop visible at the system level (i.e., identifiable in the representative network) can act as a formal cause. However, another loop acting at a much smaller scale (i.e., incorporated into a single node) would probably be classified as an efficient cause. Conversely, if some of the pathways in the observed system constitute links in a feedback loop that lies partially outside the given system (that is, the entire loop exists only at a higher level), then its agency will be exercised upon the system in question via the driving boundary conditions. It appears as an agent exerting its influence in the guise of a final cause acting on the object being studied from above and beyond.

In particular, a population within an ecosystem interacts with its environment. The biotic portion of that interaction is expressed in terms of its connections with the other members of the system. But we have already suggested that this biotic context seen by the population is itself evolving at the level of the whole system according to the principle of increasing ascendency. Therefore, one may interpret the agency behind increasing ascendency as a final cause affecting the given population. This is not a trivial shift in perspective. Heretofore, any processes inherent in the development of a population or a species have, by default, always been referred towards events and mechanisms at lower levels of the hierarchy. But now the possibility also exists to interpret evolutionary developments as responses to active agencies existing at higher hierarchical levels. Some will immediately object to this last statement as being rife with teleology and goal-seeking. However, those who object to teleological overtones are encouraged to read Mayr's (1974) apology for "teleonomic" behavior as a legitimate concern of biology. Furthermore, as shall presently become clear, any goal that might serve as an attractor for system ascendency is hardly immutable, but rather is nonunique, transitory and constantly changing as the system develops (see Allen, this volume).

The mathematical tools with which one studies how systems approach goals constitute the realm of optimization theory. In this discipline one typically begins with a mathematical description of the system under study and then uses elements of this description to define some quantitative system attribute to be optimized (the objective function.) A state of the system is sought that optimizes the value of the objective function subject to a specified set of constraints on the system variables. The actual mechanics of how to find the optimal states are manifold. If both the objective function and the system constraints are all linear in nature, then the global optimum can be determined in routine algorithmic fashion. The introduction of nonlinearities anywhere into the problem can seriously confound the solution. Sometimes when the nonlinear objective function possesses "regularizing" (smooth) properties, such as uniform concavity or convexity, analytical techniques (or at least efficient numerical techniques) can be applied.

In this paper ecosystems are quantified in terms of their intercompartmental transfers. The system ascendency serves as the nonlinear objective function, and mass balances around each species compartment, along with nonnegativity requirements on the flows, comprise the constraint set. Optimizing such systems is problematical on several fronts. Firstly, the ascendency is neither uniformly concave nor convex over the entire range of allowed flows. The situation is not hopeless, however, in that it is possible to hierarchically decompose the ascendency into three terms, each of which behaves uniformly in either convex or concave fashion (Cheung, 1985a). In general, even the mass balance constraints are nonlinear without the rather restrictive assumption that dissipations and exports remain fixed fractions of their corresponding throughputs (T_i's), as will be assumed here. To deal with these difficulties Cheung (1985b) and Goldman have written an algorithm that proceeds by successive linear approximations, adapting generalized network techniques (Kennington and Helgason, 1980) to solve at each stage the resulting "minimum-cost flow in a network with gains" problem.

As is often the case with nonlinear optimizations, the solution process does not necessarily converge upon a global optimum. Instead, the process takes the system to a local optimum relatively nearby the starting configuration. Beginning with a different network configuration will often result in the system approaching another local optimum. Although the existence of multiple local optima tends to detract from the definitiveness of the optimization, it adds realism to this mathematical representation of the development process. As Holling (1973) pointed out, ecosystems are unlikely to possess a single climax state, but rather are characterized by a multiplicity of adjoining domains of local attraction. It was noted earlier how the increase in system ascendency can result through the progressive dominance of one or a few feedback loops, and it seems a plausible speculation that each of the local regions of attraction corresponds to an ascendant loop. This identification accords with the observation of Richardson (1984), that in economic and social systems "'structural change' is usually little more than a term used to cover unexplainable behavior... We might better speak of shifting loop dominance,...the process by which control of a system moves from one set of feedback loops to another set, often with dramatic changes in behavior."

Two examples of the application of the optimal ascendency algorithm are presented here. The first involves the familiar Cone Spring network (Tilly, 1968). As shown in Figure 2a, this simple lotic system is depicted as a simple network of transfers among 5 functional compartments. The key element is the detritus, which allows for five separate pathways of recycle (Ulanowicz, 1983). All flows are in Kcal $m^{-2}y^{-1}$, and the ascendency of this configuration is 56,725 kcal-bits $m^{-2}y^{-1}$. The nearby local optimum is the one-tree shown in Figure 2b, and it possesses an ascendency of 64,291 kcal-bits $m^{-2}y^{-1}$. One is immediately faced with the question of how to interpret the local optimum. Invariably, the optima resulting from maximizing the ascendency will possess unrealistic features. For example, it is clearly impossible for detritivores to consume only bacteria, as shown in the figure.

Configurations of maximal ascendency are virtually never realized, because well before the system can reach such a state, perturbations or other constraints intervene to slow or rescind progress towards higher efficiencies. In fact, Figure 2a was already the result of a natural standoff between those tendencies towards efficiency and those in the direction of chaos. Unless the algorithm leaves such a climax input unchanged, the result is bound to deviate from reality, because the set of mathematical constraints used is always a gross simplification of those actually limiting the system.

Further steps in this program will, of course, include refining and expanding the constraints to better approximate the real system limits. (We emphasize that such normal scientific steps, of successive testing and improving the theory, are made possible here only through the embodiment of each stage of that theory in an explicit mathematical model with an operational optimization algorithm to generate its "predictions".) Such a project would significantly deepen one's understanding of the community dynamics. However, even before going to such lengths it is possible to gain useful insights into the particular system dynamics (and sometimes to learn about ecosystems in general) by carefully studying how the local optimum differs from the observed network. One need only remember that a nearby optimum represents how the system might reconfigure if many of its natural constraints were released. The optimum emphasizes the most efficient pathways available among the components, assuming that each element maintains a constant rate of respiratory loss for each unit of throughput.

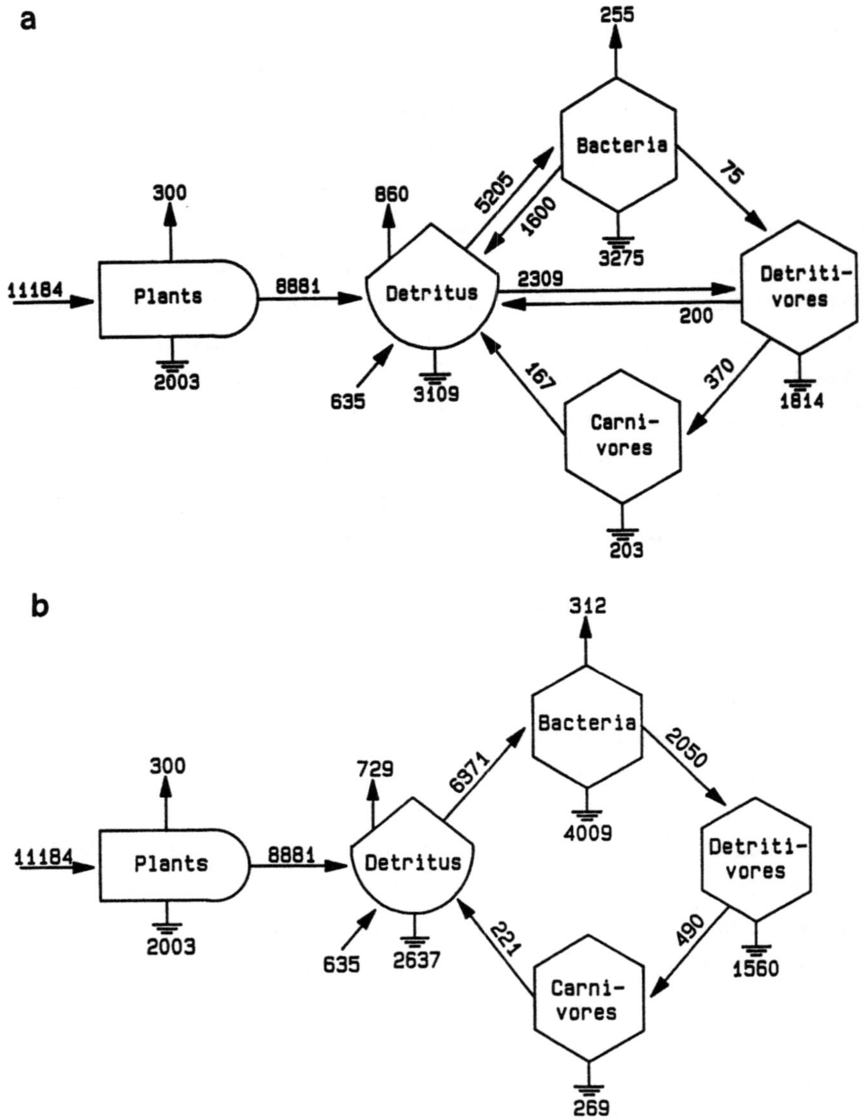

Figure 2: (a) Energy flow among the 5 functional compartments of the Cone Spring ecosystem according to Tilly (1968). Flows are in Kcal $m^{-2}y^{-1}$. A=56 725 Kcal-bits $m^{-2}y^{-1}$. (b) Optimal reconfiguration of flows in (a) maintaining constant respiration rates. A=64 291 Kcal-bits $m^{-2}y^{-1}$.

One notes that the role of the carnivores is hardly changed in the optimal configuration--save that slightly more energy passes through that component. The major difference in the optimal network is the rerouting of all the detrital carbon through the bacteria before it reaches the detritivores. Other things being equal, detritivores certainly would grow more efficiently on a diet of pure bacteria. However, such a feeding strategy would require that the detritivores separate the bacteria from the detritus to which these microorganisms are attached. In a thermodynamic sense, this separation would entail considerable work on the part of the detritivores (Cousins, 1985), and such overhead would probably be expressed as a higher rate of dissipation by these feeders (assumed constant in this optimization). One notes secondarily that the losses to detritus from the bacteria and the detritivores have disappeared in the optimal network. All indications point towards thermodynamic limitations related to particle size, dispersal and adhesion that all converge to make detritivory a "sloppy" process in comparison to predator-prey interactions.

This conclusion is underscored by the results from a second optimization. Figure 3a shows the transfers among the seven major functional units (in mg $Cm^{-2}d^{-1}$) of a tidal marsh gut in Crystal River, Florida (Homer and Kemp, unpublished; Hirata and Ulanowicz, 1985). The adjacent optimal configuration is shown in Figure 3b. As expected, the network is a one-tree (albeit a degenerate one). The pelagic grazing chain appears to be more efficient than the parallel benthic route, dependent as the latter is upon inefficient detritivory. The relative efficiency of the grazing chain over that of the detrital pathway has also been observed by Wulff and Ulanowicz (1988), and is possibly a general feature of aquatic and marine ecosystems. In oligotrophic, or low productivity systems, the grazing chain is usually dominant; but as productivity increases beyond the capacity of the herbivores to crop the additional production, the less efficient detrital pathways become more prominent.

(The careful reader may have noticed that the link from detritus to microphytes in Figure 3b did not exist in the starting network. The appearance of this new connection was because the optimization routine allows one to specify at the outset a number of links that are not present in the initial graph but which might appear under subsequent development. If this particular and somewhat problematical process had not been included as a potential transfer, the optimal configuration would have resembled that given in Figure 3b, except that the detritus would have fed into the pelagic herbivores.)

Thus, an optimal network can be considered as a virtual state towards which the observed system would tend, if existing, but unspecified constraints were relaxed. Conversely, those pathways present in the observed system but missing from the optimal configuration are those less efficient processes likely to diminish if the environment should become less erratic.

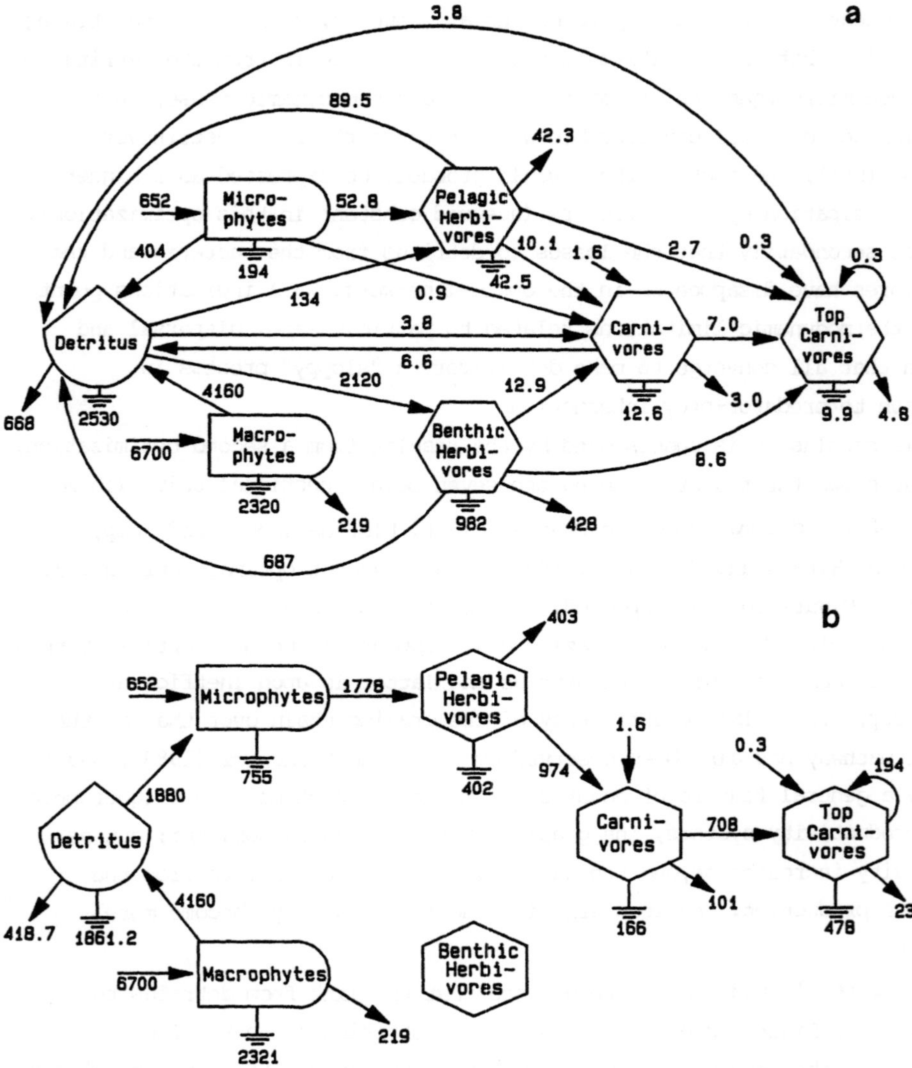

Figure 3: (a) Carbon flow among the seven major functional compartments of a Crystal River, Florida march gut according to Hirata and Ulanowicz (1985). Flows are in mg Carbon $m^{-2}d^{-1}$. A=28 279 mg C-bits $m^{-1}d^{-1}$. (b) Reconfiguration of (a) yielding locally maximal value of ascendency. A=41 163 mg C-bits $m^{-2}d^{-1}$.

Conclusions

While ecosystems remain contingent upon lower-level biological phenomena, their dynamical structures are not entirely determined by such smaller-scale events. The structure of ecosystems derives in part from formal agencies that can be described and quantified only at larger scales. Conversely, the fate of particular ecosystem elements will be influenced by mechanisms not included in the neo-Darwinian scheme of evolutionary change. How well a particular population fares is also a matter of how well it fits into the larger context of ecosystem trophic relationships, and we can now aim to quantify the influences of the whole community upon single species through the use of network, information and optimization theories.

The realm of ecodynamics transcends the scope of classical dynamics, and a more thorough quantification of such complex behavior is almost certain to lead to a deeper understanding of the living world and possibly to significant new insights into the related fields of economics, sociology and developmental biology.

References

Boulding, K. E. 1978. Ecodynamics: A New Theory of Societal Evolution. Sage Publications, Beverly Hills, California. 368 p.

Cheung, A. K-T. 1985a. Network Optimization in Ecosystem Development. Doctoral Dissertation, Department of Mathematical Sciences, The Johns Hopkins University, Baltimore, Maryland, 163 p.

Cheung, A. K-T. 1985b: ECONET: Algorithms for network optimization in ecosystem development analysis. Technical Report No. 423, Department of Mathematical Sciences, The Johns Hopkins University, Baltimore, Maryland. 63 p.

Cousins, S. H. 1985. The trophic continuum in marine ecosystems: structure and equations for a predictive model. pp76-93. IN: R. E. Ulanowicz and T. Platt (eds.) Ecosystem Theory for Biological Oceanography. Canadian Bulletin of Fisheries and Aquatic Sciences. 213.

Goedel, K. 1931. Ueber formal unentscheidbare saetze der Principia Mathematica und verwandter systeme. Monatshefte fuer Mathematik und Physik 38:173-198.

Hirata, H. and R. E. Ulanowicz. 1985. Information Theoretical analysis of the aggregation and hierarchical structure of ecological networks. J. theor. Biol. 116:321-341.

Holling, C. S. 1973. Resilience and stability of ecological systems. Ann. Rev. Ecol. Syst. 4:1-23.

Kennington, J. L. and R. V. Helgason. 1980. Algorithms for Network Programming. John Wiley and Sons, New York. 291 p.

Mayr, E. 1974. Teleological and teleonomic: A new analysis. Boston Studies in the Philosophy of Science 14:91-117.

McEliece, R. J. 1977. The Theory of Information and Coding. Addison-Wesley, Reading, Massachusetts. 302 p.

McLuhan, M. 1973. Understanding Media: The Extensions of Man. Mentor Books, New York.

Odum,E. P. 1969. The strategy of ecosystem development. Science 164:262-270.

Platt, T. and K. H. Mann and R. E. Ulanowicz. 1981. Mathematical Models in Biological Oceanography. Unesco Press, Paris. 157 p.

Richardson, G. P. 1984. The evolution of the feedback concept in American social science. Ph.D. thesis. Massachusetts Institute of Technology, Cambridge, Massachusetts. p.

Rosen, R. 1985. Information and complexity. pp221-233. IN: R. E. Ulanowicz and T. Platt (eds.) Ecosystem Theory for Biological Oceanography. Canadian Bulletin of Fisheries and Aquatic Sciences. 213 p.

Tilly, L. J. 1968. The structure and dynamics of Cone Spring. Ecol. Monographs 38:169-197.

Tribus, M. and E. C. McIrvine. 1971. Energy and information. Sci. Am. 225(3):179-188.

Ulanowicz, R. E. 1980. An hypothesis on the development of natural communities. J. theor. Biol. 85:223-245.

Ulanowicz, R. E. 1983. Identifying the structure of cycling in ecosystems. Math. Biosci. 65:219-237.

Ulanowicz, R. E. 1986. Growth and Development: Ecosystems Phenomenology. Springer-Verlag, New York. 203p.

Wills, G. 1978. Inventing America. Doubleday, Garden City, New York. 398 p.

Avoidance of Ecological Risk in Optimal Exploitation of Biological Resources

C. Wissel

Fachbereich Physik und Fachbereich Biologie,
Philipps-Universität Marburg, Renthof 6, D-3550 Marburg, Fed. Rep. of Germany

1. INTRODUCTION

There exists a vast number of mathematical models dealing with problems of optimal exploitation of biological resources [1-5]. There appear different types of models. This is due to the fact that people pursue different aims and objects by using mathematical models. The purpose of the paper in hand is to find some general trends for some ecological problems which arise in the exploitation of populations. Therefore simple models of population dynamics are used. Because of their abstractions and idealisations they reveal qualitative results which can be generalized. However they are not suitable to make quantitatively correct predictions for a specific situation.

In this paper the following problems are considered:
It is well known that biological populations can be endangered by human exploitation. There are several examples where fish stocks collapsed or have been extinguished by overfishing. As men continue harvesting natural populations it is an important ecological problem to assess the risk that populations may be extinguished by exploitation [1,2,6]. So far rather little attention has been paid to this problem. Only very rough measures of this risk of extinction have been given [6-8]. Thus one aim of this paper is to give a reasonable quantification of the risk of extinction.

For this purpose a stochastic description of the population dynamics is necessary. It is trivial that harvesting strategies of exploitation exist which extinguish the exploited population in a deterministic way [9]. This happens when the harvesting rate exceeds the maximal growth rate of the population. But also if one does not apply this crude action the probability of extintion is enhanced by the exploitation of a population because unpredictable random events affect the population growth. Unfavourable random influences together with the exploitation can cause so small individual numbers that the population runs a high risk of extinction. In this case only a probabilistic description of this risk can be given [10,11].

The common stochastic description of a population dynamics with random influences is [10-14]

$$dN/dt = f(N) + g(N)\,\xi(t) \; . \tag{1}$$

The mean growth of the individual number N is determined by the growth rate f(N). We use the logistic growth

$$f(N) = rN(1-N/K) \tag{2}$$

as the simplest example. But also other functions for the growth rate f(N), for instance with Allee-effect [16], will be used in this paper in order to find general results. The second term on the right hand side of (1) is the fluctuating part of the growth rate. Usually one considers the randomly fluctuating quantity $\xi(t)$ as Gaussian white noise [10-14]. It is called environmental or external noise because it describes the reaction of the growth rate to randomly changing environmental conditions. The function g(N) describes the dependence of the random influences on the population size. One often [12-14] assumes that the randomly fluctuating part of the individual growth rate is $\sigma\,\xi(t)$ independent of the population size N. In order to get the growth rate of the total population one has to multiply by N. Therefore one gets

$$g(N) = \sigma N \tag{3}$$

where σ is the strength of the environmental noise.

The exploitation of a population is described by [1,4,5]

$$f(N) \longrightarrow f(N) - Y(t) . \tag{4}$$

That means that the growth rate f(N) is diminished by the harvesting rate Y(t), i.e. the number of individuals caught per unit time. Most of the models of exploitation try to determine an optimal harvesting strategy Y^*, i.e. this harvesting rate in the course of time which maximize the total yield or the total economic profit. This is a problem of optimal control theory [17-19]. There exists a conventional stochastic optimal control theory [18] which performs the maximization in the presence of environmental noise. It yields the optimal harvesting rate Y^* in feed-back (closed loop) form

$$Y^* = Y^*(N(t)). \tag{5}$$

That means Y is not given as an explicit function of time but as a function of the population size N. If one wants to apply such strategies it would be necessary to know the population size N(t) exactly, continuously all the time. This is highly unrealistic and unpracticable. The best one can achieve is an information at discrete instants of time, for instance each year. In addition these informations are subject to uncertainties and errors. Thus one aim of this paper is to introduce an approach which <u>avoids</u> this <u>unpracticable feed-back control</u> and gives the optimal harvesting strategy in an applicable form. In addition it should open the possibility to include partial information on the population size which is obtained at discrete instants of time (discrete monitoring).

It is well known that the probability density P(N,t) for the individual number fulfills a Fokker-Planck (Kolmogorov) equation [10-15]

$$dP/dt = -\partial(PA)/\partial N + 1/2\ \partial^2(PB)/\partial N^2 \tag{6}$$

where the coefficients are given by

$$A(N) = f(N) \tag{7}$$

$$B(N) = g^2(N) \tag{8}$$

if one uses the Ito calculus. Using (2) and (3) one finds a so-called built-in boundary [10] at N=0. It is easy to show that the individual number N cannot reach the value N=0 in a finite time. Thus in this approach there is no chance that the population may be extinguished even if the individual number N is very small. This unrealistic result remains true for other growth functions f(N) and for functions g(N) which contain higher powers of N or corrections at higher values of N. It is obvious that the continuous description (1) of the discrete individual number N is incorrect especially at low values of N whereas at high N it may be a good approximation. Therefore another aim of this paper is to give a different approach which is correct also at low individual numbers N and which allows for the possibility of extinction of the exploited population.

Having solved these methodological problems this paper handles with the interdependence of the economic demand for a high profit and the ecological demand for a low risk of extinction. One may expect an antagonism between these two demands. In this case one should search for a compromise. Finally the influence of discrete monitoring on the profit and the risk of extinction will be investigated.

2. STOCHASTIC POPULATION DYNAMICS

A stochastic description of a population dynamics which is suitable also at low individual numbers is the master equation of the birth and death type [10,20]

$$dP_n(t)/dt = \lambda_{n-1}P_{n-1} + \mu_{n+1}P_{n+1} - (\lambda_n+\mu_n)P_n \quad . \tag{9}$$

Here the probability $P_n(t)$ of having n individuals at the time t is given by the birth rates λ_n and death rates μ_n. They are the probabilities that per time one individual is born or dies, respectively, when the individual number is n . It is well known that extinction, i.e. n=0, is possible in this description.

In order to find the birth and death rates one can use the fact that the Fokker-Planck equation (6) may be a correct description at high individual numbers. Indeed one finds a master equation of the type (9) if one discretizes the right hand side of (6) as usually done in numerical solutions [10]. It can be shown that this approach is correct at higher individual numbers. For this purpose one compares the stationary distributions and mean first passage times of both equations (6) and (9). There exist analytic expressions for these quantities [10,15]. In this way one finds that the birth and death rates are given by the coefficients of the Fokker-Planck equation (6):

$$\lambda_n = 1/2\ [B(n)+A(n)] \tag{10}$$

$$\mu_n = 1/2\ [B(n)-A(n)]. \tag{11}$$

Using (2), (3), (7), (8), (10) and (11) we can determine the high-n-part of μ_n and λ_n which are given by the first two terms of the following equations:

$$\lambda_n = 1/2\ \sigma^2 n^2 - 1/2\ r/K\ n^2 + \lambda_o n \tag{12}$$

$$\mu_n = 1/2\ \sigma^2 n^2 + 1/2\ r/K\ n^2 + \mu_o n \tag{13}$$

That (6) is an unsuitable description at low individual numbers is reflected by the fact that one would get a negative value for μ_n by (11).

Therefore special considerations for low individual numbers are necessary. There intraspecific competition does not play any role. Thus one expects that the birth and death rates per individual are constants λ_o and μ_o independent of the population size n [12-14]. The birth and death rates of the total population are obtained by multiplication with n. The corresponding low-n-parts are given in (12) and (13). The difference of (12) and (13) gives the mean growth rate as can be seen from (10), (11) and (7). The term proportional to n^2 is correct indeed. The comparison with (2) shows that

$$\lambda_o - \mu_o = r \tag{14}$$

is the maximal individual growth rate. The sum of (12) and (13) gives the stochastic noise. Taking into account (3) and (8) one finds that $\sigma^2 n^2$ represents the environmental noise. Correspondingly

$$Rn := (\lambda_o + \mu_o)n \tag{15}$$

describes a different noise, which is called demographic or internal noise. It results from the variability of the individuals in the population which has the consequence that birth and death processes are not deterministic but are described by probability rates.

One may ask if both of these noise sources, the environmental and the demographic noise, are necessary for the simplest description of a population dynamics which allows for the possibility of extinction. But indeed we need both of them as the demographic noise R gives a vanishingly small contribution to the birth and death rates at high individual numbers n whereas the environmental noise contributes considerably. At low individual numbers n the situation is just reverse. There only the demographic noise is essential. Thus in a population of higher individual number n the environmental noise causes a finite probability that the number n is occasionally small. In this case the demographic noise becomes important which results in a finite probability that the population may be extinguished. The interplay of both noise sources are necessary to find a finite chance that a larger population may die out.

This approach has been tested in a stochastic model for the island theory. With a modification of (12) and (13) which takes into account immigration and interspecific competition the well known relation between the number S of species of a certain taxonomic group and the area A of an island could be deduced:

$$S = c\,A^z\,.$$

As found from field data values of the exponent z around z=.25 have been calculated by this model. By this success the confidence in the stochastic description (12) and (13) is corroborated.

The problem to find a reasonable quantification for the risk of extinction is also solved by this approach. It is simply given by the probability $P_0(T)$ that the population is extinguished, i.e. that the individual number is n=0 at the end of a time interval T under consideration.

Finally the exploitation of the population is modelled by adding the harvesting rate Y(t) to the death rate:

$$\mu_n \longrightarrow \mu_n + Y(t) \qquad (16)$$

In this way it is already considered that harvesting is a stochastic process, i.e. that the catch is now larger, now smaller [21]. In the following two harvesting strategies will be distinguished. In the quota strategy the harvesting rate Y(t) is chosen independent of the population size n. That means that the fishing is performed so strongly and so long that this number of individuals is caught per time interval how it is prescribed by Y(t). In an effort strategy one takes

$$Y(t) = n \cdot E(t) \qquad (17)$$

Here the harvesting rate Y(t) is proportional to the individual number n, i.e. proportional to the densitiy of individuals. This may be realized if a certain number of fishing ships are sent out to sea for a certain time. The total catch is expected to be proportional to the density of the fish. The proportionality factor E(t) is called the effort. It should be proportional to the number of used ships. It includes the catchability coefficient [1,4].

3. OPTIMIZATION

In order to find a optimization procedure which avoids feed-back control one has first to appoint an expression for the profit which should be maximized. As usual in economics the so-called present value of the economic return is calculated by [1,4]

$$J = \int_0^T dt \exp(-\delta t) \; [p\sum_n nP_n(t)E(t) - cE(t)\} \qquad (18)$$

In the first term in the bracket the mean individual number is multiplied by the effort E(t) which gives the mean harvesting rate because of (17). Multiplied by the price p one gets for one sold individual on the market gives the profit per time. One has to subtract the current cost per time cE(t) of the effort. Then this net profit per time is summed up, i.e. integrated over the time interval T under consideration. For this planning horizon one takes 5 to

10 years in economics. Finally it is taken into account that it is better to earn the money today instead of tomorrow. Therefrom the exponential factor in (18) results where δ is the discount rate of the money. The modification of (18) for a quota strategy is trivial.

The obvious aim of the economics is to maximize this profit J. But there is a constraint for the effort:

$$0 \leq E(t) \leq E_{max} \tag{19}$$

If there is no exploitation the effort is $E(t)=0$. The maximal feasible effort E_{max} is determined by the number of available ships for instance.

Therefore the problem is to maximize the objective functional J in (18) with the constraint (19) where $P_n(t)$ is given by the dynamical equation (9). This is a problem quite common in conventional deterministic optimal control theory [17-19]. The stochastic property of the present model is reflected in the high number of components of $P_n(t)$ only. The difference to the conventional stochastic optimal control theory lies in the fact that $P_n(t)$ is used as the dynamical variable to be controlled instead of the stochastic quantity $N(t)$. By this way we avoid a feed-back control of the form (5).

In order to show the general applicability of this approach (for quota and efforts strategies etc.) we rewrite the objective functional J in (18) and the master equation (9):

$$J = \int_0^T dt \, \{\sum_n P_n(t)[C_n(t)E(t)+D_n(t)]+F(t)E(t)\} \tag{20}$$

$$dP_n/dt = \sum_m [A_{nm} + B_{nm}E(t)]P_m \tag{21}$$

The maximum principle [17-19] says that the maximization of J in (20) with (19) and (21) is equivalent to the maximization of the Hamiltonian H with respect to $E(t)$.

$$H = \sum_{nm} \psi_n (A_{nm} + B_{nm}E)P_m + \sum_n P_n (C_n E+D_n) + FE \tag{22}$$

The costate vector ψ_n is determined by

$$d\psi_n(t)/dt = -\partial H/\partial P_n \tag{23}$$

with the end condition $\psi_n(T)=0$. The Hamiltonian H depends linearly on E(t) by the switching function

$$\phi(t) = \sum_{nm} \psi_n B_{nm} P_m + \sum_n P_n C_n(t) + F(t) \tag{24}$$

Thus the maximization of H gives

$$E(t) = 0 \quad \text{for} \quad \phi(t) < 0$$

$$\tag{25}$$

$$E(t) = E_{max} \quad \text{for} \quad \phi(t) > 0$$

Therefore the dynamical equations (21) and (23) have to be solved for piecewise constant E(t). The other remaining problem is the determination of the switching times τ_i where E(t) switches over between 0 and E_{max}.

The solution of (21) and (23) for piecewise constant E(t) is performed solving the eigenvalue problem of the matrices A_{nm} and $A_{nm}+B_{nm}E_{max}$ [22]. For this purpose the transformation [23]

$$W_{nm} = P_n^{*-1/2} A_{nm} P_m^{*1/2} \tag{26}$$

is performed with the definition

$$P^*_{n+1} = P^*_n \lambda_n / \mu_{n+1} \tag{27}$$

The matrix W_{nm} is symmetric tridiagonal. Its eigenvalues can be determined by the well-behaved numerical bisection method [24]. The eigenvectors are obtained by the inverse iteration method. It should be emphasized that this solution of the eigenvalue problem has to be done at the beginning. Then the solutions of (21) and (23) and the switching function $\phi(t)$ can be written down immediately for all possible switching times. In this way it is possible to perform the following numerical calculations in a reasonable time.

The next problem is the determination of the optimal switching times τ_i. One starts with a trial set $\{\tau_1, \tau_2, \ldots \tau_q\}$ which is comprehended in a q-dimensional vector $\underline{\tau}$. Consider small changes $\delta\underline{\tau}$ which results in small changes $\delta E(t)$ and $\delta P_n(t)$. Using (20) the change of the profit J in first order is

$$\delta J = \int_0^T dt [\sum_n \delta P_n (C_n E + D_n) + \delta E (\sum_n P_n C_n + F)] \quad (28)$$

In [25] it is shown using (21) and (23) that (28) gives

$$\delta J = \underline{\Phi}(\underline{\tau}) \cdot \delta\underline{\tau} \quad (29)$$

where the vector $\underline{\Phi}(\underline{\tau})$ is defined by

$$\phi_i(\underline{\tau}) = E_{max} \; s_i \phi(\tau_i) \quad (30)$$

It is $s_i=-1$ if $E(t)$ switches from 0 to E_{max} at τ_i and $s_i=1$ if from E_{max} to 0. Thus J can be enlarged by choosing

$$\delta\tau_i = \sum_j K_{ij} \phi_j(\underline{\tau}) \quad (31)$$

where K_{ij} is a positive definite matrix. By iteration of this procedure J can be enlarged until the situation $\phi_i(\underline{\tau})=0$ is reached which is optimal because of (25). Using the Davidon-Fletcher-Powell algorithm [17] only few iteration steps are necessary. In this procedure one has to allow for the possibility that the number q of switching times τ_i may be reduced by disappearence at t=0 or t=T or by the fusion of two switching times. If the number q of the τ_i in the trial set is too small the switching function $\phi(t)$ will be zero at some values of t in addition to the τ_i.

Instead of the bang-bang control (25) there may exist a singular solution [17] of the optimization problem (19)-(21) with $\phi(t)\equiv 0$ for a certain time interval. In this case it would be necessary to solve (21) with a time dependent E(t). There is no chance to perform this solution. Therefore an approximation of this singular solution by a bang-bang control (25) with many switching times τ_i is tried. This approach has been tested for a case where the singular solution could be determined analytically [26]. As shown in [25] the values of J and $P_o(T)$ calculated with bang-bang control approach the correct values very rapidly when the number q of switching times is increased. Thus we increase the number q of the switching times until the procedure stops at a finite number of switching times (pure bang-bang control) or until asymptotic values of J and $P_o(T)$ can be determined sufficiently accurately.

4. PROFIT AND RISK

In this paragraph some typical results for the interdependence between risk of extinction and optimal profit are shown. Different modifications of the growth rate (2) have been considered which may include Allee effect (critical depensation) [16]. As a different measure for the risk the probability P_c that the population size n falls below a certain limit n_c has been used, too [25]

$$P_c = \sum_{n<n_c} P_n(T) .$$

A wide range of parameter values which seem biological reasonable has been investigated. In this way we have found the following general trends.

The principle difference between a quota and an effort strategy can be investigated quite simply without solving the stochastic optimal control problem (19)-(21). In a deterministic model there is no difference between an effort and a quota strategy. There the individual number N is a known deterministic quantity. In this case one can use (17) to determine the harvesting rate Y(t) from the effort E(t) and vice versa. The deterministic optimal harvesting [1,4], i.e.

N=K/2, E=r/2 and Y=rK/4, is applied and the influence of the stochastics investigated. In the stochastic case the individual number n is an unknown stochastic quantity. Therefore in spite of (17) there is a difference between a quota and an effort strategy.

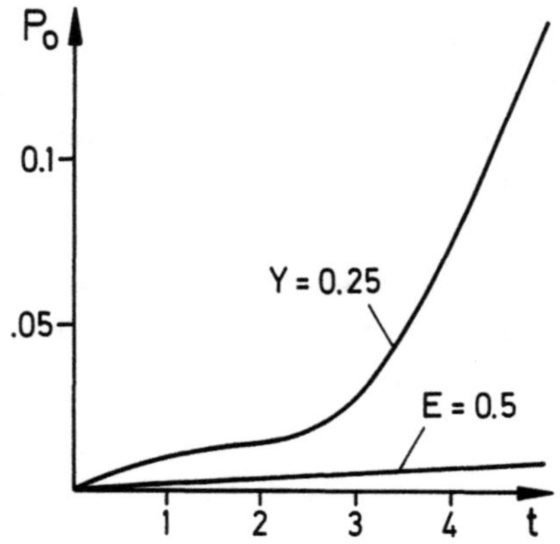

FIG.1 Probability of extinction P_0 versus time t for deterministically optimal constant quota Y and effort E strategy. Logistic growth. K=100, σ^2=.001r, R=r. Time t in units of 1/r.

This is shown in Fig.1. Already for small noise strengths σ^2=0.001 r and R=r the risk of extinction $P_0(t)$ is by order of magnitude higher for the quota strategy after a short time t. This by simple stability arguments expected effect has been quantified by the method shown at (26). If one applies the stochastic optimization (19)-(21) the same general trend is found. The quota strategies result in about the same profit J as the effort strategies (see Fig.4 and Fig.5) but cause a much higher risk of extinction $P_0(T)$.

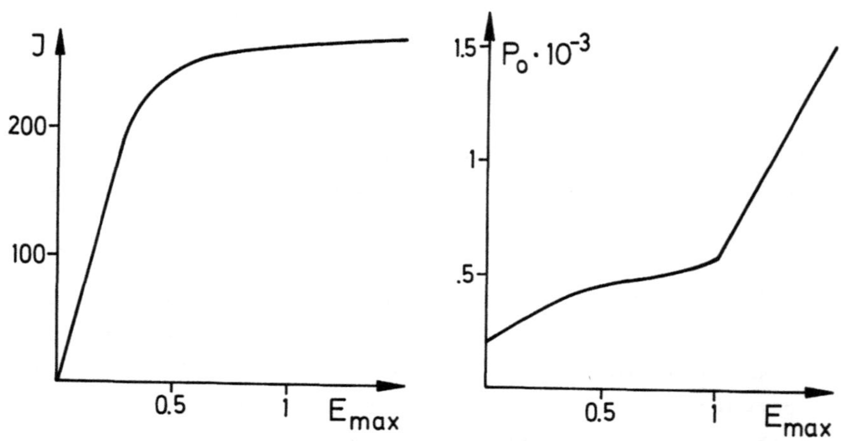

FIG.2 Optimal profit J (in units of pK) and risk P_0 versus maximal harvesting effort E_{max} (in units of r) for a fast growing population. Logistic growth. K=100, σ^2=.1r, R=4r, Tr=10, δ=.02r, c=0, N(0)=50.

The most decisive parameter for the trend of the results is the ratio of the discount rate of the money δ and the maximal growth rate r of the population. For an effort harvesting of a fast growing population in Fig.2 the optimal profit J and the resulting risk $P_0(T)$ are shown versus the strength of the exploitation, i.e. the maximal harvesting effort E_{max} which may be determined by the maximal numbers of ships in the case of a fishery. The values of the risk $P_0(T)$ are rather low. There is a saturation behaviour of the profit J. Thus very high values of the effort E_{max} are economically not worthwhile. Without a substantial loss of the profit one can reduce E_{max} so far (in Fig.2 $E_{max} \approx .6$) that the risk P_0 is of the same order of magnitude as it is in an undisturbed population at $E_{max}=0$. Therefore in a fast growing population there is no antagonism between the demand for a high profit and the demand for a low risk.

For slowly growing populations with maximal growth rate r of the same magnitude as the discount rate δ the situation is changed. A

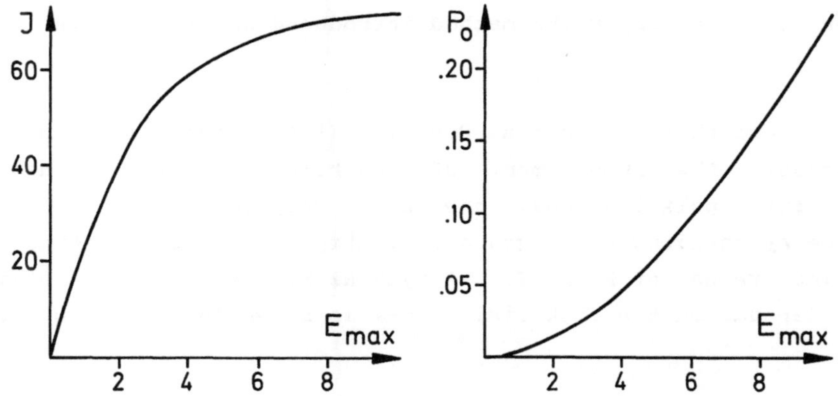

FIG.3 Optimal profit J and risk of P versus maximal harvesting effort E_{max} (in units of r). Logistic growth. K=100 σ^2 .001r, R=r, Tr=1, δ=r, c=0. N(0)=50.

typical result is shown in Fig.3. Although the strengths of the noise σ^2 and R and the planning horizon T has been decreased compared to Fig.2 the risk of extinction P_0 is larger by order of magnitude. If one wants to reduce P_0 to such low values as in Fig.2 one has to diminish E_{max} so strongly that the optimal profit would be lowered considerably. Here an antagonism between the demand for a high profit and the demand for a low risk appears. That means if one exploits a slowly growing population in an economically optimal way one runs a high risk of extinguishing this population.

In this situation one may search for a compromise for these two demands. Probably the best way is to perform the maximization of the profit J subject to the constraint that the risk of extinction does not exceed a given limit P_L. This constraint can be transformed by

$$P_L \geq P_0(T) = \int_0^T dt \; dP_n(t)/dt \qquad (32)$$

Using (21) for the derivative in the integral of (32) one finds an isoperimetric condition [17]. This can easily be taken into account by multiplying this integral by a Lagrangian multiplier and adding it to the integral in (20). The general form of (20) is not changed thereby [25]. Thus one can apply the method introduced above also in this case.

It is obvious that one obtains $P_L = P_0(T)$ if the constraint (32) is active. Because the appointment of the risk limit P_L is not a scientific but a political one, one should vary P_L in the results. This is done by changing the Lagrangian multiplier and calculating the corresponding value of $P_L = P_0(T)$. A typical result how the optimal profit J depends on the risk limit $P_L = P_0$ is shown in Fig.4. Without

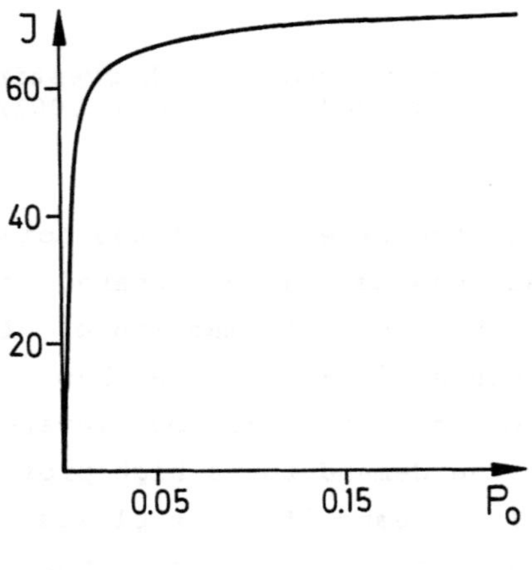

FIG.4 Optimal profit J (in units of pK) versus risk (limit) P_0 $(=P_L)$. $E_{max}=10r$. Other parameters as in Fig.3.

any constraint (32) for the risk one would be at the upper end of the curve where the risk would be $P_0 \approx .25$ as can be seen from Fig.3 with $E_{max}=10$. Performing the optimization with the constraint (32) which results in a different harvesting strategy E(t) Fig.4 shows that one can diminish the risk of extinction P_0 considerably (in Fig.4 by a factor 10) without a substantial loss of the profit J. That means slowly growing populations are endangered unnecessarily if they are exploited simply in an economically optimal way. As shown in Fig.4 one can always change the harvesting strategy E(t) in a way that the risk of extinction is reduced considerably and the profit remains almost unchanged.

One may ask if one can find the same compromise for a quota strategy. In this case the dependence of J on P_0 shows indeed the same saturation behaviour in Fig.5 as before. The comparison with

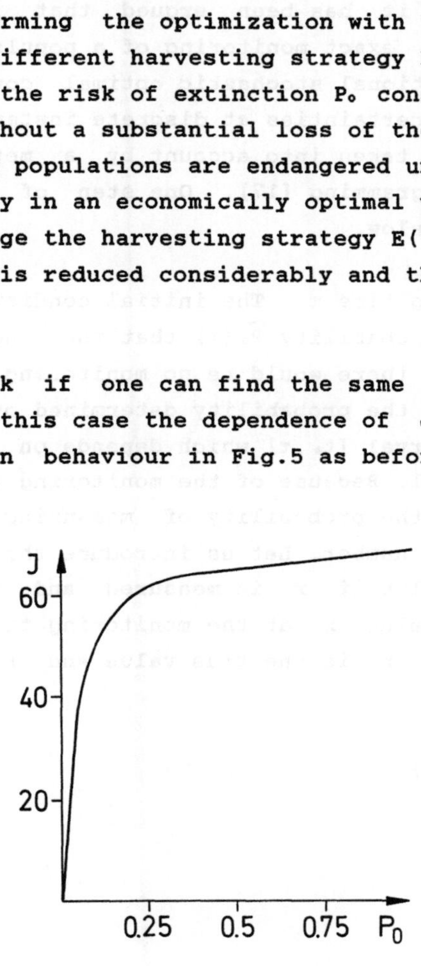

FIG.5 Optimal profit J versus risk (limit) $P_0 (=P_L)$. Maximal harvesting rate $Y_{max}=10rK$. Other parameters as in fig.3.

Fig.4 shows that the maximal profit is about the same in both of these cases. But for the quota strategy much higher values of the risk P_0 appear. If one wants to reduce the risk P_0 to the same low values as in the case of effort harvesting in Fig.4 one has to sacrifice a huge amount of the profit. Thus for the quota strategy no compromise between the demand for a low risk and the demand for a high profit exists.

5. DISCRETE MONITORING

In the introduction it has been argued that instead of the unpracticable, permanent exact monitoring of a population which is demanded by the conventional stochastic optimal control theory a monitoring subject to uncertainties at discrete instants of time is feasible. This will be taken into account by a method, which is similar to dynamic programming [17]. One step of this iterative procedure will be shown below.

Consider one monitoring time τ. The initial condition for the time interval $[\tau,T]$ is the probability $\widetilde{P}_n(\tau)$ that the individual number is n at the time τ. If there would be no monitoring $\widetilde{P}_n(\tau)$ would be equal to $P_n(\tau,E_2)$, i.e. the probability determined by the stochastic process in the time interval $[t_0,\tau]$ which depends on the harvesting effort E_2 in this interval. Because of the monitoring this probability is changed. Let M_{rn} be the probability of measuring r presupposed n is the true individual number. Let us introduce the probability T_{nr} that n is the true value if r is measured and the probability $W_r(\tau)$ of measuring the value r at the monitoring time τ. Then the combined probability that n is the true value and r is measured at τ is:

$$M_{rn}P_n(\tau,E_2) = T_{nr}W_r(\tau) \qquad (33)$$

From this follows

$$W_r(\tau) = \sum_n M_{rn}P_n(\tau,E_2) \qquad (34)$$

$$T_{nr} = M_{rn}P_n(\tau,E_2)W_r^{-1}(\tau) \qquad (35)$$

If the uncertainty of the measurement is described by M_{rn} the other probabilities T_{nr} and $W_r(\tau)$ can be determined by (34) and (35).

In order to perform the optimization one determines the maximal profit $J^*_1(\{\widetilde{P}_n(\tau)\})$ for the time interval $[\tau,T]$ with the initial condition $\widetilde{P}_n(\tau)$ first. This can be done by the method shown in paragraph 3. If the value r is measured at τ one has to take $\widetilde{P}_n(\tau) = T_{nr}$ as initial condition. Because $W_r(\tau)$ is the probability of measuring r the expe. .atıon value of the optimal profit for the time interval $[\tau,T]$ is

$$\sum_{r} W_r(\tau) J^*_1(\{\widetilde{P}_n(\tau) = T_{nr}\}) \tag{36}$$

The profit $J_2(E_2)$ of the time interval $[t_o,\tau]$ for an effort E_2 in this interval is described by an integral (18) with corresponding limits. Thus one can determine the optimal profit J^*_2 of the time interval $[t_o,T]$ by

$$J^*_2 = \max_{E_2} \{J_2(E_2) + \sum_{r} W_r(\tau) J^*_1(\{\widetilde{P}_n(\tau) = T_{nr}\})\} \tag{37}$$

Remember that $W_r(\tau)$ and T_{nr} depends on E_2 because of (34) and (35). The decisive point is that (37) is of the form (20). Thus the methods of paragraph 3 can be applied. The next iteration step with a monitoring time at t_o can be performed if J^*_2 plays the role of J^*_1. This backward iteration is similar to dynamic programming [17].

One expects the strongest influence of the monitoring on the results if the measurement at the monitoring times can be performed exactly. In this case it is

$$M_{nr} = T_{rn} = \delta_{nr} \tag{38}$$

For this case in Fig.6 it is shown how the optimal profit J and the risk of extinction P_o depends on the position of the monitoring time $\tau=t_m$. From an economical point of view one should perform the monitoring as early as possible whereas by the conservation point of view a later time t_m is preferred.

An information on the population size is often achieved routinely every year. Therefore in Fig.7 it is shown how the profit J and the risk P_o change if one introduces a monitoring at the beginning of the first year, then in addition in the second year and so on. We see that the profit J increases and reaches a saturation level soon. If one takes the costs of the monitoring into account then only a small number N_m of monitoring times are worthwhile. The risk of extinction

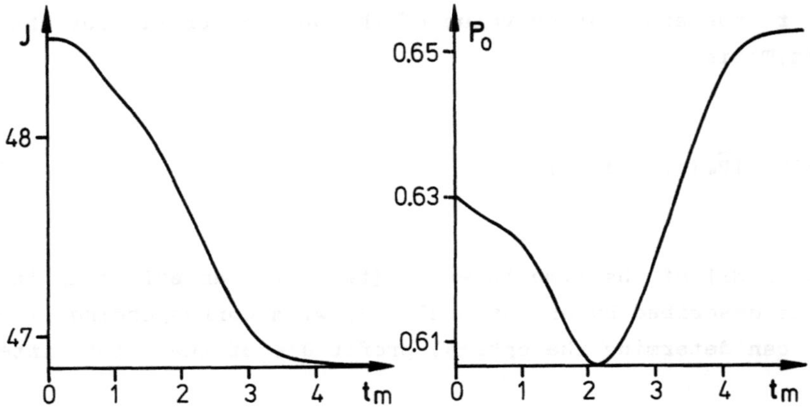

FIG.6 Optimal profit J (in units of pK) and risk P_0 versus the position one monitoring time t_m. The quasistationary distribution is taken for the inital condition $P_n(0)$. Logistic growth $\sigma^2=.5r$. $R=2r$, $E_{max}=r$, $T=5r$, $\delta=0$, $c=0$.

P_0 decreases with increasing number N_m of the monitoring time. Thus here is no antagonism between the demand for a high profit and the demand for low risk. For both demands monitoring is advantageous. But the Table shows that the relative increase of the profit J and the

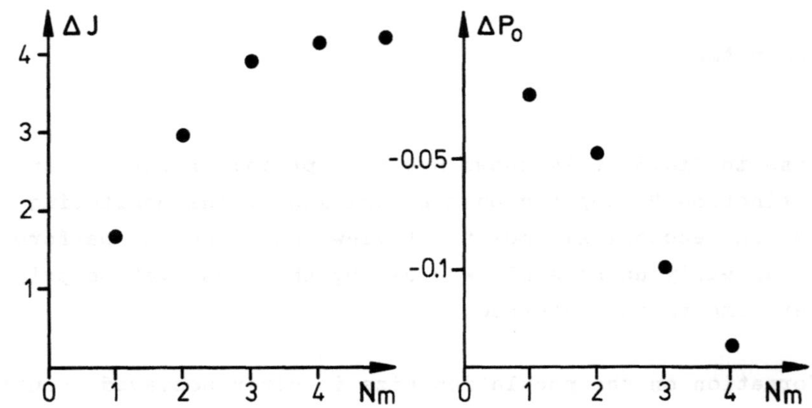

FIG.7 Change of the optimal profit J and the risk P_0 versus the number N_m of monitoring times τ_i. Parameters as in Fig.6.

relative decrease of the risk P_0 by the monitoring is rather small although rather high values of the noise strengths σ^2 and R and of the planning horizon T are used. Therefore the general trends of the last paragraph should remain true when monitoring is included especially when uncertainties of the measurements are taken into account.

Nm	J	J(%)	Po	Po(%)
1	1.65	3.5	.022	3.4
2	2.86	6.1	.049	7.5
3	3.92	8.4	.100	15.3
4	4.16	8.9	.136	20.8
5	4.22	9.0	.147	22.5

TABLE Absolute and relative change of the optimal profit J and risk P_0 versus the number N_m of the monitoring times τ_i as shown in Fig.7.

6. SUMMARY

In this paper a stochastic description of population dynamics is introduced which allows for the extinction of the population. Demographic and environmental noise are both important and are both included in the description by a master equation. A reasonable quantification of the risk of extinction is obtained by the probability that the individual number is zero at the end of the considered time interval. A stochastic optimization approach is introduced which corresponds to the deterministic optimal control theory. It differs from the conventional stochastic optimal control theory by the fact that the probability is taken as the dynamical variable to be controlled instead of the individual number. In this way the feed-back control of the conventional approach is avoided which would demand the permanent exact knowledge of population size - an unpracticable demand. In this introduced approach it is possible to include information on the population size obtained at discrete instants of time which may be subject to errors. From a conservation point of view it is essential that this approach can perform the maximization of the profit subject to the constraint that the risk of extinction does not exceed a given limit.

Using this method the interdependence of the ecological demand for a low risk and the economical demand for a high profit is investigated. Quota strategies give a much higher risk of extinction than effort strategies. For fast growing population with effort strategy the risk is rather low. But if for slowly growing populations the maximal growth rate is of the same order of magnitude as the discount rate of the money the populations are strongly endangered by the exploitation. But a compromise between the economical and ecological demand can be found. There exist always harvesting strategies which lower the risk of extinction considerably without a substantial loss of the profit. For quota strategies this compromise does not exist. Monitoring at discrete instants of time increases the profit and lowers the risk. The general trends mentioned above are not changed by this monitoring.

ACKNOWLEDGEMENT

This research was supported by the Deutsche Forschungsgemeinschaft

REFERENCES

1) CLARK,C.W. 1976, Mathematical Bioeconomics. Wiley, New York.

2) VINCENT,T.T. and SKOWRONSKI, J.M. (Editors), 1981. Renewable Resource Management. Lecture Notes in Biomathematics, Vol.40, Springer, Berlin.

3) MAY,R.M.(Ed.), (1984). Exploitation of Marine Communities, Report of the Dahlem Workshop 1984, Springer, Berlin.

4) CLARK,C.W. (1985). Bioeconomic. Modelling and Fisheries Management. Wiley, New York.

5) GOH,B.S., 1980. Management and analysis of biological populations. Elsevier, Amsterdam.

6) MAY,R.M., BEDDINGTON,J.R., HORWOOD,J.W. and SHEPHERD,J.G., 1978. Exploiting natural populations in an uncertain world. Math.Biosci., 42: 219-252.

7) KIRKWOOD,G.P., 1981. Allowing for risk in setting catch limits based on MSY. Math.Biosci., 53: 119-129.

8) LUDWIG D., 1981. Harvesting strategies for a randomly fluctuating resource. SIAM J.Appl.Math., 37: 116-174.

9) SILVERT W., (1977). The economics of over-fishing. Trans.Amer.Fish.Soc. 106, 121-130.

10) GOEL,N.S. and RICHTER-DYN,N., 1974. Stochastics models in biology, Academic Press, London.

11) KLEIDING,N., 1975. Extinction and growth in random environments., Theor.Pop.Biol., 8: 49-63.

12) TURELLI,M., 1977. Random environments and stochastic calculus., Theor.Pop.Biol., 12: 140-178.

13) TUCKWELL, H.C., 1974. A study of some diffusion models of population growth., Theor.Pop.Biol., 5: 345-357.

14) CAPOCELLI,R.M. and RICCIARDI,L.M., 1971. A diffusion model for population growth in random environment., Theor.Pop.Biol., 5: 28-41.

15) GARDINER,C.W., 1983. Handbook of stochastic methods., Springer, New York.

16) JACOBS,J., 1984. Cooperation, optimal density and low density thresholds: yet another modification of the logistic model, Oecologia, 64: 389-395.

17) PIERRE,D., 1969. Optimization theory with applications., Wiley, New York.

18) FLEMING,W.H. and RISHEL,R.W., 1975. Deterministic and stochastic optimal control. Springer, New York.

19) LEE,E.B. and MARKUS,L., 1967. Foundations of optimal control theory. MacGraw Hill, New York.

20) NISBET,R.M. and GURNEY,W.S.C., 1982. Modelling fluctuating populations. Wiley, New York.

21) HUTCHINSON,C.E. and FISCHER,T.R., 1979. Stochastic control theory applied to fishery management. IEEE Trans.Sys. Man and Cybern. SMC-9, 5: 253-259.

22) WISSEL,C., 1985. Solution of the master equation of a bistable reaction system. Physica 128 A, 150-163.

23) HAKEN,H., 1983. Synergetics. An introduction. Springer, Berlin.

24) WILKINSON,J.H., 1965. The algebraic eigenvalue problem. Clarendon Press, Oxford.

25) WISSEL,C., and SCHMITT,T., (1987). How to avoid extinction of populations optimally exploited. Math.Biosci. 84, 127-138

26) SCHMITT,T. and WISSEL,C., 1985. Interdependence of ecological risk and economic profit in the exploitation of renewable resources. Ecol.Mod. 28: 201-215

Deterministic Modelling of the Combined Action of Light and Heat Stress on Microbial Growth

E. Fiolitakis, J.U. Grobbelaar[1], *C.J. Soeder, and E. Hegewald*

Institut für Biotechnologie der KFA, Postfach 1913,
D-5170 Jülich, Fed. Rep. of Germany
[1]Present address: Unit for Limnology, P.O. Box 339,
Bloemfontein 9300, South Africa

Abstract

Transient behavior, which is typical of living systems, requires that short term processes be formulated as dynamic reactions and that these take long term adaptations into account. It has been possible to include these requirements in a mathematical model, where the growth of the green alga *Scenedesmus obliquus* was analysed in terms of a combined action of light and heat stress. Michaelis-Menten kinetics were solved and the activation energy, entropy and enthalpy were in agreement with the known properties of enzymatic reactions. An activity state variable V was derived, which enabled the modelling of adaptation kinetics due to available light.

Introduction

The modelling of microbial growth is mostly limited to the description of steady states, such as e.g. Monod kinetics. Transient conditions are, furthermore, usually modeled as a straight forward extension of some stationary relationship, which had been validated over a limited range of parametric variations. The complexity of transient behaviors which is typical of living systems adapting themselves to altered external conditions, demands a kinetic expression which can handle transient data over a large range of parametric variations. Such expressions should describe short term dynamics, as well as the adaptation to long term effects.

Several models have been presented which can predict micro-algal growth as a function of temperature and light (e.g. Grobbelaar, 1981; Hill and Lincoln, 1981; and Shelef, 1968). However, these models are simplistic in the sense that they do not describe the complex relationships between productivity, light energy input and prevailing temperatures adequately. Furthermore, they mostly contain certain restrictions, whereby reduced growth is predicted at superoptimal conditions (Grobbelaar et al., 1984).

Goldman and Carpenter (1974) showed that the temperature dependence of the specific growth rate (μ) of microalgae can be described by an Arrhenius equation, if the light intensity was held constant. A distinct optimum was found, above which negative activation energies were calculated when an Arrhenius equation was applied to this section of the temperature curve (Krüger and Eloff, 1979; and Soeder et al., 1984). Soeder et al. (1984) found that the activation and deactivation energies differed for the same organism (*Scenedesmus obliquus*), depending on relative light intensities (I). According to these authors the Q_{10} values (temperature coefficient expressing the reaction velocity over a 10 °C difference) ranged from 1.66 to 3.35 for suboptimal temperatures. Moreover the Q_{10} values followed an inverse hyperbolic relationship, when plotted against relative light intensity. Soeder et al. (1984) also showed that the growth intensity (GI) increased with an increase in available light, subject to certain temperature restrictions, and that productivity (P) displayed typical hyperbolic relations (i.e. P versus I curves; cf. also Harris, 1978).

In order to include possible adaptation phenomena in the growth of microbial populations, under different environmental conditions, Fiolitakis et al. (1987) in analysing the data of Soeder et al. (1984), were able to describe the temperature response of *Scenedesmus obliquus* over the entire physiological range by introducing a damage-and-repair mechanism. By assuming such a mechanism, it was possible to interpret the growth activities of the alga at different temperatures and light intensities. The experimental data, however, on which Fiolitakis et al. (1987) had based their considerations, consisted of integral productivities and GI's over 24 hour periods, i.e. 16:8 hours light and dark cycles. In reality the concentration of algal biomass changed with time in the semi-batch cultures, rendering growth rate coefficients which were dependent on the light intensity. As only long term adaptation kinetics were considered by Fiolitakis et al. (1987), use of the integral values were justified in their analysis . However, in order to resolve the growth kinetics of a population, where the biomass concentration changes with time and to obtain kinetic parameters which are independent of light , the differential rate dependence of growth has to be considered.

In this paper a dimensionless activity state variable V is introduced, which accounts for long time scale adaptation reactions, and enables the combined action of light and heat stress to be modeled. The light dependence of short time scale growth kinetics, is expressed in terms of a Michaelis-Menten relationship between the growth rate and the available light energy. For the suboptimal temperature range a linear dependence of the state variable V against the available light energy is postulated.

Material and methods

Scenedesmus obliquus (Turp.) Kütz. (Strain SAG 276-3a of the Göttingen culture collection) was grown axenically using the mineral nutrient solution N8 in a light:dark cycle of 16:8 (Soeder et el., 1984). Measurements of optical density (O.D.), growth intensity (GI) and dry biomass were done according to Soeder et al. (1984). The cultures were grown at 20 °C and after obtaining 2 to 4 representative growth values at these quasi-steady-state conditions, the temperature was increased to the next level (Table 1). The cultures were illuminated from the sides in a thermoconstant water bath with six fluorescent lamps (OSRAM 1/32, 40W), which gave an irradiance of about 78 μEinst. m^{-2} s^{-1} at the surface of the culture tubes. The experimentally imposed different biomass concentrations implied that the effective light energy per cell decreased with increasing biomass (i.e. O.D.). The relation 1/O.D. and percentage relative light intensity was used to express the overall light availability (Soeder et al., 1984).

Mathematical considerations and structure of the model

A differential kinetic analysis served as the basis for this approach, where the light dependence of algal growth is formulated as a Michaelis-Menten function. GI-values were calculated from the following equation:

$$GI\ [d^{-1}] = ((X/X_o) - 1) \qquad \ldots(1)$$

where X is the biomass concentration at time 24h, X_o is the initial biomass, and d is the time in days (see also Soeder et al., 1984). This renders an integral rate over a day cycle. In a generalized form the rate of change in terms of available light energy (I) is:

$$dX/dt = \mu(I).X \qquad \ldots(2)$$

where t is the time in days and μ the specific growth rate. The specific growth rate according to the Michaelis-Menten kinetics is:

$$\mu(I) = \hat{\mu}(I).(I/(K_I + I)) \qquad \ldots(3)$$

where K_I is the half-saturating light intensity (analogous to a Michaelis-Menten constant) and $\hat{\mu}(I)$ the light dependent maximum specific growth rate, as established in the course of adaption to the combined action of light and heat stress.

Thus, in the context of the model, the rate constant $\hat{\mu}$ is both temperature and light dependent:

$$\hat{\mu}(I) = \varkappa(\theta).V(I,\theta) \qquad ...(4)$$

where $\varkappa$ is the Arrhenius frequency factor (d^{-1}), θ the temperature, and the state variable V, a dimensionless term, which is the activity coefficient of adaption due to the action of light and temperature stress.

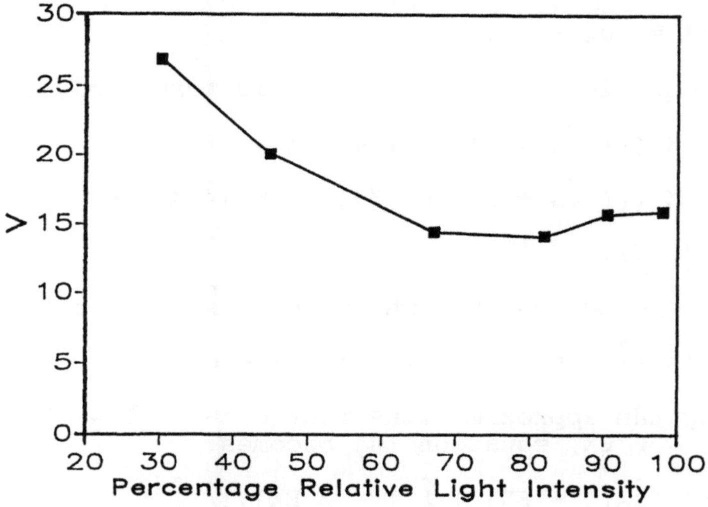

Figure 1.

The activity coefficient of adaptation due to temperature stress (V) at temperature from 20 to 30°C, against the percentage relative light intensity.

The adaption term $V(I,\theta)$ was recalculated using the model and parametric values from Fiolitakis et al. (1987). The light dependence of the adaption term V at temperatures up to 31 °C is shown in Fig. 1, which suggests a complex nonlinear relation between the action of light, heat stress and growth intensity, which could be related to enzymatic reactions. From this it is clear that at the suboptimal temperature range, i.e. up to 31 °C the adaption term is only light dependent.

As only the Michaelis-Menten light dependence of growth is considered here, the analysis can be limited to the data from optical densities of 0.02 to 0.4 (Soeder et al., 1984), which would represent the least dense cultures and therefore adequate light for photosynthesis. For this range the adaption term can be approximated by the following expression:

$$V = V_o + \alpha I \qquad \ldots(5)$$

where it has previously been shown that $V_o = 12.0$ and $\alpha = 4.744.10^4$ Einst. m^{-2} s^{-1} (see also Soeder et al., 1984). It is also postulated that this linear dependence should be valid for the entire light energy range. Under these conditions,

$$\hat{\mu}(I) = \varkappa . V_o + \alpha \ I$$

$$\text{or} \quad \hat{\mu}(I) = \hat{\mu}_o + \hat{\alpha} I \qquad \ldots (6)$$

Rewriting equation (3) to include this term yields:

$$dX/dt = X.(((\hat{\mu}_o + \hat{\alpha}I)I)/(K_I + I))$$

$$\text{or } dX/dt = X.(((\hat{\mu}_o + \hat{\alpha}I_o\, e^{-\epsilon x})\, e^{-\epsilon x})/(\hat{K} + e^{-\epsilon x}) \qquad \ldots(7)$$

where $\hat{K} = K_I/I_o$.

Equation 7 can be restated where $\hat{\delta} = \hat{\alpha}I_o$, as :

$$dX/dt = (X/(\hat{K} + e^{-\epsilon x})\{\hat{\mu}_o\, e^{-\epsilon x} + \delta\, e^{-2\epsilon x}\} \qquad \ldots(8)$$

By assuming the approximations that $e^{-\epsilon x} \approx 1 - EX$ and $e^{-2\epsilon x} \approx 1 - 2\,EX$, equation (8) becomes:

$$dX/dt = (X\,[\hat{\mu}_o(1 - EX) + \hat{\delta}(1 - 2\,EX)])/(\hat{K} + 1) - EX) \quad ..(9)$$

which can be integrated analytically. The solution becomes straightforward, after some transformations of the exponential function, rendering the following hyperbolic dependence on the initial biomass concentration:

$$GI = A/(1 + BX_o)\ , \text{ or}$$

$$1/GI = 1/A + (B/E)/A\ (O.D._o) \qquad \ldots(10)$$

where the following abbreviations were used:

$$A = e^{(\hat{\mu}o + \hat{\delta})t} - 1,$$

$$\tau = (\hat{\mu}_o + 2\hat{\delta})/(\hat{\mu}_o + \hat{\delta}),$$

$$B = E[(K + 1)\tau - 1]\, e^{(\hat{\mu}_o + \hat{\alpha})t/(\hat{K} + 1)} \quad \text{, and}$$

$$t = 1\ d.$$

Table 1.

Fitting equation (10) to the GI-data for *Scenedesmus obliquus* grown at optical densities from 0.02 to 0.4 and temperatures below 31°C (r = correlation coefficient).

Temp. °C	Regression line	r	A	B/E	$(\hat{\mu}_o + \hat{\alpha})$
20	y = 1.319x + 0.152	0.999	6.601	8.706	2.028
24	y = 1.193x + 0.078	0.999	12.804	15.275	2.625
28	y = 1.051x + 0.034	0.996	29.070	30.552	3.404
30	y = 0.988x + 0.019	0.994	51.546	50.928	3.962

Results of fitting the integral to the data presented by Soeder et al. (1984) are given in Table 1. The values for κ and $\hat{K}$ are given in Table 2, where it can be seen that $\hat{K}$ decreases with increasing temperature, which is typical of exothermic reactions. The values of κ and $\hat{K}$ are plotted against the inverse of the absolute temperatures (Fig. 2), where it follows that the slopes represents the activation energy (E) and the activation enthalpy (ΔH^*) of the reactions. An E of 49.005 kJ mol^{-1} and a ΔH^* of 98.129 kJ mol^{-1}, were derived from these plots.

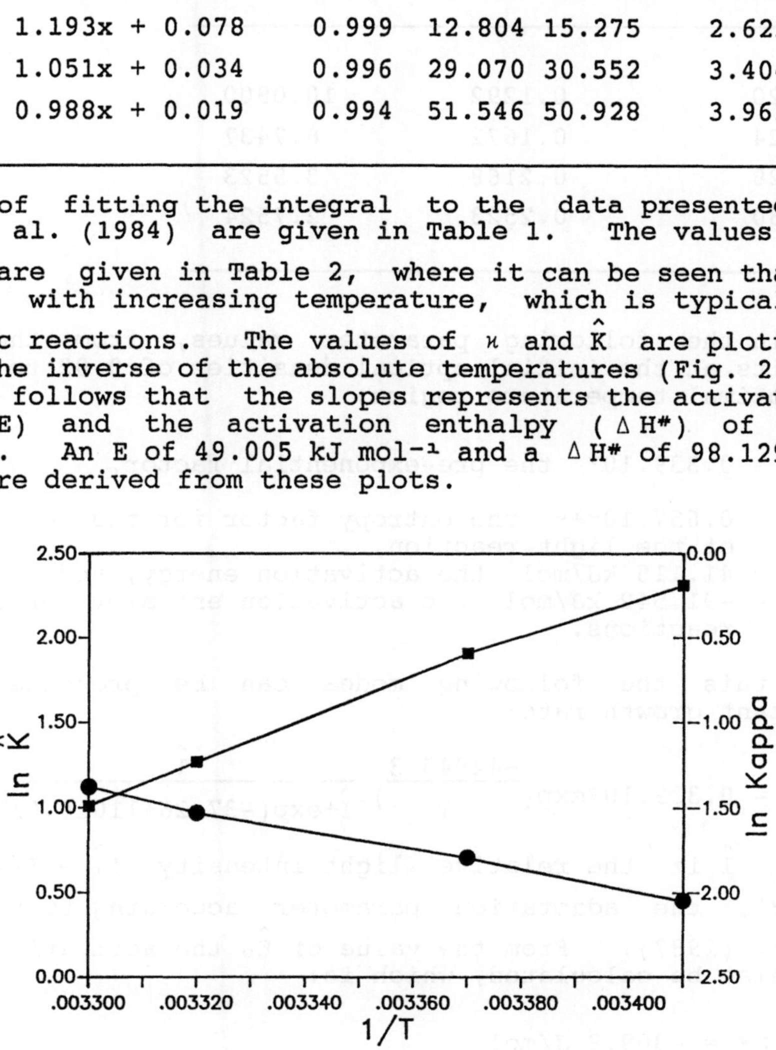

Figure 2.

The relationship of $\hat{K}$ (●) and κ (■) against temperature (1/T).

Equations (2) to (5) were verified with numerical analysis and a nonlinear fitting routine (Marquardt) was used to

Table 2.

Values for the variables $\varkappa$ and $\hat{K}$ at the various temperatures.

Temperature °C	$\varkappa$	$\hat{K}$
20	0.1292	10.0900
24	0.1672	6.7437
28	0.2168	3.5523
30	0.2523	2.7524

obtain the following parameter values, from the growth results at the initial optical densities of 0.02 to 1.20 and suboptimal temperature regime:

$\varkappa_o$ = $0.339.10^7$ the pre-exponential factor,

$\hat{K}_o$ = $0.657.10^{-16}$ the entropy factor for the activation of the light reaction,

E = 41.115 kJ/mol the activation energy, and

$\Delta H^{\#}$ = -91.542 kJ/mol the activation enthalpy for the light reactions.

From this the following model can be presented for a constant growth rate:

$$\mu(I) = 0.339.10^7 \exp\left(\frac{-44945.3}{T}\right) \cdot \frac{\hat{I}}{\hat{I}+\exp(-37.26+11010.6/T)} V(I,\theta)$$

where $\hat{I}$ is the relative light intensity ($\hat{I} = I/I_o$) and $V(\hat{I},\theta)$, the adaptation parameter according to Fiolitakis et al. (1987). From the value of $\hat{K}_o$ the activation entropy can also be calculated, which is:

$\Delta S^{\#}$ = -309.8 J/mol.

Discussion and conclusions

From the above analysis it is clear that the growth intensity (GI) of *Scenedesmus obliquus* as affected by temperature and available light, can be described mathematically. A differential solution is presented for the combined influence of temperature and light, where an Arrhenius relationship was used for the temperature response and Michaelis-Menten kinetics for light. Using a damage-and-repair model (Fiolitakis et al., 1987) was able to solve the adaptation parameter (V) which represented the combined action of heat and light stress.

It has also been possible to resolve the dependence of the rate constants with regards to available light, in the context of Michaelis-Menten kinetics. For the light reactions the activation enthalpy and the activation entropy gave reasonable values, whilst the activation energy and pre-exponential factor were in agreement with those of normal enzymatic reactions (Raison, 1980).

In the analysis of the data by Fiolitakis et al. (1987) constant initial optical densities (i.e. light availability) were considered and the adaptation kinetics were not related to actual light intensities. The fact that adaptation processes requires long time scales should be kept in mind, which implies that the adaptation parameter V, as derived in this paper (equation 5), would be representative for the suboptimal temperature range and low initial optical densities (i.e. high relative light intensities).

The results presented in this paper is, therefore, a first approach to model the dependence of adaptation kinetics on light availability, which offers a solution of the response of algae to real world conditions. It was possible to determine the Arrhenius parameters and Michaelis-Menten equilibrium constant from the short time scale growth kinetics, by taking the non-linear dependence of growth on available light into consideration. The next step in this analysis would be to model the adaptation kinetics at superoptimal temperatures.

References

Fiolitakis, E., Grobbelaar, J.U., Hegewald, E., and Soeder, C.J. (1987). Deterministic interpretation of the temperature response of microbial growth. Biotecnh. Bioengin., 30, 541 - 547.

Grobbelaar, J.U. (1981). Deterministic production model for describing algal growth in large outdoor mass algal cultures. In: Wastewater for aquaculture, Grobbelaar, J.U., Soeder, C.J. and Toerien, D.F. (eds.). University of the OFS Publ., Series C., 3, 173 - 181.

Grobbelaar, J.U., Soeder, C.J. and Stengel, E. (1984). Modeling algal productivity and oxygen production in large outdoor cultures. Jül - Spez - 282, Jülich, FRG, 49 pp.

Harris, G.P. (1978). Photosynthesis, productivity, and growth: The physiological ecology of phytoplankton. Arch. Hydrobiol. Beih., Ergebn. Limnol., 10, 1 - 171.

Hill, D.T. and Lincoln, E.P. (1981). Development and validation of a comprehensive model of large-scale production of microalgae. Agricultural Wastes, 3, 43 - 64.

Krüger, G.H.J. and Eloff, J.N. (1978). The effect of temperature on the specific growth rate and activation energy of *Microcystis* and *Synechococcus* isolates relevant to the onset of natural blooms. J. Limnol. sth. Afr., 5, 9 - 20.

Raison, J.K. (1980). Effect of low temperature on respiration. In: The biochemistry of plants, D.D. Davies (ed.). Academic Press, New York. 614 - 620.

Shelef, G. (1968). Ph.D dissertation, University of California, Berkley, USA.

Soeder, C.J., Hegewald, E., Fiolitakis, E., and J.U. Grobbelaar (1984). Temperature dependence of population growth in a green microalgae: Thermodynamic characteristics of growth intensity and influence of cell concentration. Z. Naturforsch., 40c, 227 - 233.

Different Responses in Ecosystems to Environmental Stress

E. Gutiérrez[1], *R. Guardans*[2], *and M.A. Canela*[3]
[1]Dept. Ecology, University of Barcelona, E-08071 Barcelona, Spain
[2]Environmental Protection Agency, Aragó 244, E-08007 Barcelona, Spain
[3]Dept. Applied Mathematics, University of Barcelona, E-08071 Barcelona, Spain

1. Introduction

The mechanistic paradigm is gone, so the scientific description of natural phenomena is shifting to a point of view where every event or situation is thought as the concrete realization of a set of possible events or situations, and consequently what is relevant is to describe the probabilities associated to the different events or transitions among them. The image of a stable universe composed of immutable orbits and equilibria modelled by normal distributions and linear relations between the variables is fading away, diluted in the colossal flow of incoming data.

Biogeophysical signals present stochastic variability over space and time, revealing an active and connected structure over a wide range of scales. Most of this universe appears as asymmetric and discontinuous. This paper deals with the argument that asymmetries over space and time are a very relevant part of the pattern and structure of natural phenomena, and, in particular, constitute a basic feature of ecosystems. The usual mechanical analog of an ecosystem in terms of flows and particles navigating in a deterministically defined landscape can be improved. Asymmetries and intermittencies must be accounted for and can be traced in what usually is perceived as stress, in the sense of approaching irreversibility, unpredictability, of leaving equilibrium.

The problem of achieving a macroscopical or phenomenological description keeping the required consistency with the immense microscopic complexity has been the core of many essential developments in mathematical physics over the last hundred years. Two main steps in the way to the solution are the Langevin equation and the Boltzman approximation, where the dynamics of the microscopical description translates into probabilistic statements that are valid and explain satisfactorily the dynamics of large systems.

In a certain sense, models of living organisms seem to follow contemporary technology. The 18th century dealt with cold mechanics, levers, pulleys, compartments. In the 19th and early 20th, heat and chemistry were the front. The chemistry of the living world was discovered to be driven by photosynthesis and respiration. From primitive reducing environments, life has evolved to an oxidizing medium, in a bewildering complexity of closed membranes that probably recapitulate the historical displacement from -400 mV to 800 mV.

The deterministic and symmetric models constructed with differential equations and limit cycles in state space precede and support a very simple but still abhorred statement: in biological theory we can be almost certain about the unpredictability of most relevant phenomena over a wide range of scales. The asymmetry over time was in a first approach associated to, or identified with, loss, leak, or noise. Later, a new approach through information theory and statistical physics (assisted by the new probabilistic tools introduced by mathematicians) has been succesful in its effort to assimilate the irreversibility of most biological phenomena to a possible source of new arrangements and combinations of the units undergoing the changes. This contribution was further developed by including the idea that the dissipated energy could be stored by a living system and could change the way in which the system would use the next input of energy. In other words, stating that under certain conditions the irreversibility of physical systems might be coupled with an increase in complexity.

After the electromagnetic explosion about 1940, the prevailing mode of the organism became that of an information handling system, a network of cells, transferring data from nucleic acid molecules to peptides, and communicating with other cells by transmitting specific peptidic messages. From the point of view of efficiency, the termophysical and the electromagnetic models present quite different requirements. It would be advisable to run a steam engine at high temperature, to get a gradient Δt, but a computer close to the absolute zero.

As models mimic and predict technology, our purpose is to emphasize the relevance of working in terms of algorithmical, numerical or computational paradigms, thereby drifting from the mechanics of the basic forces to the algebraic description.

The recent views of ecosystem structure and function have changed in several ways over the last three decades. Three main features of this change can be singled out.

a) The displacement of the mechanistic paradigm in which biological structure was mapped to particles and fundamental forces is a basic conceptual change that has recently taken place in the professional biological thinking. It is probably wiser now to focus on the fact that the biological variability is present in every scale, and is organized asymmetrically over space and time, leaning on membranes and phase transitions. To keep track of biological changes, particularly those taking place at scales far from our centimeter-second scale, it is useful and relevant to describe the numerical procedures applied to the extant data that underlie inevitably any interpretation or forecast (Hutchinson *et al.*, 1959; Pahaut, 1986).

b) The increase in computational power and information retrieval capabilities, which has radically changed the balance between the qualitative and the quantitative approaches to the description of complex systems (Margalef, 1968, 1981; De Angelis *et al.*, 1987). Most information concerning the living world is retrieved and stored in numerical or digital form. Therefore the direct and explicit mention of reproductible procedures and manipulations being carried out on the recorded signals is probably more efficient and operative than forcing the results to fit in a mechanistic model. It could be thus said that *digital ecology* joins *analogical ecology*, and, as a consequence of this expansion the new basic paradigms must be completed, recasted, in order to include structure and regulation patterns not only different but remote.

c) The relevance of microscopic, procariotic and subcellular processes to the functioning of large scale patterns. Microbial ecology and molecular biology have introduced essential novelties in the picture of life from the very small scale to the present global metabolism, like the evolution of mitosis, eucariotic cells and metazoa, metaphyta and fungi from procariotes, or the ontogeny and phylogeny of photosynthesis and respiration (Margulis *et al.*, 1976).

Notices about asymmetries and intermittency can be found in size distributions of organisms, mass distributions inside organisms and growth patterns. Therefore we sketch a model for the analysis of stress, based on the three points mentioned above, to be applied to data in search of the relevant asymmetries. Thus, we try to illustrate in practical terms some of the questions which have been outlined here, by presenting a framework to analyse quantitatively the different responses of ecosystems to environmental stress.

2. A model for stress

a) It has been often suggested (Margalef, 1968; Tansky, 1976; May, 1986), that there are fundamental asymmetries in the processes organizing life at any scale. The asymmetries are related to the irreversibility of time and to the inertia and hysteresis of material systems. Purely symmetric waves are excluded in ecosystems and organisms. Moreover, the asymmetry in the speed of change remains essential.

Let us drive our attention towards two variables which deserve careful consideration. The first one is the *net primary productivity* (NPP), which can be defined as the increase in biomass per unit area and unit time ($gr/m^{-2}y^{-1}$, wm^{-2}). At ecosystem scale, this variable can be directly related to the amount of tissue actually undergoing photosynthesis. It has been measured by book-keeping weighted samples of different compartments around the world. At smaller scales, the NPP can be indirectly related to photosynthesis, as in the case of secondary growth in trees, where the cambial tissue is related to photosynthesis by a complex network of syncronized displacements and time delays.

The second magnitude to be considered is the *gross turnover rate*, i.e. the quotient between the NPP and the total biomass per unit area, which decreases intermittently (Margalef, 1968). These intermittencies can be supposed to account for the production of the hard parts of the living organisms.

b) As long as asymmetries and intermittencies, time delays and temporary accumulations of mass are essential in ecosystems, it seems that a reasonably safe way to approach the subject is following the steps of the mathematical physics of turbulence, where the relevant variables are measurable and measured.

In the case of turbulence, the variables that the model tries to interpret and predict can be the velocity components in a cartesian reference. Assuming homogeneous fluid composition, x stands for the velocity along the main flow, y for the orthogonal component in the horizontal plane, and z for the vertical component. The velocity components can be given as a function of time, or simultaneously at many points. In the first case we deal with the fluctuations of the velocity of a given particle in time, and we can evaluate the Lagrangian spectrum as a function of frequency.

In the second, we can evaluate the Eulerian spectrum as a function of wave number. The power spectrum describes the contributions of oscillations with particular frequencies or wave numbers to the total variance (or power) of a variable. Most spectral estimates are calculated from data sampled at fixed points in space and represent Eulerian spectra (Frisch, 1978; Markatos, 1986).

For years, spectral analysis has been applied to problems of telecommunications, where the power spectrum is physically meaningful. In environmental research, spectral analysis is performed mainly to estimate parameters describing a complex signal. An updating of the theoretical ecology toolbox must include the search for relevant asymmetries associated to natural variability, carried on by spectral analysis methods. A recent contribution with such a perspective has been made by the authors (Guardans *et al.*, 1988). The estimation of the relative power at different frequencies over a given wave length range is a necessary step to develop operative methods for data analysis for such fields as tree growth, meteorology or remote sensing of vegetation (Fritts, 1965; Nicolis, 1986).

c) Ecosystems expose the present state of an historical process, buildt upon cultural and exosomatic memories, enjoying the irreversibility of time as the most nourrishing factor. About 2500 million years of procariotic activity resulted on a vast transformation of the surface of the planet and its atmosphere. Biological patterns and processes operate simultaneously at microscopic and global scale. Modern genetic systems, ubiquitous over the last 1000 million years, are based on mitosis taking place in cells which include organelles like plastids and mitochondria. Symbiosis and the deposition of hard parts underlie the Cambrian explosion of metazoa, metaphyta and fungi (Margulis *et al.*, 1976; Coope, 1970).

The configuration in a dense web of space and time of asymmetries of the support and transport in vascular plants, fungi, and multicellular plants in general, as well as in colonial procariotes or in animals, constitutes a form of memory posed at a scale larger than the size or life span of the individuals that retrieve it. Spiderwebs, anthills, songs and poisons are cultural superstructures, like highways and towns, libraries, migrations and sinks.

The quotation of a modern classic seems suitable here:

"Life has a range of magnitude narrow indeed compared to that with which physical science deals, but it is wide enough to include three such discrepant conditions as those in which a man, an insect and a bacillus have their being and play their several roles. Man is ruled by gravitation, and rests on mother earth. A water-beetle finds the surface of a pool a matter of life and death, a perilous entanglement or an indispensable support. In a third world, where the bacillus lives, gravitation is forgotten and the viscosity of the liquid, the resistance defined by Stokes' law, the molecular shocks of the Brownian movement, doubtless also the electric charges of the ionized medium, make up the physical environment and have their patent and immediate influence on the organism. The predominant factors are no longer those of our scale; we have come to the edge of a world of which we have no experience, and where all our preconceptions must be recast" (D'Arcy Thompson, 1961).

3. Results and exhibits

As an illustration of what has been suggested above, we drive our attention now towards two lines of research, which fit in the framework proposed in this paper.

We start by an example at human scale. Tree-ring widths, sampled in forest, suggest the study of how trees and their ecosystems have evolved under a certain type of external conditions. This example is suitable here, because the study of tree-rings is formally close to any other study where one considers a variable changing over time, recorded over a spatial grid.

Today, the amount of data available allows a frequent use of large sample statistical techniques. Among these techniques, two have been used by the authors in a recent research in order to examine how sensitive the width of the tree-rings might be to environmental fluctuations, and at which time scales the climatic signal, as detected in tree-rings, is more significant.

First, multivariate regression analysis, performed after having extracted the principal components, is used to study how the yearly climatic variability affects the habitat of trees and modulates the growth controlling process. Tree-ring width variations due to the climatic factors (precipitation and temperature), are described by the set of significant coefficients, usually known as response function. Samples of response functions are displayed in figure 1.

Second, spectral analysis is used to estimate the responses over a range of frequencies (cycles per year). Different species differ in their responses. We describe below these responses for two particular species, *Fagus sylvatica* and *Pinus sylvestris*, living close together under a Mediterranean climate, in an ecosystem located at 1000m of height (coordinates 42 20' 06" N, 01 47' 06" E). The statistical analysis shows that both species react strongly to the hydric stress, but using different strategies.

The environmental changes which alter the growth do not do so directly, but rather indirectly through their influence on rates and balances among photosynthesis, respiration, assimilation, hormone synthesis, absorption of water and minerals, and translocation of growth requirements. More than two thirds of the dry weight of trees consist of transformed sugars. Hence the growth of trees depends fundamentally on the synthesis of carbohydrates and their conversion to storage forms or hard parts. The stem tissue through which carbohydrates move between the crown and the roots is a layer of inner bark that is only a fraction of a milimeter thick (Kozlowski, 1979). This layer of bark increases 50% in a few weeks in June and stops growing by the end of August. 85% of the variation in the diameter of trees is due to the hydric stress.

The anatomy of the xerophytic leaves of *Pinus* as well as their stomatal responses are different to those from *Fagus* which are mesophytic and have little hydric regulation. *Pinus* stores less water in trunk and canopy than *Fagus*. Confronted with extreme drought, *Pinus* stops growing and waits for favorable conditions. Frequently, at the end of the season, or even in autumn, it is active again. That intermittency confers to *Pinus* a pioneering character. The growth season in *Fagus* is not interrupted, and consequently the rings coincide precisely with the seasonal forcing. Under drought conditions, *Fagus* is able to use fog as a source of water, and keeps active its cambium even if very slow.

Finally, we present a recent research, which can open new perspectives. As was mentioned when this magnitude was introduced, the NPP can be related to the amount of tissue performing

FIGURE 1

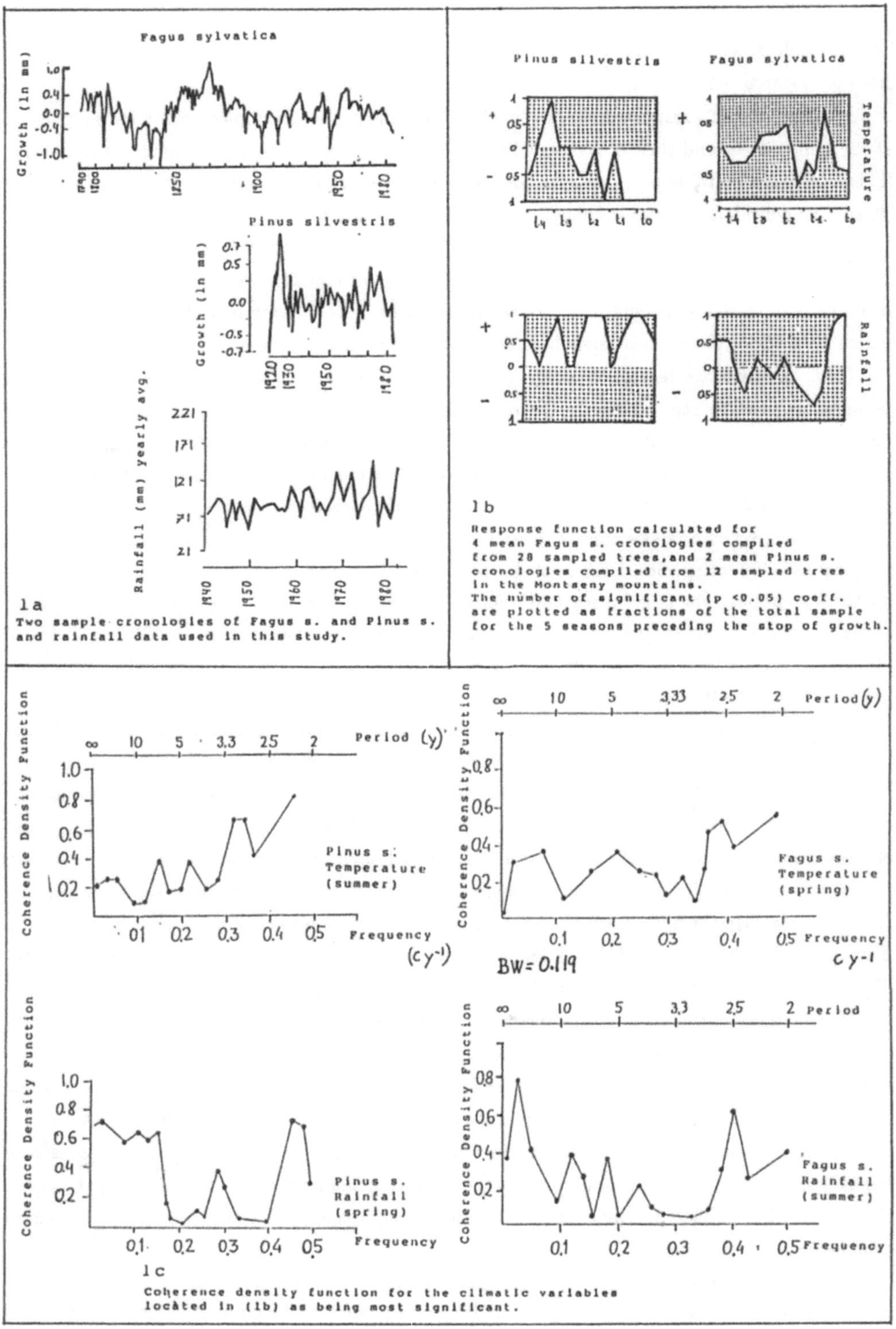

1a
Two sample cronologies of Fagus s. and Pinus s. and rainfall data used in this study.

1b
Response function calculated for 4 mean Fagus s. cronologies compiled from 28 sampled trees, and 2 mean Pinus s. cronologies compiled from 12 sampled trees in the Montseny mountains.
The number of significant ($p < 0.05$) coeff. are plotted as fractions of the total sample for the 5 seasons preceding the stop of growth.

1c
Coherence density function for the climatic variables located in (1b) as being most significant.

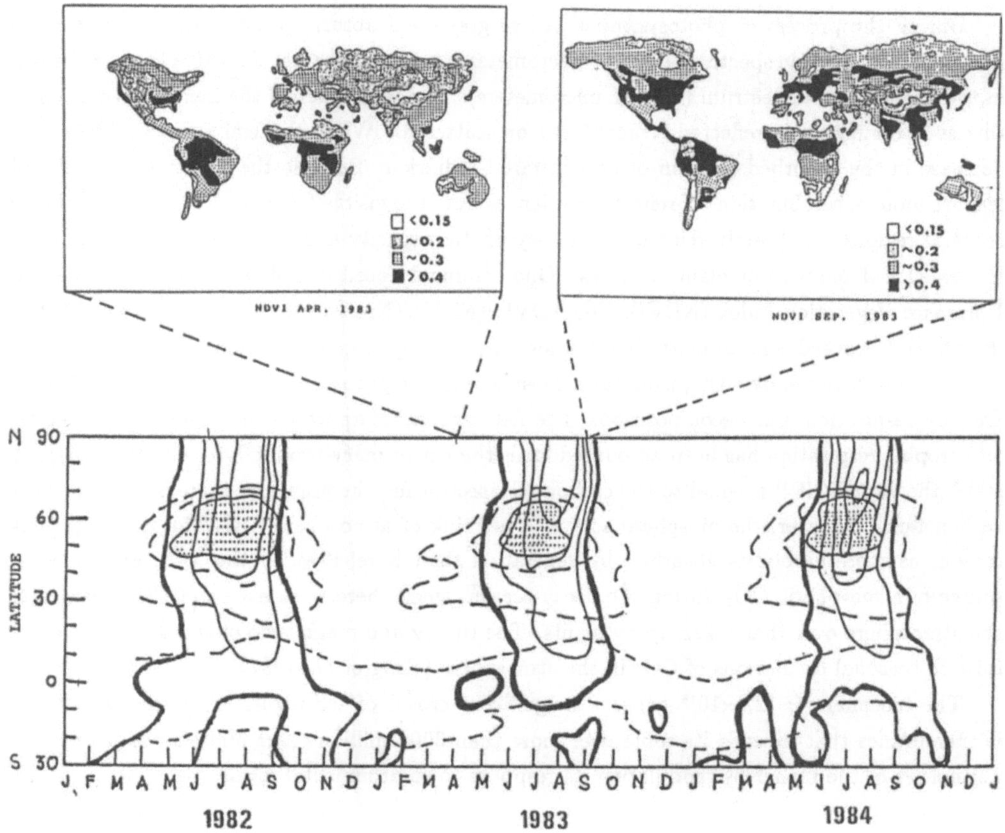

FIGURE 2 Comparison of variations of NDVI with variations of the drawdown rates of atmospheric CO2. NDVI averaged over Longitude in 4 zones, are shown by dashed lines denoting NDVI>.05, shading denotes NDVI >.15. Atmospheric CO2 drawdown rates (negative time derivatives of the CO2 concentrations) are shown as solid lines. The heavy solid line is 0 ppm/month and the contour interval is -2 ppm/month. (simplified from Fung et al 1987).

photosynthesis. At global scale, the amount of biomass carrying on this task can be estimated from an analysis of the radiation balance in the infrared range.

The radiative balance of the lower boundary layer of the atmosphere has been documented by calibrated photography at planetary scale with a georeferenced resolution of about 10 Km. A few daily pictures of the whole atmpsphere including its lower boundary have been taken by a handful of remarkable masterpieces orbitting in outer space (900–36000 Km) over the last ten years.

A very interesting contribution has recently appeared (Fung *et al.*, 1987). Instead of trying to explain the ideas involved, it is wiser to quote (almost literally) two selected paragraphs from this paper, which deserve a careful lecture for their clearity and compactness. We also present in figure 2 a simplyfied version of a picture showing the changes in the global vegetation cover.

During the process of photosynthesis, a live green leaf absorbs about 85% of the incident radiation in the visible spectrum (0.4–0.7 micrometers), and absorbs less than 10% of the radiation in the near-infrared spectrum (0.7–1.1 micrometers). The remainder of the incident radiation in this spectral region is reflected, transmitted or scattered. When the leaf senesces, there is a decrease in the absorbed fraction of near-infrared radiation. Because the spectral properties of leaf are unique, combinations of reflected radiances from the visible (V) and the near-infrared (Ni) spectral regions are directly related to photosynthetic capacity and have been used to estimate the green leaf biomass of plant canopies. One commonly used combination is the Normalized Difference Vegetation Index (NDVI). The NDVI=(Ni–V)/(Ni+V) ranges from less than 0.02 for deserts and exposed soils to more than 0.5 for fully developed green canopies.

The biosphere removes CO_2 from the atmosphere by the process of photosynthesis and releases CO_2 by respiration and decomposition. The net carbon incorporated into the biosphere, after autotrophic respiration has been accounted for is the net primary productivity (NPP). At steady state, the annual NPP is equal to the carbon released during the year, and there is no annual net carbon flux. However, the biosphere acts as a net sink of atmospheric CO_2 during the growing season, as more carbon is absorbed by vegetation than is released by the soils, and as a net source of atmospheric CO_2 during the decay season, when there is an excess of CO_2 released to the atmosphere over that taken up by plants. The timing and magnitude of this source-sink give raise to seasonal oscillations of CO_2 in the atmosphere (Fung *et al.*, 1987).

The biosphere ($\approx 12.9 \times 10^{18}$ gr) is a magnificent crowd of self-regulated processes included in membranes that extends its roots over more than 3000 million years and shapes the present physiology of the planet by modulating its response to environmental stress.

Un vent pénétrant, sans ombre, ni forme
Souffle aux quatre saisons, dispersant tout;
Rampant à terre, il soulève les feuilles jaunes
Et des montagnes tire d'informes nuages ...
—JIN PING MEI—

References

1. Coope, G. R., 1970. Interpretation of quaternary insect fossils. *Annual Review of Entomology* **15**, 997–120.

2. D'Arcy W. Thompson, 1961. *On Growth and Form*. Cambridge University Press, Cambridge.

3. De Angelis, D. L. and Waterhouse, J. C., 1987. Equilibrium and nonequilibrium concepts in ecological models. *Ecological Monographs* **57**, 1–21.

4. Frish, U., Sulem, P. L. and Nelkin, M., 1978. A simple dynamical model of intermittent fully developed turbulence. *Journal of Fluid Mechanics* **87**, 719–736.

5. Fritts, H. C., Smith, D. G., Cardis, J. W. and Budelsky, C. A., 1965. Tree-ring characteristics along a vegetational gradient in Northern Arizona. *Ecology* **46**, 393–401.

6. Fung, I. Y., Tucker, C. J. and Prentice, K. C., 1987. Application of AVHRR to study atmosphere-biosphere exchange of CO_2. *Journal of Geophysical Research* **92**, 2999–3015.

7. Guardans, R., Canela, M. A. and Gutiérrez, E., 1988. Spectral analysis of secondary growth patterns in *Pinus uncinata*. In *Biomathematics and Related Computational Problems,* edited by M. Ricciardi. D. Reidel, Dordrecht.

8. Hutchinson, G. E. and MacArthur, R. H., 1959. A theoretical ecological model of size distributions among species of animals. *The American Naturalist* **93**, 1117–125.

9. *Jin Ping Mei Cihua* (translated from the Chinese by A. Levy), 1985. Gallimard, Paris.

10. Kozlowski, T. T., 1979. *Tree Growth and Environmental Stress.* University of Washington Press, Seattle.

11. Margalef, R., 1968. *Perspectives in Ecological Theory.* University of Chicago Press, Chicago.

12. Margalef, R., 1981. Asimetrías inducidas por la operación de la energía externa en secuencias de sedimentos y de poblaciones. *Acta Geologica Hispanica* **16**, 35–38.

13. Margulis, L., Walker, J. C. G. and Rambler, M., 1976. Reassessment of roles of oxigen and ultraviolet light in Precambian evolution. *Nature* **264**, 620–624.

14. Markatos, N. C., 1986. The mathematical modelling of turbulent flows. *Applied Mathematical Modelling* **10**, 190–220.

15. May, R. M. 1986. The search for patterns in the balance of nature: advances and retreats. *Ecology* **67**, 1115–1126.

16. Nicolis, G. and Prigogine, I., 1986. *Exploring Complexity*. Freeman, San Francisco.

17. Pahaut, S., 1986. Tempo irreversibile e instabilità. *Nuova Civiltà Delle Machine* **14**, 18–26.

18. Tansky, M., 1976. Structure, stability and efficiency of ecosystems. In *Progress in Theoretical Biology,* vol. **4**, edited by F. Snell. Academic Press, New York.

An Example of Modelling in Ecotoxicology

K. Mathes, V.M. Schulz-Berendt, and G. Weidemann
Universität Bremen, FB 2, (Biologie:Ökologie),
P.O. Box 330440, D-2800 Bremen 33, Fed. Rep. of Germany

ABSTRACT:
Field studies, laboratory experiments and simulation modelling were conducted in order to judge pollutant-caused alterations at the ecosystem level. The data gained by an intensive field study (1982/1984) of the nitrogen-cycle during the initial phase of secondary succession of a ruderal plot were used to construct and evaluate a simulation model. By this model the effect of a biocide application on the N-budget was simulated by introducing results of experiments on the reaction of nitrifying bacteria on an insecticidal stress. Parallelly the same biocide was applied in the same concentration to a similar field plot like that of 1982, and the reactions of the different components of the nitrogen-cycle were studied. By comparison of field and laboratory results it turned out, that the biocide effects measureable in the field were either underestimated or overestimated by the laboratory experiments due to the actual conditions. Furthermore, the model proved to correctly simulate the field findings if used with the respective weather conditions. It is concluded, that a combination of field, laboratory and simulation experiments is necessary to get an impression of the impact of toxic chemicals at the ecosystem level.

1. INTRODUCTION

The aim of ecotoxicology is to analyze and judge the effects of environmental chemicals on ecosystems (LEVIN & KIMBALL 1984, WEIDEMANN et al. 1985, CONNELL 1986), and its main subject is the detection of pollutant-caused alterations of populations and communities and their function in their natural environments. However, no reliable ecotoxicological methods exist. These have yet to be developed.

Ecotoxicological studies require an ecosystem approach. This means not only to allow for the organisms and their populations but also for their interactions and the resulting functions. One of these functions tying together biotic and abiotic components is the nitrogen-cycle (N-cycle). Our study, which is part of a multidisciplinary investigation of vegetation, soil fauna and soil microflora (WEIDEMANN et al. 1985, WEIDEMANN et al. 1987), examines the effects of the biocide Aldicarb, a systemic insecticide, on different components of the N-cycle of a ruderal ecosystem in the initial phase of secondary succession. Three approaches are used for the ecotoxicological risk assessment of an Aldicarb application: 1. laboratory testing of the effects of the biocide on nitrate production and the kinetics of nitrification; 2. field trials with Aldicarb application and examination of changes caused in nitrate production and 3. simulation of the consequences of the measured Aldicarb effects for the NH_4^+-N- and NO_3^--N-contents of the soil solution by a dynamic simulation model of the N-cycle.

2. MATERIAL and METHODS

2.1 Study Site

The research site was the soil cover of a former rubble dump near Bremen (F.R.G.). The experimental plots were prepared by removing the topsoil (1 m), picking out the larger stones, mixing and replacing the soil.
The ecosystem development during the initial phase of secondary succession was investigated from May 1982 - May 1984 to quantify and model the essential N-pathways. In May 1985 two further sites were established in the same manner as in 1982, but were treated differently: one plot was left undisturbed and one plot was Aldicarb contaminated. Aldicarb was applied as TEMIK 5G at 2.5 g (a.i.) * m^{-2}, which resulted in a concentration of Aldicarb of 38.5 mg*kg^{-1} in the upper 5 cm of the soil. During the initial phases of succession (week 0-20) of both years, 1982 and 1985, a ruderal vegetation dominated by Chenopodium album and Sinapis arvensis appeared very quickly and produced a high standing crop. In 1985 the first plants germinated 3 weeks after the beginning of the experiments and reached a coverage of 100% 6 weeks later. From this time until the end of the investigation the coverage did not change. During this study period (May 1985 to Oct. 1985) the weather was cool and wet. The soil temperature (5 cm soil depth) ranged from 11°C to 22°C and the water content of the soil (0-8 cm soil depth) was generally higher than 15% H_2O (d.w.). Only during the first 3 weeks of succession and the last 4 weeks of the investigation period did the water content decrease to values below 10% H_2O (d.w.). For further information see WEIDEMANN et al.(1987).

2.2 Measurements

Samples were taken immediately after Aldicarb application and after 1,2,3,4,6,8,10,12,16 and 20 weeks of succession from the upper 0-4 and 4-8 cm of the soil. Six 100 cm^3 samples were taken from each site and mixed, then put into polyethylene bags and stored at 4°C.
The N_{min} contents (NH_4^+-N, NO_2^--N and NO_3^--N) were measured by extracting the soil with 1% $KAl(SO_4)_2$ (2.5:1) and analyzing the extract with an automatic analyzer (SKALAR). Net-nitrogen-mineralization was determined:

1) by incubating soil samples in polyethylene bags at the site for three weeks and
2) by incubating aliquots of the soil samples in Erlenmeyer flasks in the laboratory at 25°C and a water content of 90% of water holding capacity for three weeks.

After terminating the incubation, the N_{min}-content was measured again. The difference in N_{min}-content before and after incubation yields the quantity of net-mineralization. The NO_3^--N percentage of N_{min} after incubation is called degree of nitrification (BÜCKING 1972). Depending on incubation conditions two degrees of nitrification could be calculated for each sampling date: one from field incubation = NG_{fld} and one from laboratory incubation = NG_{lab}. In addition, parts of the soil samples were air dried and sieved (< 2 mm). 10 g of dry soil were weighed into 100 ml Erlenmeyer flasks and wet with $(NH_4)_2SO_4$ solution to produce a concentration of 330 mg NH_4^+-N • kg^{-1} of soil as a substrate for nitrifying bacteria. The nitrification kinetics were studied by measuring N_{min}-contents weekly as above until all of the NH_4^+-N had been nitrified. The kinetic parameters v_m and $(t_{1/2})^{-1}$, which had been used to describe the influence of pesticides on nitrification in pure cultures (RATNAYAKE & AUDUS 1978) or soil samples (TENA et al. 1984), were calculated from these results and served as measures for maximum activity (v_m) and division rate ($(t_{1/2})^{-1}$) of the nitrifying populations. The degree of nitrification = NG_{add} was calculated from the percentage of NO_3^--N formed after 3 weeks of incubation. Each aggregated sample was measured in dublicate at each site and for each date. The two values were nearly identical for all measurements and the mean value is represented in the figures. Detailed information about the methods used are given by SCHULZ-BERENDT (1986).

2.3 Model

A simulation model was constructed on the basis of a model of the N-cycle as given by MATHES & SCHULZ-BERENDT (1987). Its structure is shown in Figure 1. The fluxes of the N-cycle were quantified by statistically analyzing the field data of the 1982-experimental plot from May 1982 to May 1984. The NH_4^+-N- and NO_2^--N-leaching was not modelled, because of the vanishingly small quantities leached. We assumed the plants to use only nitrate, due to the characteristics of the study site. The nitrogen content of the soil fauna was not modelled as a state variable, because in relation to the other N_{org}-pools it was very small.

The model reproduces the N-cycle at a soil depth of 0-8 cm and is described as a deterministic, time-dependent model in ASS (Allgemeines Simulationssystem, SIMON 1983). State variables are ammonium and nitrate of the soil solution, organically bound nitrogen (N_{org}) of microorganisms and N_{org} of two fractions of dead organic matter (active fraction and stabilized fraction, PAUL & JUMA 1981). These state variables are connected by the processes N-gross-mineralization, N-immobilisation and nitrification. The inputs into the system are N_{min} of precipitation and the N_{org} of the litter, outputs are nitrate-uptake by plants and NO_3^--N-leaching from the soil, the temperature and water-content of the soil are incorporated as driving variables. The Euler-Cauchy-Method with a time step of 1/2 hour was found to be numerically stable. Therefore, this algorithm was used to solve the differential equations. The complete model is discussed in detail by MATHES (1987). The simulation model was evaluated by statistically comparing the simulated results for the NO_3^--N- and NH_4^+-N-content of the soil solution with those independently measured in the field. As Table 1 shows there is a fairly good correspondence.

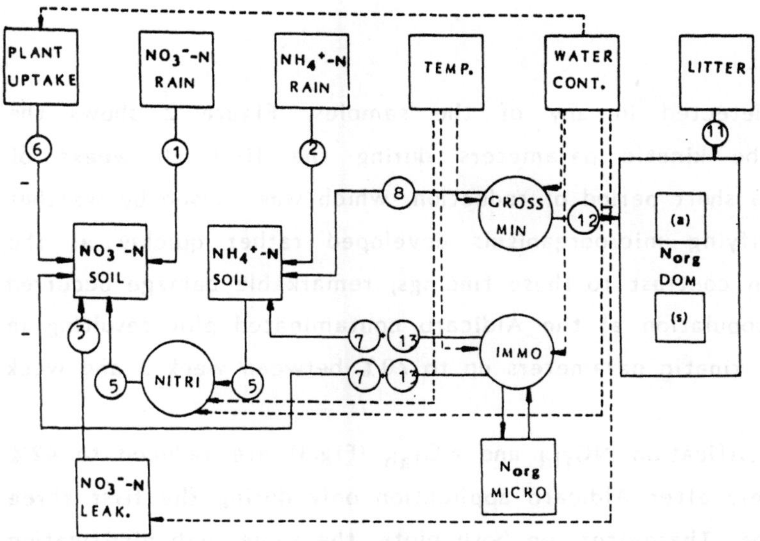

Fig.1: Structure of the simulation model
PLANT UPTAKE = nitrate nitrogen uptake by plants, NO_3^--N RAIN = nitrate nitrogen input by precipitation, NH_4^+-N RAIN = ammonium nitrogen input by precipitation, TEMP. = temperature in -5 cm of the soil, WATER CONTENT = water content in 0-8 cm soil depth, LITTER = N_{org}-N input by litter, NO_3^--N SOIL = nitrate nitrogen content in the soil solution (0-8 cm soil depth), NH_4^+-N SOIL = ammonium nitrogen content in the soil solution (0-8 cm soil depth), GROSS MIN = N-gross-mineralization, N_{org} DOM = N_{org} of dead organic matter (0-8 cm soil depth; (a):"active fraction", (b):"stabilized fraction", NITRI = nitrification, IMMO = N-immobilization, NO_3^--N LEAK. = nitrate nitrogen output by leaching, N_{org} MICRO = N_{org} content of microorganisms (0-8 cm soil depth).

Table 1.
Comparison of the simulation results and the measurements (n=17).
Measurement - mean = slope * (simulation - mean).
*: The correlation coefficient is significant (p = 0.1).

state variable	Spearman's correlation coefficient	mean measurement	mean simulation	slope
nitrate content	0.4093*	931	755	0.8
ammonium content	0.3313*	145	66	0.78

3. RESULTS

3.1 Experiments

No nitrite was detected in any of the samples. Figure 2 shows the development of the kinetic parameters during the first 20 weeks of succession. After a short period of reduction, which was caused by weather conditions,the nitrifying microorganisms developed rather quickly at the undisturbed plot. In contrast to these findings, remarkable damage occurred to the nitrifying population at the Aldicarb contaminated plot resulting in a reduction of the kinetic parameters up to 70% between week 3 and week 16.

The degrees of nitrification NG_{fld} and NG_{lab} (Fig.3) are reduced to 42% and 55% respectively after Aldicarb application only during the first three weeks of succession. Thereafter, on both plots, the same high nitrification degrees of about 90% were detected. This means that almost all of the mineralized N was immediately oxidized to NO_3^-. However, the initial reduction of NG_{lab} was less than that of NG_{fld}. On the other hand, NG_{add} shows extremely low values of about 20% up to week 12 (Fig.3).

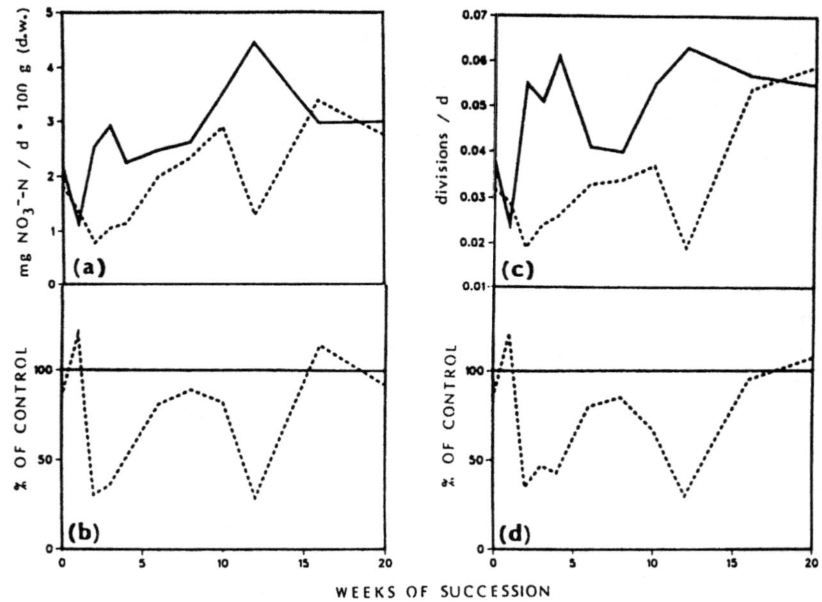

Fig.2: Kinetic parameters of nitrification. Uncontaminated plot (CONTROL ——) vs. Aldicarb contaminated plot (----).
(a)+(b) = maximum nitrifying activity (v_m).
(c)+(d) = average division rate ($(t_{1/2})^{-1}$).

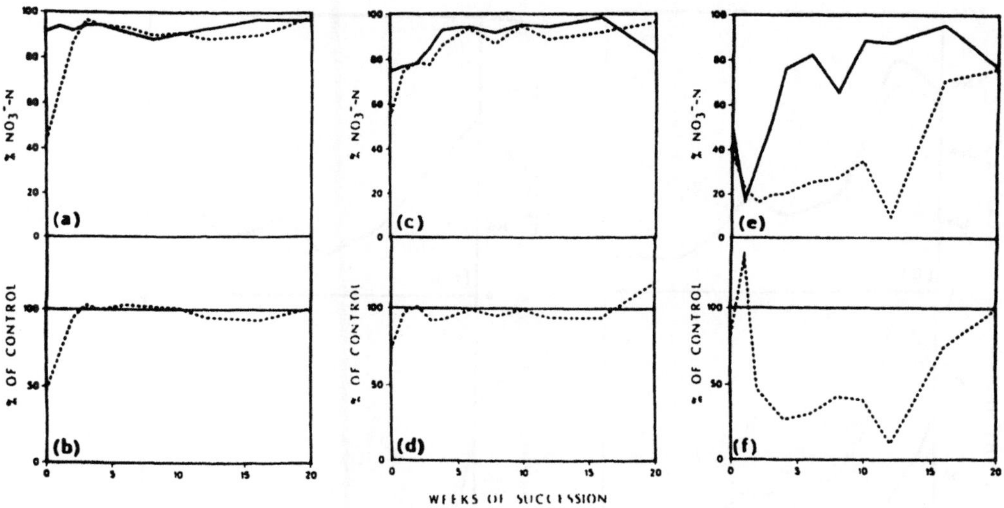

Fig.3: Degree of nitrification. Uncontaminated plot (CONTROL ——) vs. Aldicarb contaminated plot (----).
(a)+(b) = field incubation (NG_{fld}),
(c)+(d) = laboratory incubation (NG_{lab}),
(e)+(f) = laboratory incubation with addition of $(NH_4)_2SO_4$ (NG_{add}).

3.2 Simulation

The N-cycle was simulated using different sets of parameters to assess the consequences of the actual effect on nitrification after Aldicarb application for the nitrate- and ammonium-contents of the soil solution. This was done as follows. At first the impact of Aldicarb, which was measured in the field (NG_{fld}), was incorporated into the model and therefore a reduction of nitrification degree to 42% was assumed during the first 3 weeks of succession. The simulated consequences of this primary effect were studied under different weather conditions. The simulation results for a wet year (1985) are shown in Fig.4a,b and those for a dry year (1982) are shown in Fig.4c,d. This shows that the biocide leads to an increase of ammonium-content and to a reduction of nitrate-content in the soil solution. These consequences are especially apparent if the N-cycle is simulated using the weather conditions of 1985 (Fig.4a,b).

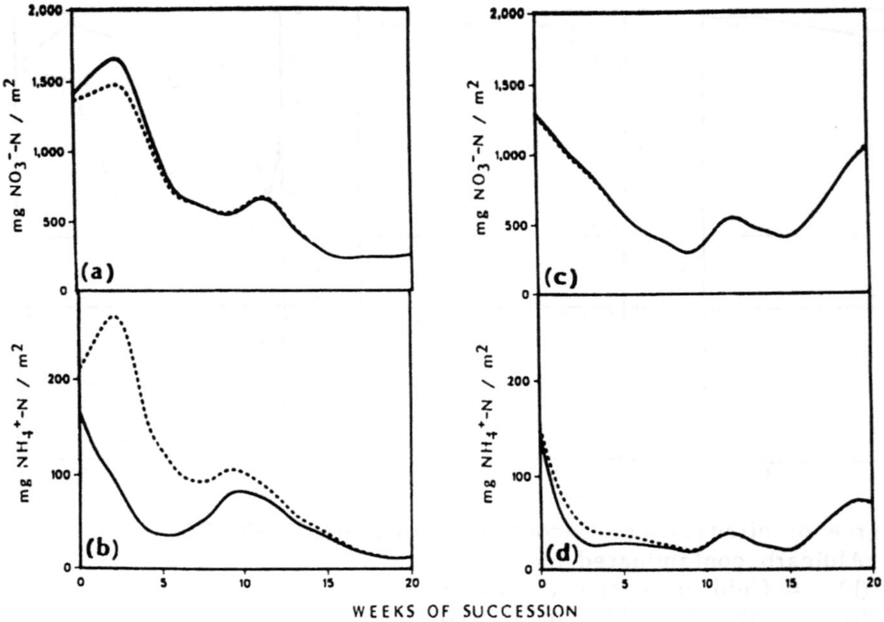

Fig.4: Simulation results. Undisturbed model (——) vs. stressed model (----).
(a) nitrate and (b) ammonium content of the soil solution (0-8 cm soil depth). Weather conditions of 1985.
(c) nitrate and (d) ammonium content of the soil solution (0-8 cm soil depth). Weather conditions of 1982.

4.DISCUSSION

While the kinetic parameters of nitrification showed marked negative influence after Aldicarb application for 16 weeks (Fig.2), the degree of nitrification was only reduced during the first 3 weeks under field and laboratory conditions (Fig.3 a-d). On the other hand the degree of nitrification, which is a measure of the actually nitrified amount of ammonified nitrogen, showed a lower Aldicarb effect under laboratory conditions (NG_{lab}, Fig.3 c,d) than under field conditions (NG_{fld}, Fig. 3 a,b) in the beginning of the experiment, while with substrate added in the Aldicarb contaminated laboratory trial the degree of nitrification was reduced for 16 weeks. These results indicate, that under field conditions the damaged microflora is still able to nitrify the same low amount of ammonium to the same degree as the undamaged one. However, higher substrate concentrations, which could possibly occur after manuring, can not be oxidized completely as a consequence of the alterations by

Aldicarb. Thus with higher NH_4^+-N concentrations in the soil solution the degree of nitrification in the field would be influenced to the same extent as observed for NG_{add} (MATHES & SCHULZ-BERENDT 1988). If this is true, the assessment of the effect of Aldicarb on nitrification depends on ecosystem conditions. Furtheron, the comparison of the results show that by laboratory experiments field effects of the biocide application are either underestimated or overestimated. In each case the ecological consequences of the biocide induced damage to the nitrifying microflora cannot be derived from the data. To make this possible the simulation model of the N-cycle is used. However, this is only admissible, if the model has been proved appropriate, as has been done statistically (Table 1) and by sensitivity analysis (MATHES 1987). If the changes in the nitrate production induced by Aldicarb as shown above were incorporated into the model the resulting consequences for ammonium- and nitrate concentrations in the soil solution became apparent. The results of the simulation indicate that the intensity and the timing of the changes in NH_4^+-N and NO_3^--N concentrations in the soil solution depend on weather conditions. It is quite clear, that these ecosystem effects could not be predicted without this form of analysis. It must be kept in mind, however, that for the present these simulation results are only valid for the model. For the case represented in Fig. 4a,b field experiments were independently conducted, and they induced the same effects of Aldicarb at the ecosystem level as the simulated ones (MATHES & SCHULZ-BERENDT 1987). All simulations presented here may be used to judge the qualitative impact of Aldicarb on the N-cycle of the ruderal site. The calculations of the "undisturbed model" should be compared with those of the "stressed model". In such a comparative analysis, simplifying assumptions that do not bias the results may be acceptable for an ecological interpretation whereas they may be unacceptable in the assesssment of an absolute risk (O'NEILL et al. 1982).

Acknowledgements:

This research was supported by the Ministry of Research (PTU 037281A7).

5. REFERENCES

Bücking, W. 1972. Zur Stickstoffversorgung von südwestdeutschen Waldgesellschaften. Flora 161:383-400.

Connell, D. 1986. Ecotoxicology - A new approach to understanding hazardous chemicals in the environment. Search 17:27-31.

Levin, S.A., and K.D. Kimball (eds.). 1984. New perspectives in ecotoxicology. Environm. Managem. 8:375-442.

Mathes, K. 1987. Computersimulation: Eine Methode zur Beurteilung von Umweltbelastungen -Untersucht am Beispiel des Stickstoff-Haushalts eines Ruderal-Ökosystems. Dissertation, Universität Bremen, FB 2, 174 S.

Mathes, K., and V.M. Schulz-Berendt. 1987. Ein Modell des N-Haushalts zur ökotoxikologischen Analyse und Beurteilung von Chemikalienwirkungen. Verh. Ges. Ökol. 16 (i. press).

Mathes, K., and V.M. Schulz-Berendt. 1988. Ecotoxicological risk assessment of chemicals by measurements of nitrification combined with a computer simulation model of the N-cycle. Toxicity Assessment - An international Quarterly - (Valencia 1987), i. press.

O'Neill, R.V., R.H. Gardener, L.W. Barnthouse, G.W. Suter, S.G. Hildebrand, and C.W. Gehrs. 1982. Ecosystem risk analysis: a new methodology. Environm. Toxicol. Chem. 1:167-177.

Paul, E.A., and N.G. Juma. 1981. Mineralization and immobilization of soil nitrogen by microorganisms, P. 179-194. In F.E. Clark, and T. Rosswall (eds.), Terrestrial nitrogen cycles, Ecol. Bull. Vol. 33.

Ratnayake, M. and L.J. Audus. 1978. Studies on the effects of herbicides on soil nitrification II. Pesticide Biochem. Physiol. 8: 170-185.

Schulz-Berendt, V. 1986. Der Stickstoff-Haushalt eines Ruderalstandorts als Grundlage der Beurteilung von Ökosystem-Veränderungen. Dissertation, Universität Bremen, FB2, 142 S.

Simon, K.-H. 1983. ASS-Benutzer-Anleitung. GHS Kassel, Fb 19.

Tena, M., R. Garrido, and M. Magallanes. 1984. Sugar beet herbicides and soil nitrification. Soil Biol. Biochem. 16:223-226.

Weidemann, G., D. Bolte, H. Koehler, K. Mathes, J. Müller, V.M. Schulz, and G. Vollmer. 1985. Auffindung von Indikatoren zur prospektiven Bewertung der Belastbarkeit von Ökosystemen: Bezugsökosystem Ödland, P. 85-119. In B. Scheele, F. Führ, and E. Stüttgen (eds.), Auffindung von Indikatoren zur prospektiven Bewertung der Belastbarkeit von Ökosystemen. Vol. 7, KFA Jülich, Jül-Spez-296.

Weidemann, G., H. Koehler, K.Mathes, and V.M. Schulz-Berendt. 1987. Konzept und Anlage eines Ökosystems zur ökotoxikologischen Beurteilung von Chemikalien. Verh. Ges. Ökol. 16, i.press.

Spatial Simulation of Population Dynamics in the Evaluation of Foraging Theory in Complex Ecological Systems

F. Perez-Trejo

IFIAS-Maastricht, O.L. Vrouweplein 21, NL-6211 HE Maastricht, The Netherlands

INTRODUCTION

The distribution and persistence of species in nature is the product of very complex biotic and abiotic interactions. The resources that a species may need for food, shelter, and reproduction are non-uniformly distributed in space and time. Each species is also enmeshed in a complex web of interactions with other species that are part of its environment. It is the evolutionary process of these interactions that has produced the spatial patterns of the populations that are observed in nature. The resources that a species may require (including its food) are generally considered to have a patchy distribution. The patches themselves form patterns of more or less fragmented habitats within areas of "low" habitat value. These spatial patters may be caused by abiotic factors (such as fire, soil differences, climatic factor) interacting with the species that differentially utilize the resources in those patches [1]. The collective effect of deforestation, pollution, urbanization, agriculture, and industrialization by man has increased the level of fragmentation of natural habitats way beyond what many species can tolerate. The complexity of the effects of man's activities on processes and interactions represents a veritable challenge to theoretical ecology. Every effort to understand the very complex non-linear dynamics of these systems can help in developing more ecologically sound strategies for managing natural systems.

This article is concerned with understanding the causal relationships that govern the dynamics of spatially structured populations. A hierarchical framework is proposed for integrating processes at different levels of organization that could help explain the observed higher-level dynamics of populations without being reductionist.

CONCEPTUAL FRAMEWORK

The study of spatially structured population dynamics presupposes that lumping the spatial diversity with which these populations interact into parameters that have no spatial dimensions would be leaving out the key processes that make these populations spatially structured. Therefore, foraging behaviour was chosen as a useful integrator of interactions of processes at different hierarchical levels of the ecosystem. Since foraging behaviour of species with high mobility, such as large herbivores, is the product of an evolutionary process of species adapting to a spatially diverse environment, it must respond to the variability in the patterns of resources in space.

Behaviour patterns have evolved to deal with the complex interactions that individuals must process in an ecosystem in their struggle for survival. As species have adapted over evolutionary time to strong foraging interactions (such as predator-prey systems), at first glance one would expect that foraging strategies could be predicted using optimization theory. But foraging strategies are the result of historical accidents that have prevailed through feedback mechanisms in the ecosystem [2], and do not necessarily respond to any optimization process. In the case of predators that feed on prey that is distributed scarcely in space, and of high nutritional value, optimal foraging theory (OFT) has explained quite satisfactorily the kind of prey and the search strategies that are observed in nature [3,4].

The foraging behaviour of large herbivores, on the other hand, does not follow the behaviour patterns or diets that the OFT would predict [5, 6]. This is not surprising, considering the markedly different spacial and temporal distributions of "prey" (forage) that constitutes the food of large herbivores. The plants that a large herbivore would consider as food have a much more abundant spatial distribution, even though they may vary greatly in nutritional value and aggregation patterns.

In spite of these notable differences, large herbivore-forage interactions are as much the product of an evolutionary process as any predator-prey system, and therefore, there should exist some theory that can "explain" the observed foraging behaviour patterns.

In the study of foraging behavior it is essential to consider that the systems we are studying are the product of an evolutionary process, and that the behaviors we might observe are the product of many complex interactions. If we compare the kind of foraging behaviour that a small herbivore, such as a prairie dog, should be selecting from the available plant species, with the observed diets, we see that is it far below what the OFT would predict, assuming the animal was trying to maximize its food quality intake. What the OFT fails to consider is how the behaviour patterns of these animals have evolved to utilize the different spatial patterns in its environment to accommodate needs for food, cover form predators, reproduction, etc. The prairie dogs are "making decisions" about what, where, and how long to forage in response to many factors: the temporal and spatial distribution the forage, the relative risk of predation in different areas of the landscape, and their population density [7]. What may seem to be a sub-optimal foraging strategy in the light of OFT, may be a quite satisfactory strategy with respect to the factors that explain the patterns of use of the spatial domain in which the species exists.

MODEL DESCRIPTION

A hierarchical decision model is being developed to test the possible strategies that a species could use for selecting different patches in the landscape. The model first generates randomized spatial patterns of vegetation according to observed distribution and proportion of plant species in the field, including bare ground. The vegetation matrix that is generated is a strip 20 plants across a meter (assuming 5 cm diameter plants) and can have any desired length. The model representation of the landscape can then be expanded by generating several 1 meter strips next to each other. Associated with each plant is an array of plant characteristics that currently includes height and nutritional value (represented by crude protein level) and varies for each plant according to a random normal distribution. The program is then given a set of "rules" that a hypothetical herbivore uses for moving across the computer-generated landscape and selecting plant species to eat within different patches. The program is very flexible about the kinds of rules that can be tested. They include how many plants to evaluate before each bite, what height, crude protein level, minimum (or maximum) level for selection, the threshold level for switching to another patch, etc. After each selection the plant community is updated for height and nutritional value to simulate how the perceived status of the forage changes due to the effect of grazing itself. The hypothetical herbivore evaluates the possible plants available in a patch of the landscape (Figure 1) and selects those plants that would fit the rules until no remaining plants are above 3% protein. The animal then moves on to another patch, utilizing rules for patch selection (Figure 2), eg: the closest patch with plants containing more than 3% protein, and begins plant selection all over again.
In this manner the hypothetical herbivore moves across the computer-generated landscape and a cumulative diet is compiled by the program.

Several kinds of foraging strategies are being tested to determine what kinds of diets they produce from a series of randomized runs. The resulting diets and patch selection patterns can then be compared against empirical observations and measurements of diets that have been made in the field.

RESULTS AND DISCUSSION

Model results for cattle and sheep show that simple strategies, such as:"graze everything that is fairly nutritious (over 3% crude protein)", are able to explain the complex foraging patterns that have been reported in grazing experiments [8]. This would indicate that in landscapes with contiguous patches, such as the short grass prairie simple rules at the patch level can explain the general foraging patterns that are observed at the community level. These are termed "deductive" rules. A deductive foraging rule is defined as a rule that applies to selelction of diet itmes among a few palnts, that when integrated over a large enough temporal or spatial dimension explains the observed resposes of the system at a higher

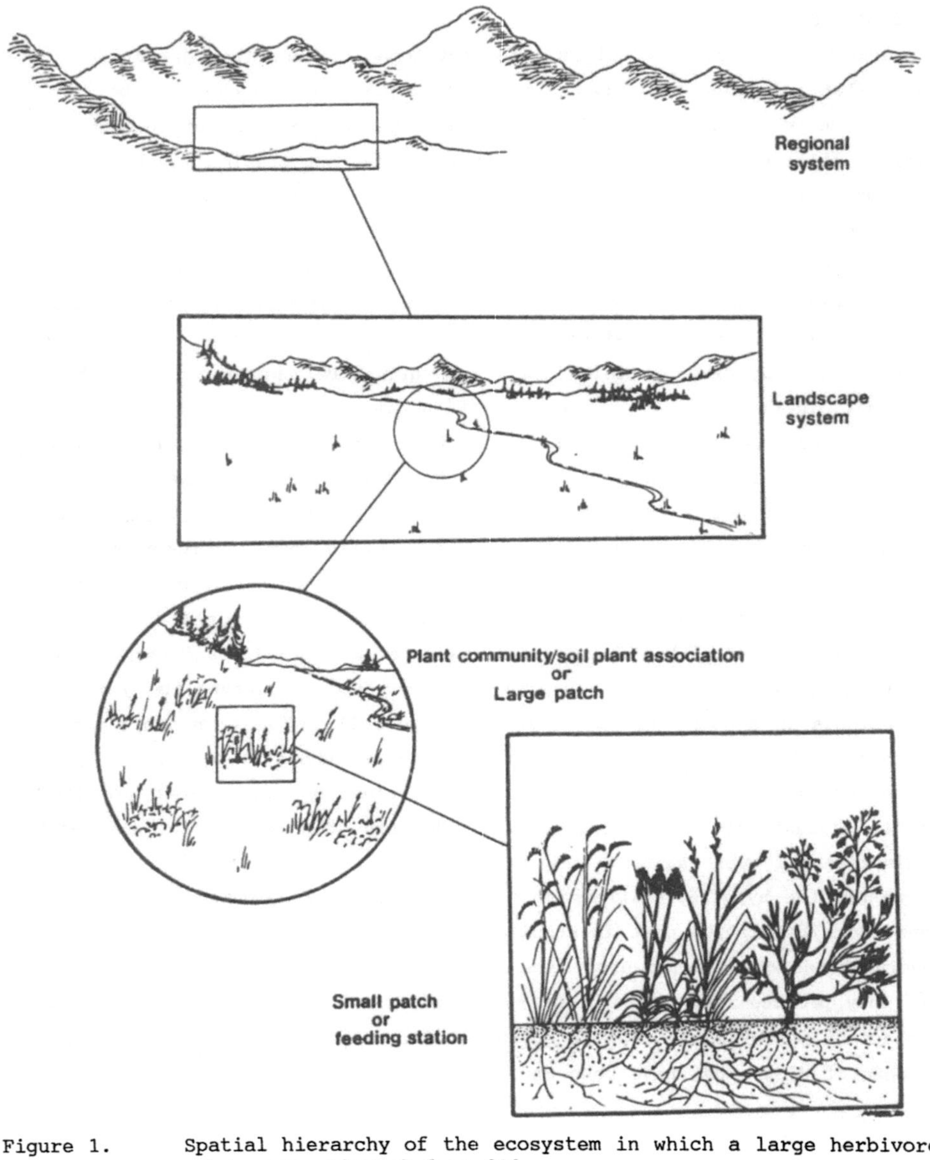

Figure 1. Spatial hierarchy of the ecosystem in which a large herbivore can interact. Adapted from [9].

organizational level (at the patch or landscape level). Other kinds of rules at lower organizational levels would be expected not to be able to explain the general behaviour patterns observed higher organizational levels. Such patterns that do not respond directly to any plant selection behaviour, but to other higher level rules at the landscape or regional plane, would be termed "inductive" rules. An example of such higher level behaviors are the rules that might govern migration patterns, such as climate.

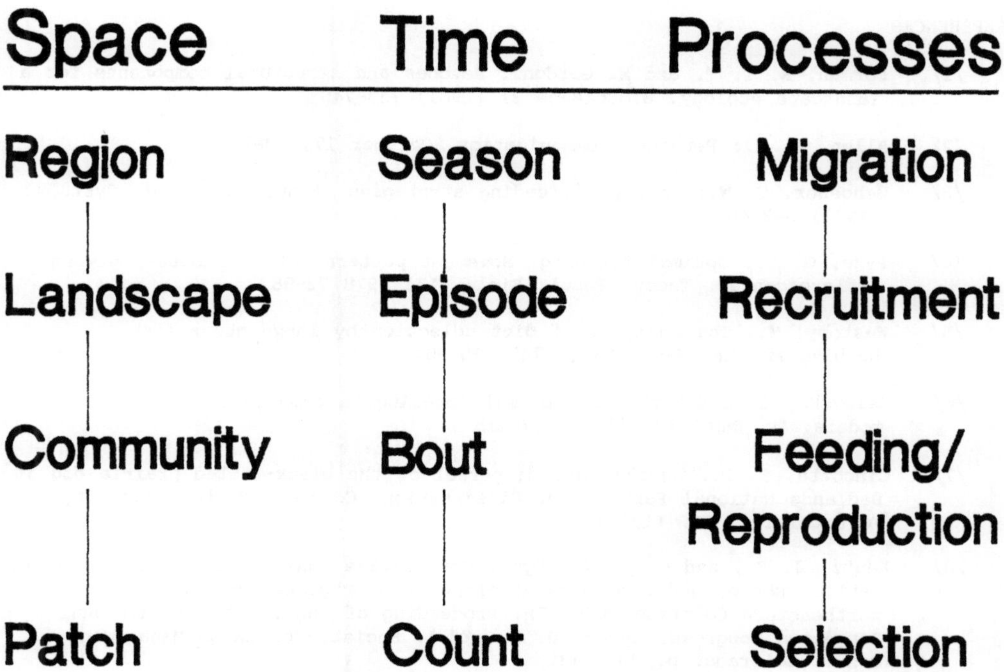

Figure 2. Relationships between levels of space, time, and process hierarchies.

One would expect the rules to become much more complex as the number of possible interactions increases; for example, when an animal must select in which area to graze, and for how long. But if we consider these decisions to be simple strategies at different hierarchical levels of the ecosystem, the model does not have to include the overwhelming complexity of the system to provide meaningful explanations of numerous behaviour patterns that have little to do with optimization.

These kinds of models can provide valuable learning tools for conducting "experiments" within a computer that can give us some insight about the theories that explain the evolution and behaviour of such complex systems. The promising results from such a model are helping in the formulation of theory that would explain the complex causal relationships that govern the divers range of foraging behaviors that are observed in nature. As we are able to explore the causal links between organizational levels of the ecosystem, the theory will help in the understanding of the community structure that drives population dynamics.

REFERENCES

/1/ Forman, R. T. T. and M. Gordon.: Patches and structural components for a landscape ecology. BioScience 31 (1981) 733-740

/2/ Allen, P. M.: Personal communication. October 17, 1987

/3/ Schoener, T. W.: Theory of feeding strategies. Annu. Rev. Ecol. Syst. 11 (1971) 369-404

/4/ Pyke, G. K.: Optimal foraging: Movement patterns of bumblebees between inflorescences. Theor. Popul. Biol. (13) 1978 72-98

/5/ Westoby, M.: The analysis of diet selection by large generalist herbivores. Am. Nat. 108 (1974) 290-304

/6/ Belovski, G. E.: Herbivore optimal foraging: a comparative test of three models. Am. Nat. 124 (1984) 97-115

/7/ Cincotta, R. P.: Habitat and dispersal of the black-tailed prairie dog in Badlands National Park. Ph.D. Dissertation. Colorado State University, Fort Collins, CO. (1985)

/8/ Kautz, J. E., and G. M. Van Dyne: Comparative analyses of diets of bison, cattle, sheep, and pronghorn antelope on shortgrass prairie in northeastern Colorado, USA. In: Proceeding of the First International Rangeland Congress. Hyder, D. N. (Ed.). Society for Range Management, Denver Colorado. p. 438-443.

/9/ Sneft, R. L., M. B. Coughenour, D. W. Bailey, L. R. Rittenhouse, O. Sala, and D. M. Swift: Large herbivore foraging and ecological hierarchies. In preparation. (1987)

Models for the Management of Renewable Resources Embedded in Complex Marine Systems

P.A. Shelton

Sea Fisheries Research Institute, P.Bag X2, Rogge Bay 8012,
Cape Town, South Africa

ABSTRACT

The difficulty in building useful models of the interactions between components of complex ecosystems is demonstrated with an artificial system. Problems associated with structural specification and parameterization are shown to limit the usefulness of the more complex models of the artificial system and the implications of this are considered in the context of the management of exploited animal populations in marine ecosystems.

The management of marine renewable resources has to do with populations which are embedded in complex ecosystems. It follows therefore that the dynamics of a harvested population are determined in part by the dynamics of the rest of the ecosystem. To treat harvested populations as isolated, self-generating entities may be oversimplistic and management based on such treatments may be ineffective. In this paper the building of more realistic models for complex ecosystems is considered in the context of the management of a commercially exploited pelagic fish population in the Benguela Current ecosystem. The difficulty in building useful models of the interactions between the components of such systems is demonstrated with an artificial system. Brief consideration is given to an alternative approach of using observations of general properties of whole systems to make predictions about the behaviour of individual components.

Benguela pelagic fish system

Pelagic fish, typified by anchovies and sardines, are small, fast growing, short-lived, near-surface schooling fish which are often abundant in areas of nutrient enrichment such as upwelling ecosystems. Populations of these fish yield exceptionally large commercial harvests, although often only for brief periods, before declines or collapses occur. Recovery of collapsed populations, even if fishing pressure is removed, may take place only after a time-lag, if at all.

In the Benguela Current ecosystem off the west coast of southern African, anchovy, sardine and roundherring occupy an intermediate position in the food chain, feeding on phytoplankton and zooplankton, competing with each other or preying on each others eggs and larvae, and being preyed upon by a variety of predators such as seals, seabirds, predatory fish, and not least of all, man (Figure 1). In addition to the interaction with other components of the biota, the effect of the physical environment on recruitment, by influencing the survival of the early planktonic stages, is large and has the potential to generate very variable year class sizes.

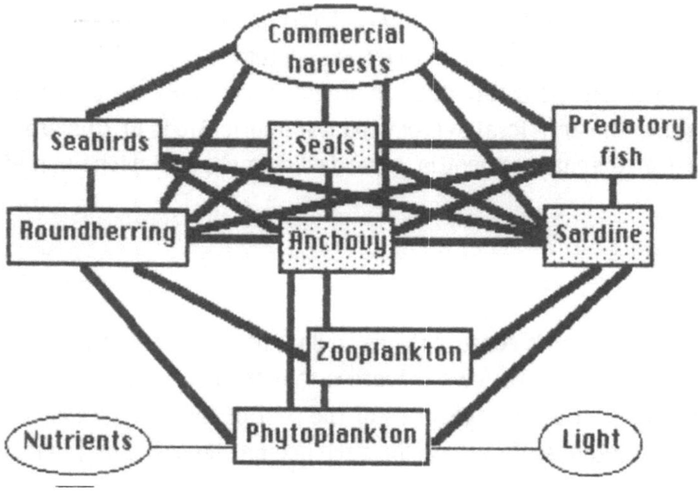

Figure 1: The major connections between anchovy and other components of the Benguela Current ecosystem. The stippled components are equivalent to those shown in the simplified, artificial system in Figure 2.

Even if the effect of the physical environment is ignored (or assumed to be random), it is intuitively obvious that the size of the anchovy population in such a system could show considerable variability, depending on changes in the size of the other interacting components. A modeler, presented with the problem of predicting changes in the size of the anchovy population with time under different levels of anchovy harvest, may try to include some of these interactions in a model. Indeed the need to take such an approach has been advocated for the management of multispecies fisheries in the Southern Ocean ecosystem [1].

An artificial system

To examine this approach, an artificial system consisting of three coupled difference equations was constructed. The system included a harvested population A, a competitor population B and a predator population C (see Figure 2).

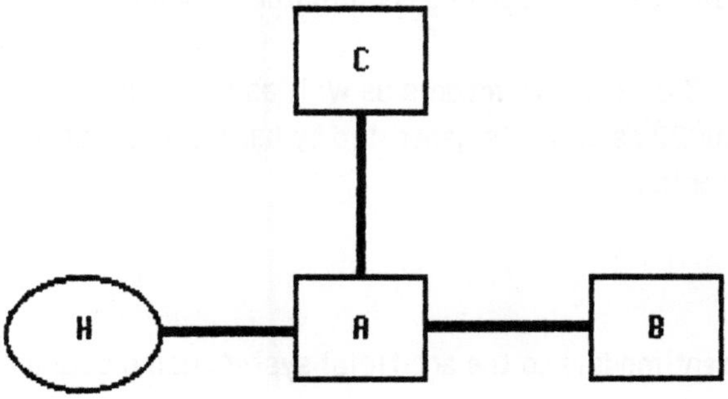

Figure 2: A simplified pelagic fish system, including a harvested population A, a single competitor B, and single predator C. H represents the harvest on A.

The basic model governing the dynamics of A was defined as

$$A_{t+1} = b_1 A_t - b_2 A_t^2 - H_t \quad (1)$$

Where A_t = the size of population A in year t

H_t = the harvest of A in year t

This model (equivalent to model f2 in Walters, p186 [2]) was expanded by including simple terms for competition with population B and predation by population C. Thus the dynamics of A within the artificial system were defined by

$$A_{t+1} = b_1 A_t - b_2 A_t^2 - b_2 b_3 B_t A_t - b_4 C_t A_t - H_t \quad (2)$$

Note that the relative growth rate of population A after harvest, $(A_{t+1} - H_t)/A_t$, is a linear function of A, B and C with parameters b_1, b_2, $b_2 b_3$ and b_4.

The dynamics of the competitor population B were defined by

$$B_{t+1} = u_1 B_t - u_2 B_t^2 - u_2 u_3 B_t A_t \quad (3)$$

and the dynamics of the predator population C by

$$C_{t+1} = w_1 b_4 C_t A_t - w_2 C_t^2 \quad (4)$$

The following parameter values were set: b_1 = 2.0, b_2 = 0.02, b_3 = 2.0, b_4 = 0.025, u_1 = 2.0, u_2 = 0.02, u_3 = 1.62, w_1 = 1.5, w_2 = 0.01. With these parameter values the system displayed single, stable, non-zero equilibria over a range in H from 0 to 9. At higher levels of harvest A and C approached 0 and B increased to a maximum.

Population A in the artificial system was fished at a wide range of H in order to provide 22 data points for A, B and C with sufficient constrast for parameter fitting by regression of $(A_{t+1} + H_t)/A_t$ against A, A and B, A and C, and A, B and C. It is important to note that a narrow range of harvests would have provided little change in the size of A, and as a consequence little information for parameterization. As a result, even fitting the correct model would have yielded poor parameter estimates. Such "adaptive probing" is considered a useful tool for learning about natural complex ecosystems [2].

The results from fitting the 3 alternative models as well as the correct underlying model (as a check), using the 22 data points generated by harvesting A at different levels, are summarized in Table I.

Table I:
Results from fitting different models to the artificial system using data generated by harvesting A.

Model	Parameters	r^2	SE
(1) $(A_{t+1} + H_t)/A_t$ vs A_t, B_t and C_t	$b_1 = 2.0$	1.0	0.0
(correct model, including	$b_2 = -0.02$		
competition and predation)	$b_3 = 2.0$		
	$b_4 = -0.025$		
(2) $(A_{t+1} + H_t)/A_t$ vs A_t and B_t	$b_1 = 4.359$	0.997	0.008
(model including only competition)	$b_2 = -0.110$		
	$b_3 = 1.9$		
(3) $(A_{t+1} + H_t)/A_t$ vs A_t and C_t	$b_1 = 1.442$	0.999	0.002
(model including only predation)	$b_2 = 0.001$		
	$b_4 = -0.031$		
(4) $(A_{t+1} + H_t)/A_t$ vs A_t	$b_1 = 1.795$	0.602	0.084
(single species model)	$b_2 = -0.022$		

As would be expected, the correct model (1) returned the original parameters, had a coefficient of determination of 1 and a standard error of 0. The predation model (3) and the competition model (2) both had high coefficients of determination and small standard errors, whereas the single species model (4) had a comparitively low coefficient of determination and a larger standard error.

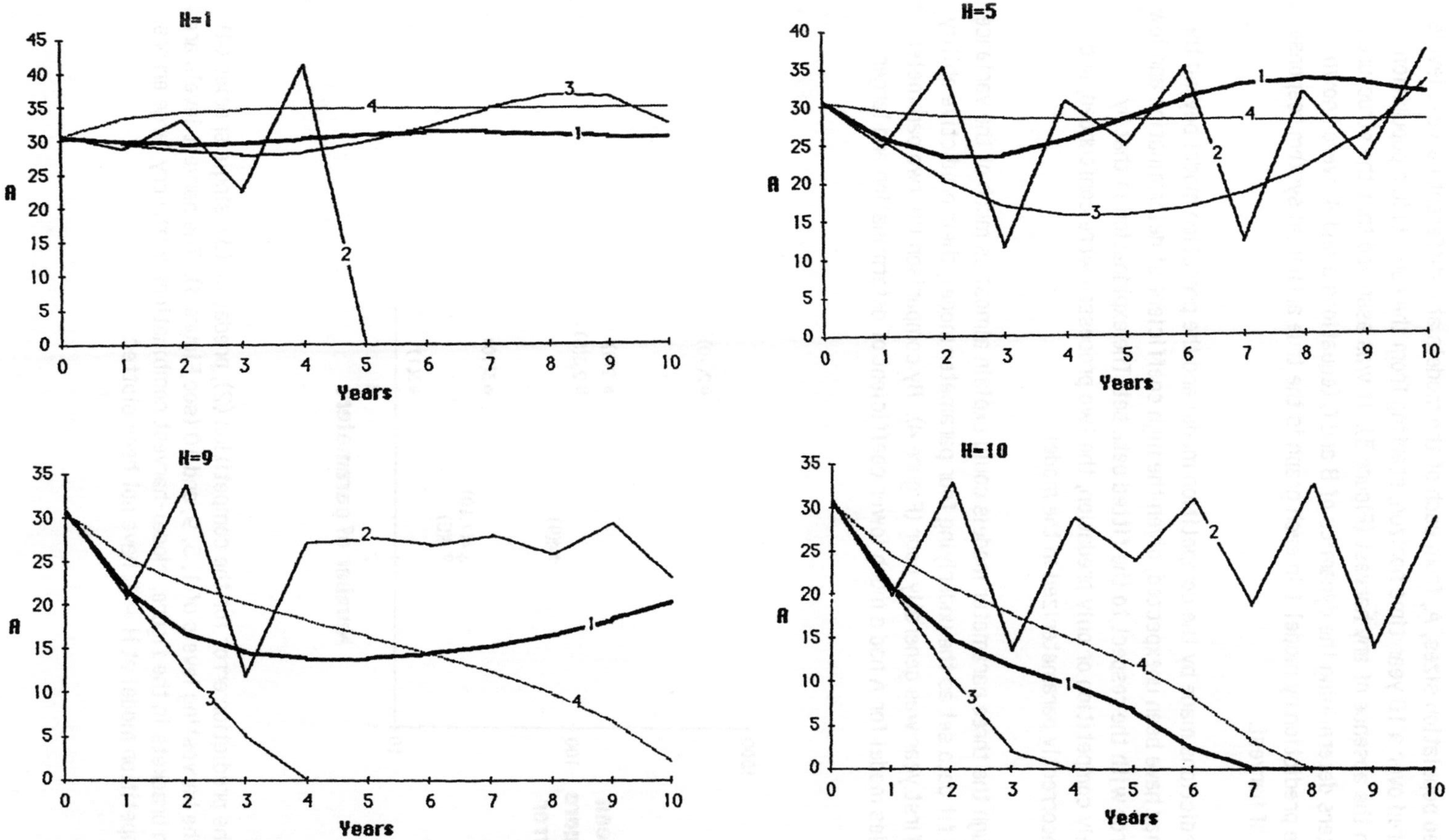

<u>Figure 3:</u> Model predictions for the population size of A over a ten year period at four different harvesting levels (H). 1= correct model incorporating both competition and predation, 2=competition model, 3=predation model, and 4=single species model predictions

The predicted population sizes, A, from each of the models at 4 different harvest levels were compared over a 10 year time horizon, starting from the equilibrium population size for A in the absence of any harvest (Figure 3). It was assumed that the structure and parameters determining the dynamics of B and C (equations 3 and 4) were known exactly. The prediction by model 1 in each graph is the true artificial system response to the level of harvest.

The poor predictions made by the competition model and the predation model beyond the first year may have been unexpected, given the high coefficient of determination and low standard error with the respect to the fitted data set. The explanation is that by including only competition or only predation, the two processes were confounded, and therefore incorrectly parameterized in the model.

Thus, although the three parameter models could explain almost as much of the variance in A for the fit data set as the underlying four parameter model, their predictive ability beyond the first year was generally poor (Figure 4). By comparison the two parameter single species model for A had a much lower coefficient of determination and larger

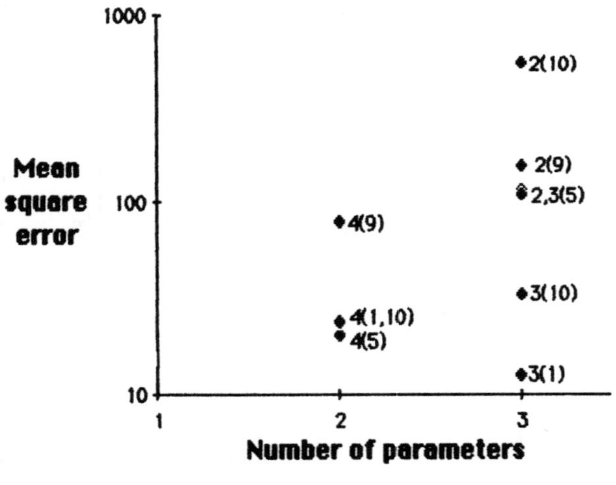

Figure 4: The prediction error for the competition (2), predation (3) single species (4) models at the harvesting levels of 1, 5, 9, and 10 (see Figure 3). The harvest levels are indicated in brackets in the figure. Model-harvest combinations with very large errors, e.g. the competition model at H = 1, have not been plotted.

standard error with respect to the fit data set, and gave poorer (over-estimates) of A for the first year, but the mean square error for the predicted data set for the 10 year period was less than for the competition and predation models in a number of cases. Thus management based on the more complex models could be less effective than if it were based on the simple model.

Coping with complexity

The above demonstration of some of the problems associated with the building of complex models of complex systems underscores the conclusions of a number of ecologists: "...there is little evidence that very complex ecosystem models have performed significantly better than simple ones...." [3], "...often the usefulness of such models, once they have been built, is disappointing."[4], and "It is a reductionist myth that models become monotonically better as the details of natural processes are more fully articulated." [2]. It is generally considered that there is an optimum complexity or "articulation" beyond which model performance deteriorates [5, 2].

In marine ecosystems, problems associated with structural specification and parameterization may cause this optimum to occur at a low model order, perhaps only two or three parameters. The utility of such simple models, for example the single species surplus production models widely used in fisheries management, needs to be carefully examined in the context of the management of populations in complex systems. Although most attempts to fit multispecies surplus production models to time series of data have resulted in interaction terms which are not statistically significant [2], competition or predation could well lead to unanticipated dynamics of exploited populations managed on the basis of single species models (see for example [6]), irrespective of how well the two or three parameters of the simple model have been estimated (see Figure 3).

If the building of complex models of complex systems is constrained by inadequate data and understanding, then alternative approaches need to be examined. One such approach may be to analyse the properties of the whole system rather than the interactions of the components of the system. Properties related to food web structure([7], see also [8]), and ataxonomic relationships between body size and abundance [9, 10] are promising in this respect, and may be able to be used to build qualitative but robust models of the general which can be exercised in making useful predictions about the particular. As an example, particle-size data have been used to predict potential pelagic fish yields for the Benguela Current ecosystem [11]. Although predictions from models of this kind may appear to be too qualitative to be useful for management, in dealing with complex systems the choice may be between being roughly correct or precisely wrong.

REFERENCES

[1] MAY, R.M., J.R. BEDDINGTON, C.W. CLARK, S.J. HOLT AND R.M. LAWS. (1979). Management of multispecies fisheries. Science 205, 267-177.

[2] WALTERS, C. (1986). Adaptive management of renewable resources. pp. 374. Macmillan Publishing Company, New York.

[3] SILVERT, W.L. (1981). Principles of ecosystem modelling. In: Analysis of Marine Ecosystems (A.R. LONGHURST, ed.), pp. 651-676. Academic Press.

[4] STARFIELD, A.M. and A.L. BLELOCH (1986). Building Models for Conservation and Wildlife Management. pp. 253. Macmillan Publishing Company, New York.

[5] CONSTANZA, R. AND F. H. SKLAR. (1985). Articulation, accuracy and effectiveness of mathematical models: a review of freshwater wetland applications. Ecol. Modelling 27, 45-68.

[6] BEDDINGTON, J.R. AND J.G. COOKE. (1982). Harvesting from a prey-predator complex. Ecol. Modelling 14, 155-177.

[7] BRIAND, F. (1983). Environmental control of food web structure. Ecology 64, 253-263.

[8] MAY, R.M. (1986). The search for patterns in the balance of nature: advances and retreats. Ecology 67, 1115-1126.

[9] SHELDON, R.W., A. PRAKASH AND W.H. SUTCLIFFE (1972). The size distribution of particles in the ocean. Limnol. Oceanogr. 17,327-340.

[10] RODRIGUEZ, J. AND M.M. MULLIN. (1986). Relationship between biomass and body weight of plankton in a steady state oceanic ecosystem. Liminol. Oceanogr. 31,361-370.

[11] MOLONEY, C.L. AND J.G. FIELD. (1985). Use of particle-size data to predict potential pelagic-fish yields of some southern African areas. S. Afr. J. mar. Sci. 3, 119-128.

Part IV

Forest Ecosystems

Ecodynamics as a Function of Ecosystem Structure: Forest Dieback as an Example

H. Bossel

Environmental Systems Analysis Group,
University of Kassel, D-3500 Kassel, Fed. Rep. of Germany

Why ecological systems analysis?

Our welfare and our future depend on the functioning of ecological systems, or, to be more precise, on the processes in a complex web of interconnected biotic and abiotic systems at different levels of organization: for example, the photosynthesis in a leaf, the growth of plants, the development of forests, the dynamics of ecosystems, the interactions of ecosystems in geographic regions. As our power to interfere with the functioning of these systems increases with the effectiveness of our technologies, our correct understanding of these systems and their behavior becomes more vital than ever.

"A system is more than the sum of its parts": This dictum is now generally accepted, but it still does not seem to prevent the spending of the bulk of scientific resources and efforts on the painstaking study of detail, to the detriment and widespread neglect of the study of systemic relationships. Both scientific approaches are necessary, but they must be in the proper balance if there is to be scientific progress.

For the study of the time behavior of ecological systems, the methods of systems analysis, modelling, and simulation are indispensible. The complexity of real systems with their rich behavioral spectrum can only be appropriately described by system models of adequate complexity. Moreover, such models must be used to properly relate and structure within the systemic context the information obtained from research of detail.

The objective of (ecological) systems analysis is the parsimonious description of a given system for a given purpose. This requires the identification of the behaviorally relevant systems elements and interconnections (structure). It therefore does not call for the indiscriminate recording of all detail, as in a photograph, but rather for the intelligent selection of essentials, as in a cartoon. If we are to understand ecodynamics, the study of system structure is essential. In this paper this point will be made by first recalling some system-theoretic results, and then looking at results from physiological models of tree development under pollution stress.

Causality - but at which level?

When dealing with dynamic systems, our sense of causality is seldom a reliable guide. In dealing with the behavior of complex ecological systems, we are always tempted to relate the observed response directly to the system's inputs, and we are confused when no clear relationship between inputs and outputs can be discerned.

The discussion of forest dieback is a case in point: While the emissions of most major air pollutants in Western Europe have - on the whole - been held constant, or have even slowly decreased over the past two decades, forest dieback has reached alarming proportions in the course of just a few years: no direct input/output correlation can be identified. As we shall see later, no such relationship can generally be expected on system-theoretic grounds. In view of this, the enormous research effort spent on identifying direct cause-effect relationships seems almost pathetic.

If the numerous ecosystem inputs which could play a role in causing forest dieback are to be related to the observed tree growth impairment and mortality, then this must include the consideration of the relevant physiological processes taking place in the tree. These will have their own characteristic dynamics (eigendynamics) which may show no direct correlation with the system input.

Descriptive models and explanatory models

There are two distinct approaches to the modelling of systems: descriptive (statistical) models and explanatory (structural) models /1/, /2/. Despite their significant differences, they are often confused in argumentation, where the objections applying to one method are inappropriately used to argue against the other.

In descriptive models (Fig. 1), no attempt is made at explaining the response from the peculiarities of the system. Instead, the system is viewed as a 'black box', and a relationship between the observed output and the known input of the system is established by postulating a convenient mathematical relationship between these quantities and then estimating the parameters of the relation by adjusting them to best fit the experimental data. Regression or correlation analysis are examples of this widely used method.

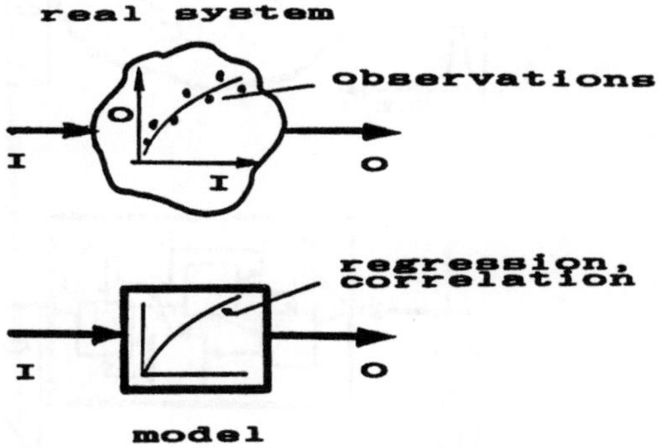

Fig. 1 - Descriptive (statistical) models describe the relationship between the input and the output of a system by a convenient mathematical relationship which is usually unrelated to the actual system structure.

While this method has its merits in cases where there is little or no knowledge of possible structural relationships within the 'black box', or where these have no significant influence on dynamics (as in elastic stress-strain relationships), there are serious objections against applying the descriptive method to the study of complex systems:

- Available structural knowledge about the system is (usually) not used. The statistical relationship is often arbitrary, structurally invalid, and necessarily based on historical data which cannot reflect possible structural changes in the system - which may lead to completely different behavioral modes.
- The parameters of the statistical model (usually) have no meaningful counterpart in the real system. Since the model results (usually) depend critically on the values of these parameters, their estimation assumes a dominant role in the analysis, even though its contribution to the understanding of the system is zero.

For these and other related reasons, descriptive (statistical) models are not well suited for the study of ecological systems, and of ecodynamics in particular. We therefore concentrate on structural (explanatory) models, where an effort is made to improve model validity (structural, behavioral, empirical, and application validity) by describing the behaviorally relevant system structure and system elements as well as possible.

Structural models (Fig. 2) have the following characteristics:

- The elements of structural models correspond to real (sometimes aggregated) elements of the real system.
- The functions of model elements are modelled after those of the real system.
- The structural connections between elements of the model correspond to identifiable relationships of the real system.
- Parameter values, initial values, exogeneous inputs etc. correspond to identifiable quantities of the real system.
- If the structure is modelled faithfully, the model may contain the same dormant structural components as the real system, which may be activated under certain conditions and may lead to different behavioral modes not observed in the past.

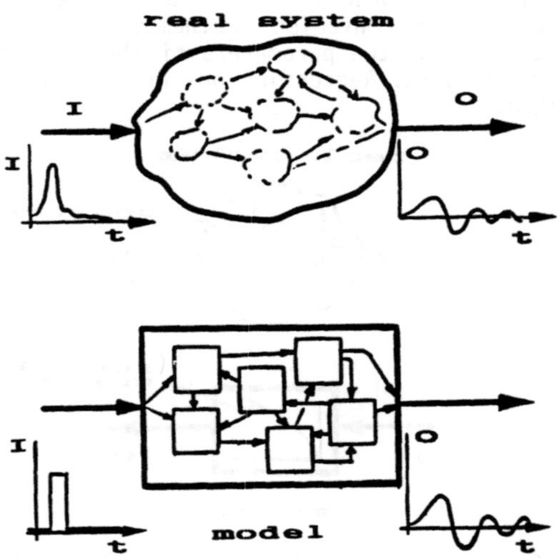

Fig. 2 - Explanatory (structural) models attempt to describe correctly the essential structure of the system, which is responsible for its behavior.

Structural models therefore have several important advantages over statistical models:

- On account of the better representation of structure, their behavioral predictions are more reliable, even for conditions not encountered in the past.
- Estimation of (physically meaningless) system parameters having a critical effect on model system behavior is not an issue, as (usually) only parameters are used which can be (or have been) directly determined in the real system or in more general experiments (e.g. gravitation constant, photosynthetic production, etc.). In most cases, valid models can be developed by using general scientific or particular experimental parameter values, with no need for statistical parameter estimation at all.
- With their emphasis on structure and the resulting behavior, structural models provide a much better understanding of the system studied than statistical models which treat the system as a 'black box'.

Inside the Black Box

At each point in time, a system is characterized by its current state, the vector of the current values of its state variables. The state variables are the memory of the system, reflecting the accumulated effects of its history. All real systems have such memories and hence state variables - or else there could not be continuity in their development. Mostly, these state variables are 'levels' of matter or energy.

The dynamics of systems come about because state variables **z** (t) change with time, for three reasons (Fig. 3):

1 - they are affected by inputs **u**(t) to the system (forcing)
2 - they are influenced by other state variables
3 - they may feed back on themselves.

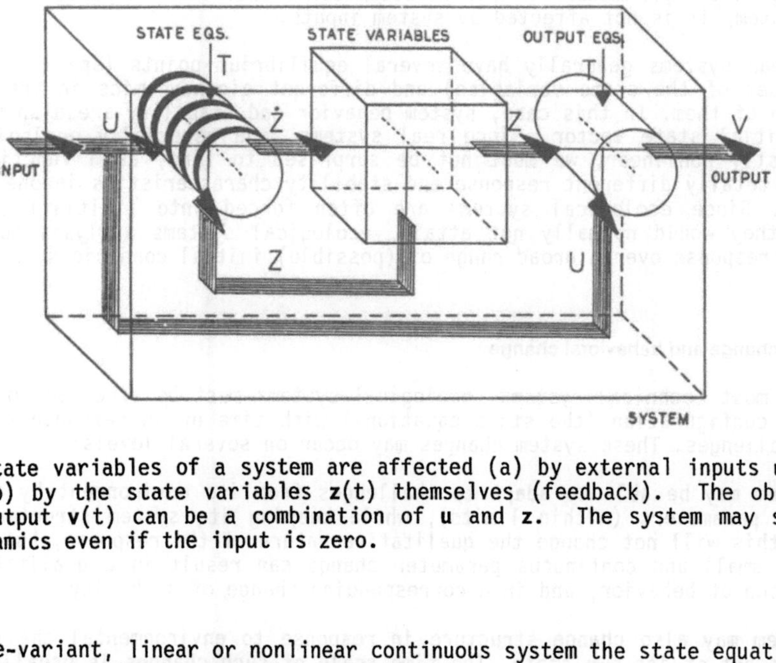

Fig. 3 - State variables of a system are affected (a) by external inputs **u**(t) and (b) by the state variables **z**(t) themselves (feedback). The observable output **v**(t) can be a combination of **u** and **z**. The system may show dynamics even if the input is zero.

For any time-variant, linear or nonlinear continuous system the state equations can therefore be written in the general form /3/, /4/

$$\mathbf{z}'(t) = \mathbf{f}\ (\mathbf{z}(t),\mathbf{u}(t),t)$$

The output will generally be some algebraic combination of input and state variables. Hence the output equations become

$$\mathbf{v}(t) = \mathbf{g}\ (\mathbf{z}(t),\mathbf{u}(t),t)$$

The important thing to note in the state equations is that the state is determined not only by the system input, but also by the state itself. This means that any system whose state variables are interconnected (including self-loops), is capable of producing an eigendynamic response, even if there is no external input (autonomous response). As a consequence, the output of such feedback systems cannot be related to the input without a full consideration of the system structure, and of its consequences for the system's dynamics. Hence the conclusion that the valid modelling of systems requires a correct representation of structure (structural validity).

Eigendynamics as a function of feedback structure

The ability of feedback systems to generate their own autonomous dynamics also means that their stability is determined by the state equations. Only in the case of a linear system are the dynamic response modes and the stability of a system independent of the initial values of the state variable. The output equations, on the other hand, merely represent algebraic recombinations of system input and system state, hence they have no effect on the dynamic characteristics of the system.

The attention therefore focusses on the state equations, i.e. the feedback structure of the system. They are responsible for the characteristic dynamics of the system.

As a reminder of the complex dynamic response which even simple systems are capable of, consider the autonomous response modes of a linear feedback system with two state variables: its response may be aperiodic or periodic, stable or unstable, depending on the parameters of the feedback structure. This response is a property of the system; it is not affected by system inputs.

Nonlinear systems generally have several equilibrium points (specific combinations of values of the state variables) and different eigendynamics in the neighborhood of each of them. In this case, system behavior and stability are also a function of the initial state vector. Since real systems, and especially ecological systems, are mostly nonlinear, we must not be surprised to find, as a function of system state, totally different response and stability characteristics in one and the same system. Since ecological systems are often forced into (initial) system states which they would normally not attain, ecological systems analysis must study the system response over a broad range of (possible) initial conditions.

System change and behavioral change

Unlike most technical systems, ecological systems must be expected to change their system configuration (the state equations) with time or in response to environmental challenges. These system changes may occur on several levels:

A system may be able to adapt to challenges from its environment by changing some of its parameters (within limits), while keeping its system structure unchanged. Often this will not change the qualitative nature of the response, but in principle even a small and continuous parameter change can result in a qualitatively different mode of behavior, and in a corresponding change of stability.

A system may also change structure in response to environmental challenges or the development of its own state. The time scale of such changes is usually longer than that for parameter adaptation. In ecological systems, this self-organization ability will be vital in periods of severe stress, and if the systems study must deal with such periods, then the systems analyst will have to be able to identify and understand the systems's capability for self-organization. Structural change is not always beneficial for the system; stress overload may actually lead to structural change which causes irreversible destabilization and destruction of the system.

Change of structure always means a change of location and nature of equilibrium points, of response modes and of stability characteristics. In ecological systems, one must be aware that different potential system structures may exist, and that the system may change structure (and hence the nature of its behavior) under certain conditions. In the following, this will be demonstrated by discussing the results of systems analysis of tree response to pollution stress using tree-physiological models.

Constructing a tree simulation model

We have argued that behavioral validity requires structural validity of a model: the behaviorally relevant structure of the real system must therefore appear in the model. Since the development of a tree is governed by the interaction of many physiological processes, these processes and their interconnections must be modelled. Fortunately, the nature and parameters of the different processes are sufficiently well known for major tree species to allow construction of tree-physiological models using known data. This is important: there is no need for dubious parameter estimation or adjustment to fit the model response to observations. If the relevant component processes and their interactions are modelled faithfully, the tree response and model response should correspond closely.

The choice of state variables depends on the level of aggregation permissible for a given task. Relevant state variables for the description of tree development dynamics are (at a high level of aggregation): leaf mass, feeder root mass, wood mass, amount of assimilate (Fig. 4). These are structurally connected via rates of change of the state variables, such as: rate of photosynthate production, rate of assimilate use for leaf and root renewal, wood increment and respiration, rate of nutrient uptake, rate of litter loss, etc. Evidently, photosynthesis, nutrient and water uptake etc. are functions of the environmental inputs such as temperature, precipitation, solar radiation, nutrient availability, etc.

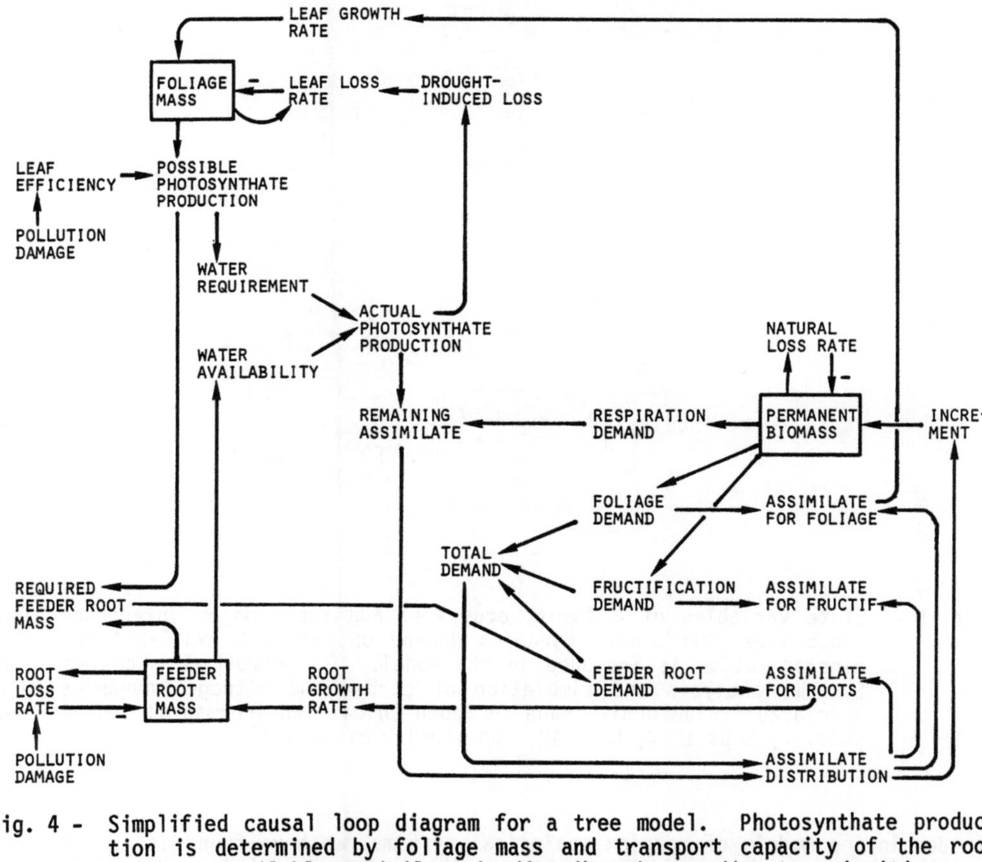

Fig. 4 - Simplified causal loop diagram for a tree model. Photosynthate production is determined by foliage mass and transport capacity of the root system. Available assimilate is distributed according to priorities: respiration needs first, renewal needs second. The remainder, if any, is used for increment. If renewal needs are only inadequately satisfied, an accelerating process of deterioration (breakdown) begins.

If the consequences of air-pollution are to be studied, their potential effects on the system must be included in the model /5/, /6/. Pollution can act on the tree only at its interfaces with the environment, mainly its leaves and its feeder roots. In principle, it can affect the functions of these vital organs in several ways: (1) loss of photosynthetic efficiency, (2) premature aging and leaf loss, (3) impairment of nutrient and water uptake and transport, (4) increased replacement needs for premature leaf and/or feeder root loss (higher turnover). These effects can be summarized into two major types of impairment of tree functions: (I) reduced production of photosynthate, and (II) increased consumption of assimilate. In the end, both amount to the same thing: a reduced availability of assimilate for maintenance and increment.

In the systems diagram of the connected tree processes (Fig. 5) it becomes clear that we are dealing with a complex feedback system where, for example, the processes in the canopy are intimately related with the feeder root processes. It is therefore not surprising to find in the simulations (and in the empirical observations) that the state of any one of the state variables is strongly affected by the states of the others.

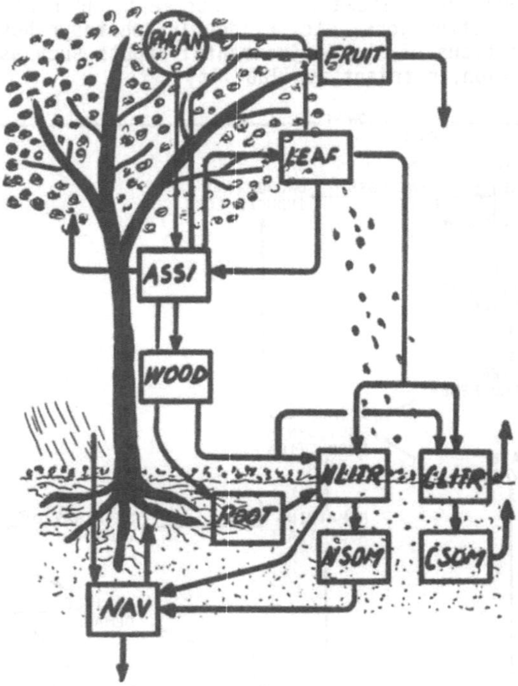

Fig. 5 - State variables of a forest ecosystem and their major interconnections. Since tree development dynamics depend on nutrient availability, litter mineralization is included in the model. (H. Bossel, H. Schäfer et al: Systems analysis and simulation of carbon and nitrogen dynamics of an Acacia auriculaeformis stand in South China. Environmental Systems Analysis, 1, Sept 1987, p. 1-42, University of Kassel).

In developing models of complex ecological systems, we have found it convenient to use the system-graphic method to represent the system structure in detail (Fig. 6). The resulting simulation diagrams can be entered directly into the computer using special software (ASS, DYSS), without requiring mathematical formalization and programming /7/, /2/.

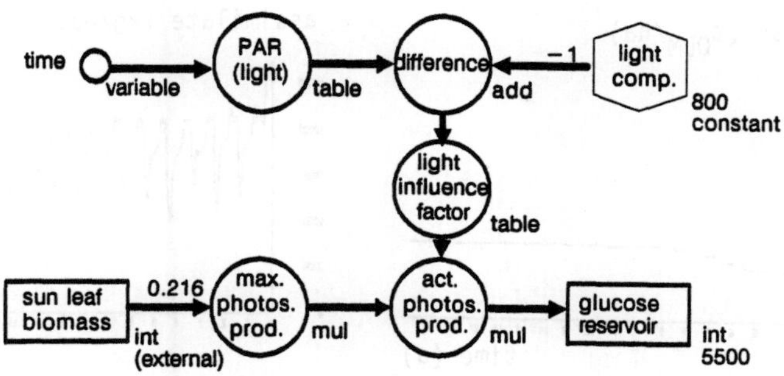

Fig. 6 - Simulation diagram for a small part of the photosynthesis submodel in the BEECH simulation model. These graphs are entered directly into the computer using special simulation software.

Results from a simulation model for beech

A tree functions as a part of an ecological system, hence more complex models such as the BEECH model /8/ consist of interrelated component models representing environmental factors, photosynthesis, respiration, growth, and transpiration, as well as vital soil parameters and processes such as soil water, soil temperature, soil chemistry, and mineralization. This model is the most complex of a series of tree models we have developed. Simulations with this model use weather service reference year data with 15 minute intervals, and actual site data and tree parameters for a Solling test site.

The simulation results for this complex model of a deciduous tree are qualitatively not different from simulation results obtained with smaller models of coniferous trees /6/ and even with very small models containing only the essential elements of the tree system /7/.

If there is no impairment of vital functions by pollution damage to either foliage or feeder roots, the tree produces a steady surplus of assimilate, which it invests in increment, assuring steady growth (Fig. 7). The assimilate reserve changes with the seasons; it decreases from a maximum near the end of the vegetation season to a mininum in the spring at the time of leaf formation, and before the onset of photoproduction by the new leaves.

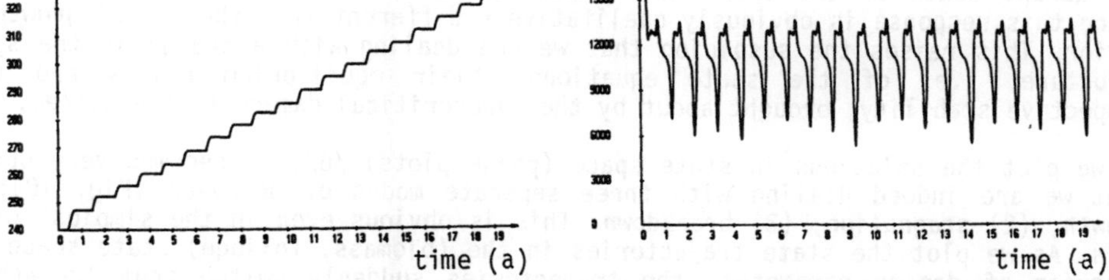

Fig. 7 - Development of wood biomass and assimilate supply under normal growth conditions (BEECH model).

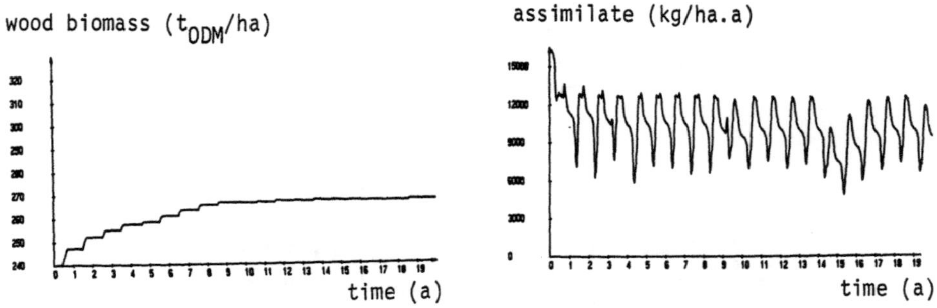

Fig. 8 - Development of wood biomass and assimilate supply under subcritical damage conditions: stagnation (BEECH model).

If the chronic pollution damage to either the foliage or the feeder roots stays below a critical level, the tree continues to function as before, although the assimilate surplus is too small to allow significant increment (Fig. 8). Additional assimilate drain, as in fructification, still does not drive the assimilate reserves to a dangerously low level.

If the pollution damage is increased only very slightly beyond a critical level, the assimilate reserves no longer allow increment, but they also drop below a level required for system maintenance by respiration and renewal (Fig. 9). A starvation process is initiated which quickly kills the tree.

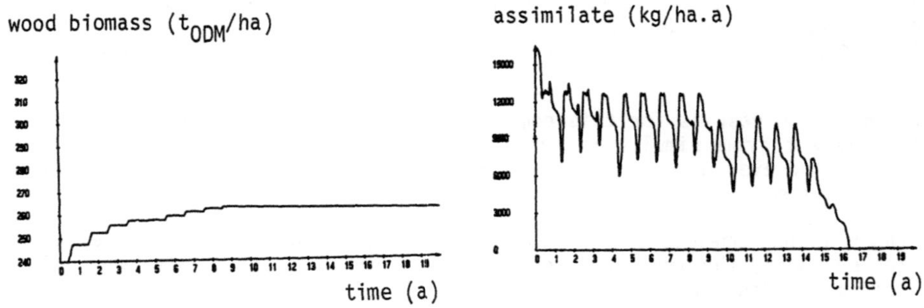

Fig. 9 - Development of wood biomass and assimilate supply under supercritical damage conditions: breakdown (BEECH model).

Three modes of tree development

The abrupt death of the tree under supercritical pollution stress is surprising, since this response is obviously qualitatively different from the normal growth behavior. This raises the suspicion that we are dealing with a change in the system structure, i.e. of the state equations, their equilibrium points and their respective stability, brought about by the supercritical damage to the system.

If we plot the solutions in state space (phase plots) /6/, it becomes very obvious that we are indeed dealing with three separate modes of behavior (Fig. 10): (1) growth, (2) stagnation, (3) breakdown. This is obvious even in the simplest of models. As we plot the state trajectories in the (biomass, foliage) state space as a function of damage parameter, the trajectories suddenly switch from 'growth' to 'stagnation' as the damage parameter increases beyond a certain value, and then again from 'stagnation' to 'breakdown', as the critical value for the damage parameter is reached.

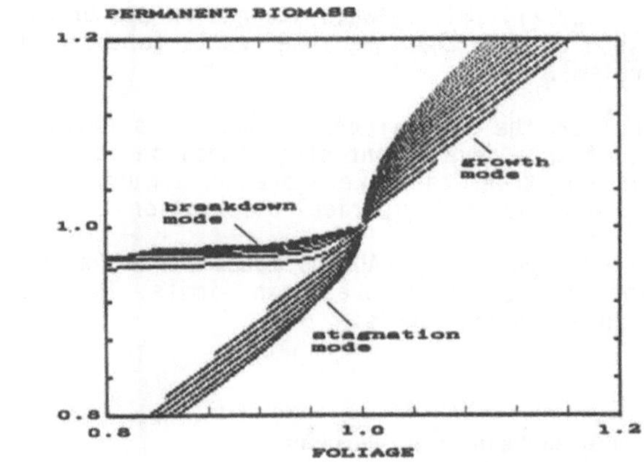

Fig. 10 - Phase plots of tree simulations for different pollution stress show three distinct behavioral modes: (1) growth, (2) stagnation, (3) breakdown.

Further investigations with all the models show that the stagnation mode is indefinitely stable, i.e. even after a long time of subcritical damage, the (model) tree will still survive.

Tree stability diagrams

The analytical stability analysis of complex nonlinear models such as the BEECH model or the SPRUCE model is not possible, but simulation experiments can provide a comprehensive picture of system stability and behavioral modes as a function of key parameters (Fig. 11) /6/.

Choosing random combinations of leaf and root damage and identical initial conditions, we can run a large number of simulations, each over long time period, noting in each case whether the system was ultimately in the growth mode, the stagnation mode, or the breakdown mode. These results can then be recorded in the two-dimensional damage parameter space. In this manner, the damage limit between growth and

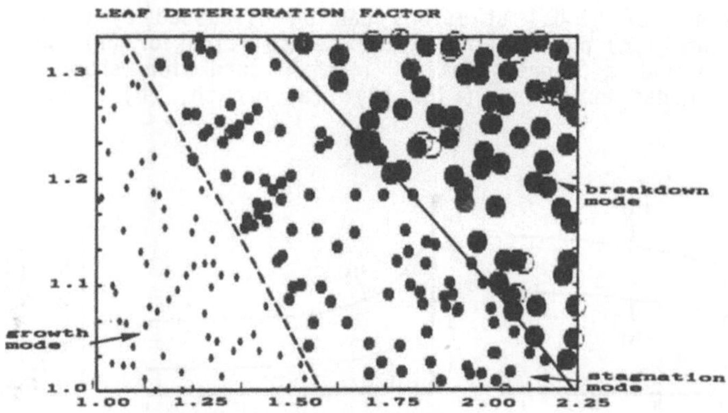

Fig. 11 - Behavioral mode diagram for young spruce, obtained from 250 simulations with the SPRUCE model using random values for the leaf deterioration factor and the feeder root turnover factor. (1) Growth mode: small dots, (2) stagnation mode: intermediate dots, (3) breakdown mode: large dots.

stagnation, and the critical limit between stagnation and breakdown, can be identified. These limits show clearly that the tree reacts to the combination of both damage stresses (synergism).

As the results depend on the initial conditions, this method can be used to investigate the effect of tree development stage (age) on its damage resilience. Such studies show that younger trees can take a greater amount of pollution damage than older trees, in agreement with the empirical observations.

Very simple tree models such as the BAUMTOD model /7/ allow an analytical study of the growth/stagnation and stagnation/breakdown limits. These confirm the observations made with the more complex models.

Mathematical systems analysis of the tree state equations

Mathematical analysis of the state equations of the full BEECH or SPRUCE models for their equilibrium points, and their respective stability and behavioral characteristics is out of the question. However, it is possible to simplify these systems significantly, while maintaining the essential structural connections /9/. In this manner, a two-dimensional, highly nonlinear system for the leaf mass l and the feeder root mass r is obtained.

$$\dot{l} = \frac{\alpha_1 \cdot l^2 + (\alpha_2 \cdot CRT + \alpha_3) \cdot l + c_1 \cdot r}{c_2 + \alpha_4 \cdot l}, \quad \dot{r} = \frac{(\alpha_5 \cdot CRT + \alpha_6) \cdot l^2 + \alpha_7 \cdot l \cdot r - c_2 \cdot r}{c_2 + \alpha_4 \cdot l} \qquad (1)$$

These state equations for l and r are not only nonlinear in l and r, they also depend (via the α_i) on the damage parameter λ, and on the limiter CRT, which is itself a function of the root-to-leaf ratio r/l and of the damage parameter: CRT = CRT(x), where $x = (1/\alpha)(r/l)$ (with $\alpha = \alpha_4$), and

$$CRT = \begin{cases} 0 & \text{for } x < 0 \\ 1 & \text{for } x > 1 \\ x & \text{for } x \text{ in } [0,1] \end{cases} \qquad (2)$$

For the relevant parameter range, existence and uniqueness of solutions of the state equations (1) for a representative initial condition can be proved locally. The existence of global solutions can be discussed by considering the direction field in the (l,r) phase plane. Numerical solutions of (1) for different damage parameters λ and a representative initial condition show again the same three behavioral modes as before (Fig. 12): (a) growth, (b) stagnation, and (c) breakdown.

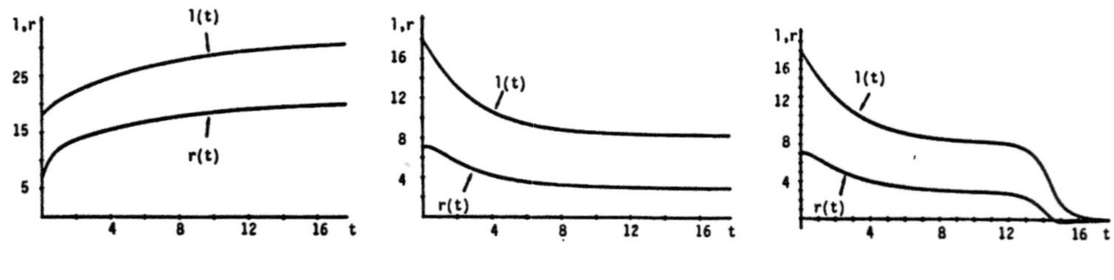

Fig. 12 - Numerical solutions for the two-dimensional tree model (Eqs. 1). This highly simplified system still exhibits the three behavioral modes: (a) normal growth, (b) stagnation, (c) breakdown /9/.

Finding equilibrium points and their stability characteristics from (1) directly is hardly possible. The stability analysis is therefore performed separately for each of the three systems of differential equations resulting from applying each of the conditions (2). It can be shown that for each of the three systems there exists the trivial solution $l_1(t)=0$, $r_1(t)=0$ as well as an additional equilibrium point $l_2(t)=l_2$, $r_2(t)=r_2$. The stability of these equilibrium points depends on the damage parameter λ.

From the definition of x and the conditions for CRT in (2) we note that the two cases CRT=1 (system I) and CRT=x (system II) are separated in the (l,r) phase plane by the straight line $r=\alpha\cdot l$ (x=1) (where α is a function of the damage parameter λ) (Fig. 13). If this line is crossed by the state trajectory, then the state equations change from system I to system II. The equilibrium points and their stability also change.

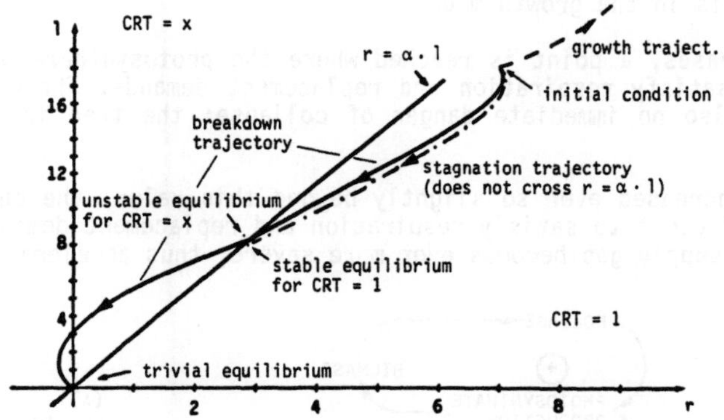

Fig. 13 - The growth and stagnation solutions have an equilibrium point with non-zero values for leaf and root mass, while the equilibrium point of the breakdown solutions is the origin. Breakdown occurs when the state trajectory crosses into the domain of attraction of the breakdown solution. The system structure changes as the separating line is crossed.

The three observed behavioral modes can now be discussed in the phase plane as follows:

System I has a nontrivial stable equilibrium point, which may be to the right of the initial state, if the damage parameter is low enough. In this case, the system will initially grow (increasing l and r) and then approach a steady state.

As the damage parameter increases, the nontrivial stable equilibrium point shifts to the left of the initial state: the system loses both leaf and root mass, but approaches a steady state (stagnation). With increasing damage parameter, the equilibrium point moves further to the left. The solution remains of the stable stagnation type until the equilibrium point crosses the shift line (separating system I and system II) when the damage parameter has reached a critical value.

When the stable equilibrium point of system I has moved to the left side of the shift line, the state trajectory will continue to be attracted by it, until it crosses the shift line. At this point, system equations I are replaced by systems equations II which have a stable equilibrium point only at the origin. The system therefore enters the breakdown mode and collapses.

The breakdown behavior observed in the tree models therefore turns out to be the consequence of a change of system structure and a corresponding switch of stable equilibrium points when the system is driven across a certain limit by a supercritical damage parameter. The collapse is not caused directly by a trivial increase of the damage parameter, but rather by the fact that it sets into motion a 'stable' self-destructive process.

The structure of dominant feedback loops

The change of system structure and the resulting change in behavioral mode can also be discussed by analyzing the dominant feedback loops for different damage conditions (Fig. 14) /6/.

If there is little or no damage, the photosynthate production will be more than sufficient to satisfy respiration and replacement demands, and to produce increment: the tree is in the growth mode.

As damage increases, a point is reached where the photosynthate production is just sufficient to satisfy respiration and replacement demands. There is no increment, but there is also no immediate danger of collapse: the tree is in the stagnation mode.

If damage is increased ever so slightly beyond this value, the photosynthate generated is insufficient to satisfy respiration and replacement demands. As production decreases, the supply gap becomes ever more severe, thus accelerating the collapse.

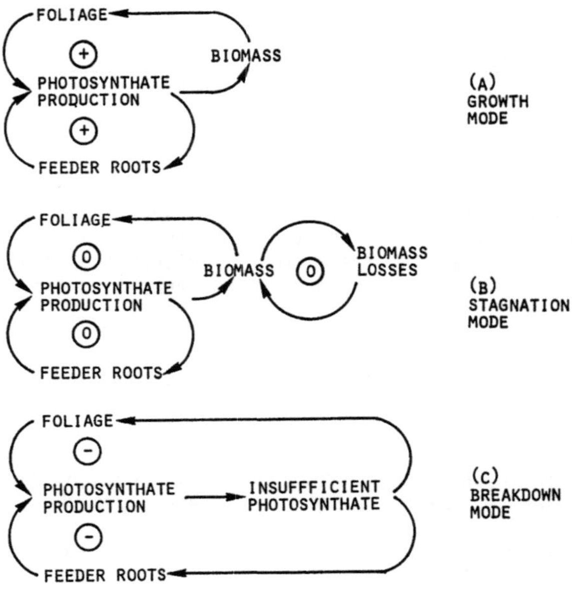

Fig. 14 - Dominant system loops determining the three modes of behavior. (A) In the growth mode, surplus photosynthate leads to increase in biomass, foliage, and feeder roots, and hence further enhancement of photosynthetic production. (B) In the stagnation mode, photosynthate production just suffices to keep up with natural losses of biomass, leaves, and feeder roots. (C) In the breakdown mode, an insufficient supply of photosynthate causes imcomplete replacement of leaves and feeder roots, leading to further accelerated reduction of photosynthetic production, and of leaf and feeder root mass.

Conclusion

We can draw a number of conclusions from this discussion:

(1) Ecological systems analysis, and especially the study of ecodynamics, must be based on explanatory, i.e. structurally valid models. Descriptive models will be acceptable only under very restricted conditions, or as aggregate summary models of component processes.

(2) The feedback structure of the interconnections of the system's state variables (formalized in the state equations) is responsible for the eigendynamics of the system. Attention should be focussed on it, rather than on attempts to explain output as a direct consequence of input.

(3) In the course of their development, ecological systems may change (a) parameters, (b) structure. These changes can change response characteristics and dynamics completely. They must be anticipated and properly represented in simulation models.

(4) The (mostly) nonlinear character of state equations of ecological systems can result in different behavior and stability characteristics for different initial states (or if the system is forced into a new state). Ecological systems analysis must deal with the whole spectrum of possible states.

(5) The system structure of tree physiological processes is the cause of rapid breakdown dynamics of trees when chronic pollution damage exceeds a certain critical value.

References

/1/ C.T. de Wit: Simulation of living systems. In: F.W.T. Penning de Vries, H.H. van Laar (eds): Simulation of plant growth and crop production. PUDOC-Centre for Agricultural Publishing and Documentation, Wageningen 1982.

/2/ H. Bossel: Systemdynamik - Grundwissen, Methoden und Programme zur Simulation dynamischer Systeme. Vieweg, Braunschweig/Wiesbaden 1987.

/3/ F. Csaki: Die Zustandsraum-Methode in der Regelungstechnik. Akademiai Kiado, Budapest 1973.

/4/ D.G. Luenberger: Introduction to Dynamic Systems-Theory, Models, and Applications. John Wiley, New York 1979.

/5/ H. Bossel, W. Metzler, H. Schäfer (Hrg.): Dynamik des Waldsterbens - Mathematisches Modell und Computersimulation. Fachberichte Simulation Bd. 4, Springer, Berlin 1985.

/6/ H. Bossel: Dynamics of Forest Dieback - Systems Analysis and Simulation. Ecological Modelling, 34, p. 259-288.

/7/ H. Bossel: Umweltdynamik - 30 Programme für kybernetische Umwelterfahrungen. TeWi Verlag, München 1985.

/8/ H. Schäfer, H. Bossel, H. Krieger, N. Trost: Systemanalyse und Simulation der Entwicklungsdynamik von Altbuchen unter Immissionsbelastung. Verh.Ges.Ökol. (Jahrestagung Gießen) 16, 1987.

/9/ H. Krieger: Stabilitätsanalyse eines parameterabhängigen Waldmodells. MS Thesis, Dept. of Mathematics, University of Kassel 1986.

Ecosystem Theory Required to Identify Future Forest Responses to Changing CO_2 and Climate

A.M. Solomon

International Institute for Applied Systems Analysis,
A-2361 Laxenburg, Austria

SUMMARY

The current anthropogenic disturbances in global environmental systems require immediate study and understanding. Yet, piecemeal and applied approaches to solving these interrelated problems hold little promise because the underlying processes of environmental change and ecological response are not understood. Development of an appropriate and valid body of theory is desperately needed. The nature of certain ecosystem theory requirements is revealed by model experiments on responses by forest ecosystems to future climate and CO_2 effects. Increasing concentrations of atmospheric CO_2 potentially could generate multiple and even opposing effects on forests. Greenhouse experiments have shown that enhanced CO_2 positively affects woody seedling growth, and that these effects may also occur in saplings and mature trees under elevated CO_2 concentrations. Yet, today's close geographic correspondence between certain climate variables and forest distributions suggests that climate changes resulting from future CO_2 increases could destroy many currently existing forests. The potential response of forests to these conflicting forces was examined using a computer model of tree growth and forest stand development. The model is composed of abstractions from current ecological theory and uses knowledge of forest dynamics to simulate simultaneous changes in CO_2 and climate, and the known responses of forest stands to these variables. Model runs of several hundred years were constrained by a climate and CO_2 scenario which changed annually, as suggested by current energy use projections. Simulation results suggest that initial forest response to both climate and CO_2 changes may be minor because of tree longevity. In the long term, the positive effects on forests, simulated as CO_2 fertilization, were overwhelmed by the negative effects, simulated as drastic changes in the climatic status quo. Yet, the value of the simulations is severely diminished by weaknesses in the theoretical underpinnings of the models. Particularly (but not exclusively) constraining is the absence of a cohesive, tested theory to predict (1) forest ecosystem behavior under chronically changing environmental regulation, and (2) tree species behavior along boundaries of geographic ranges.

INTRODUCTION

The trace gas composition of the global atmosphere continues to change in response to natural causes, such as volcanic activity, and anthropogenic ones, such as fossil fuel use. The radiatively active gases (carbon dioxide, ozone, methane, water vapor, etc.) are of particular concern. For example, carbon dioxide is transparent to short wavelengths that compose the sunlight intercepted by clouds or the earth but not to the much longer wavelengths of infrared subsequently radiated back to the sky. A "greenhouse" effect occurs when some of this infrared radiation is absorbed by CO_2 and reradiated to earth, rather than to space.

The resulting CO_2-induced climate change should generate responses in forests. Many (but not all) trees and forest communities are now, or soon may be, subjected to a different and therefore more stressful climate than that to which they were adapted as germinating seeds (Solomon and West 1985). In addition, the change in CO_2 may directly affect plant growth and forests in that enhanced atmospheric CO_2 concentrations have increased the growth of tree seedlings in greenhouse and growth chamber experiments (Lemon 1983; Oechel and Strain 1985).

Lemon (1983) and Strain and Cure (1985) have provided detailed examinations of the effects of enhanced CO_2 on photosynthesis, respiration, growth, and development of plants in greenhouse experiments. As yet, however, no research data indicate that mature trees growing in forests will be capable of taking advantage of the measured increases in dry matter production and in drought tolerance found in greenhouse herbs and woody seedlings.

Indeed, the opposite response (growth loss) could actually occur with enhanced atmospheric CO_2. Plants acclimate (cease to respond) to raised CO_2 concentrations after several days or months (Kramer and Sionit 1987; Oechel and Strain 1985), casting doubt on the long-term implications of short-term experiments with high CO_2 concentrations. Field studies measuring changes in tree growth in response to acidic precipitation and gaseous air pollutants revealed that annual tree growth has declined (Johnson 1983; McLaughlin, West, and Blasing 1983; Plochmann 1984; Nilsson and Duinker 1987), despite increases in global CO_2 of 25% to 30% since about 1850. Even the growth increases at very high altitudes (LaMarche et al. 1984), parallel with CO_2 increases, are ambiguous at best. For example, the timing of enhanced tree growth at these temperature-limited growth sites coincides more closely with the warming of the past century than with the CO_2 increases.

Forests will respond to changes in climate and CO_2 "fertilization," if at all, as a function of changing competitive advantages among species. Competition negates the simplistic view that all forest trees would benefit from increased CO_2. Instead, growth advantages conferred on one species must incur growth losses in less competitive species in a complex, but predictable, way. The fundamental problem is that there is presently no body of ecological theory which can successfully predict the outcomes of the competition at any specified time or place. Indeed, one cannot even provide an accurate scientific estimate of whether the stands will be positively or negatively affected, although opinions range from strong positive growth anomalies in all trees (e.g., Goudriaan 1986) to strong negative anomalies (e.g., Solomon 1986). The discussion below uses models to evaluate the importance of future conflicting and compensating forces which will operate on forests through the mechanism of interspecies competition. The absence of a theoretical basis for the models is illustrated in the resulting uncertainties and demonstrates the necessity for resolution of the uncertainties.

APPROACHES TO ESTIMATING FOREST COMMUNITY RESPONSE TO ENVIRONMENTAL CHANGES

The most reliable approach to projecting the states of future atmospheric CO_2 and climate would involve data from past and present environments, coupled with conceptual and mathematical models of the essential environmental, biological, and ecological processes involved in future events (see Reichle, Trabalka, and Solomon 1985). The models used to study the effects of climate and CO_2 on forests should be mechanistic to the point of being derived directly from relevant theoretical constructs. Lacking the appropriate theory, however, we must apply as much available knowledge as possible, knowing that the results will be less than satisfactory. One important result of trials with such models will be to identify gaps in theory that require scientific research.

There is no comprehensive, all-purpose model; there is, rather, a series of models and model approaches, each having strengths and weaknesses relative to the problems to which they are applied. Two general model approaches are available for projecting the responses of forest trees to climate change alone, each based on different assumptions. The first is a pragmatic approach in which correlations among data sets, describing potential causes and effects, are substituted for knowledge of cause and effect processes. The present geographic distributions of climate variables are correlated with geographic distributions of biotic variables, such as the geography of ecosystems, communities, or densities of tree species. The expected future values of these climate variables can then be replaced directly with biotic variables (Emanuel, Shugart, and Stevenson 1985; Solomon et al. 1984).

Unfortunately, empirical approaches are inherently incapable of characterizing the transient response patterns of forest ecosystems. The maximum life expectancy of trees of most species is of the same order of magnitude as the expected appearance of doubled CO_2

concentrations (Trabalka et al. 1985). Thus, the static forest ecosystems projected from empirical models probably will not be formed until many years after attainment of a climatic steady state. The number of years involved and the nature of the transient forest ecosystems simply cannot be estimated using empirical projection techniques. Yet, the short-term (i.e., 100-200 years) transient responses are of most interest in any analysis of the anthropocentric climate impacts on forests (Solomon and West 1985).

Perhaps an even more telling deficiency in the empirical approaches is their inability to deal with carbon fertilization effects. The knowledge gained from greenhouse experiments on seedlings cannot be simply projected to populations or communities composed of adult trees, each competing for light and nutrients with other trees in the forest stand. To remedy these deficiencies, a very different approach is required.

This second approach uses simulation models of the cause-and-effect relationships (along with empirical relationships when cause and effect are unknown). The objective is to apply basic data and principles (i.e., tree species natural histories, ecological and physiological processes, and environmental variables) to projections of the response of interacting and nonlinear ecosystems to climate change. Like the empirical approaches, simulations require estimates of future climate to drive the model experiments.

These models combine features of the mechanistic leaf models (scaled in minutes and millimeters) favored by physiologists (Oechel and Strain 1985) with the empirical spatial models (scaled in years and kilometers) favored by plant geographers (e.g., see Bartlein, Prentice, and Webb 1986). The models simulate tree responses that represent the summation of physiological processes, rather than dealing with the actual physiological processes underlying tree response to variables such as temperature, age, or moisture. The forest stand simulation approach (Shugart 1984) is particularly appropriate for examining potential carbon fertilization effects on forest stands, because the few experimental greenhouse data now available can be used to develop individual tree response functions (e.g., see Regehr, Bazzaz, and Boggess 1975, Figure 3).

Forest stand simulation models have been under development for over fifteen years (JABOWA, Botkin, Janak, and Wallis 1972; FORET, Shugart, and West 1977, 1980). Adapting the models to assessing forest issues in the global CO_2 problem has required a long development period (Solomon et al. 1980, 1984; Solomon, West, and Solomon 1981; Solomon and Shugart 1984; Solomon and Tharp 1985; Solomon and Webb 1985; Solomon and West 1986; Solomon 1986). The concepts and biology incorporated into the latest versions of the model are described in detail by Shugart (1984) and Solomon et al. (1984). The mathematical expressions are provided by Shugart and West (1977) and Solomon and Shugart (1984). Regional variants of the FORET models have been tested on vegetation at several locations in the United States, Canada, Australia, and other regions (see Shugart 1984; Dale, Hemstrom, and Franklin, 1986; Solomon 1986). To supplement these studies of model validity in space, FORET has also been tested using long temporal sequences of 10,000 to 20,000 years for which there are fossil pollen records of actual forest composition (see Solomon and Webb 1985).

A diagram of the model structure is presented in Figure 1. In the idealized forest, growth of each tree species at each age (response function of diameter to time, center right) would occur at the greatest rates ever measured among forest-grown trees. However, such growth rarely occurs in the model because annual growth is reduced by extrinsic (warmth and moisture response functions, upper right) and intrinsic (stand density and shading response functions, lower right) limits to growth. New trees are added to simulated plots (establishment, lower left), and established trees are removed through increased probability of death due to slow growth (suppressed trees, center left) or by increased probability of death with age (mortality, upper left).

The foregoing describes the stand simulator used recently in climate effects studies that have not included CO_2 fertilization (Solomon and Shugart 1984; Solomon et al. 1984; Solomon and Tharp 1985; Solomon and West 1985; Solomon 1986). The results from some of these studies suggest the need to include direct CO_2 effects in simulation models (Solomon and West 1985). Modifications of the model to include fertilization effects of atmospheric CO_2 would involve changes that might resemble those shown as dotted lines in Figure 1. The optimum growth with added CO_2 (center right) would increase at all ages, coincidentally enhancing the maximum age each species could attain and thereby reducing rates of mortality (center left). The complex of growth effects attributable to enhanced CO_2 is also expected to increase the photosynthetic optimum temperature and the maximum temperature at which each species can grow (warmth, upper right). In addition, tree species should become more tolerant to drought (moisture, upper right). These shifts in response functions, combined with increased photosynthesis in shade (shading, lower right), should enhance biomass per unit area (density, lower right).

Such a realistic simulation of carbon fertilization is not currently possible. Quantitative estimates of the direct CO_2 response functions are unavailable for any major species groups. Indeed, few of the functions have been measured in any tree species. What little is known about CO_2 fertilization effects was simulated to assess the implications of the effects. The model modifications included only the growth function (center right, Figure 1). Note that although only the growth curve is involved, other modeled processes change indirectly. For example, suboptimal temperature and precipitation values become less stressful because of enhanced growth. Maximum ages of each species are unchanged, but the additional

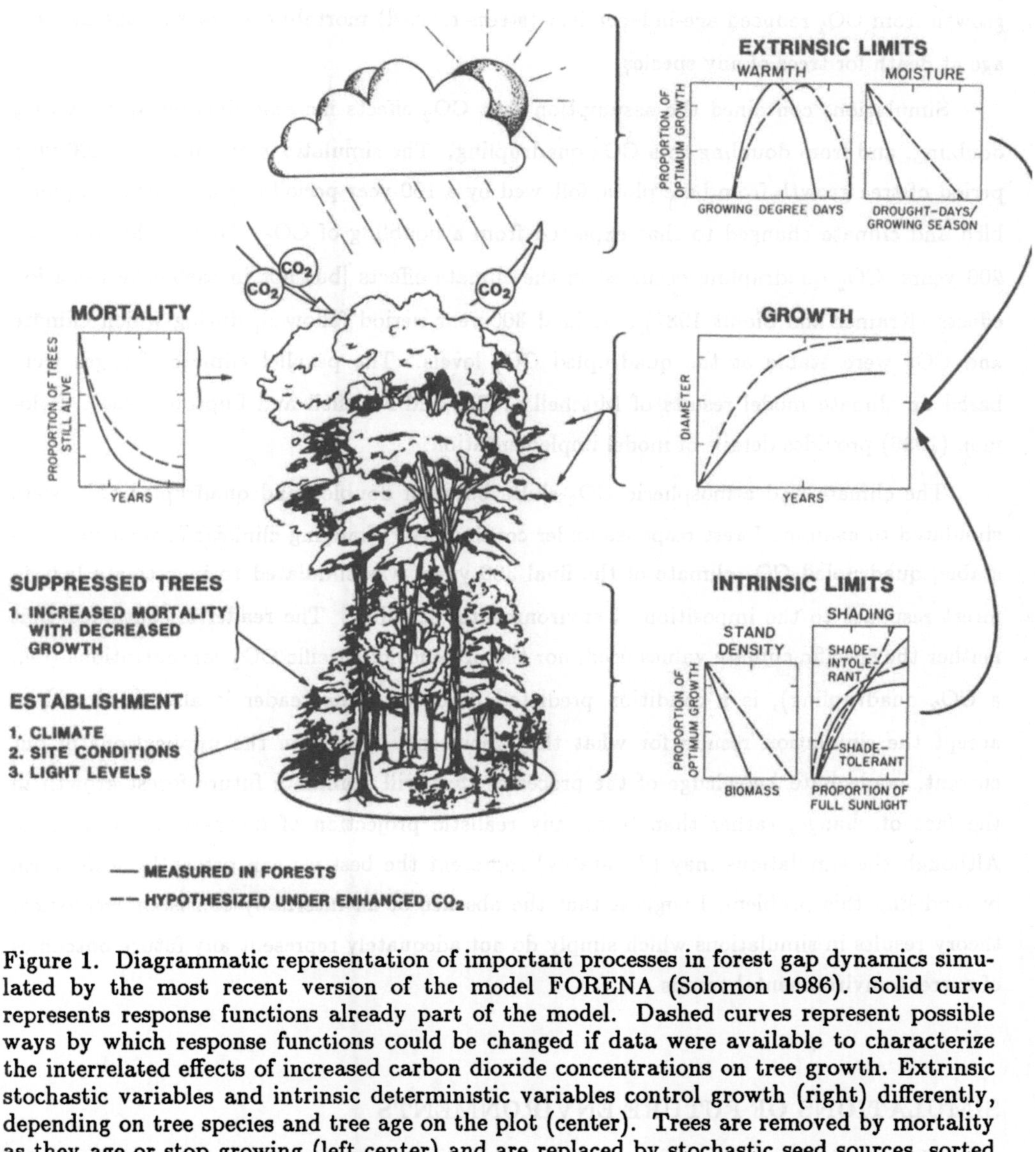

Figure 1. Diagrammatic representation of important processes in forest gap dynamics simulated by the most recent version of the model FORENA (Solomon 1986). Solid curve represents response functions already part of the model. Dashed curves represent possible ways by which response functions could be changed if data were available to characterize the interrelated effects of increased carbon dioxide concentrations on tree growth. Extrinsic stochastic variables and intrinsic deterministic variables control growth (right) differently, depending on tree species and tree age on the plot (center). Trees are removed by mortality as they age or stop growing (left center) and are replaced by stochastic seed sources, sorted by site conditions (left bottom).

growth from CO_2 reduced age-independent (stress-related) mortality, increasing the average age at death for trees of any species.

Simulations contained the assumption that CO_2 effects increase linearly up to a CO_2 doubling, and from doubling to a CO2 quadrupling. The simulations began with a 400-year period of tree growth from bare plots, followed by a 100-year period during which CO_2 doubled and climate changed to that expected from a doubling of CO_2. During the following 200 years, CO_2 quadrupling occurred in the climate effects [but not in carbon fertilization effects (Kramer and Sionit 1987)]. A final 300-year period followed, during which climate and CO_2 were stable at the quadrupled CO_2 levels. The parallel climate changes were based on climate model results of Mitchell (1983) and Mitchell and Lupton (1984). Solomon (1986) provides details of model implementation.

The climate and atmospheric CO_2 shifts between doubled and quadrupled CO_2 were simulated to examine forest response under continuously changing climatic conditions. The stable, quadrupled CO_2 climate of the final 300 years was simulated to investigate lags in forest response to the imposition of environmental stability. The reader is cautioned that neither the specific climate values used, nor the presence of specific CO_2 concentrations (i.e., a CO_2 quadrupling), is a condition predicted to occur. The reader is also cautioned to accept the simulation results for what they represent. They are the implications of our current, inadequate knowledge of the processes that will dominate future forest growth in the face of change, rather than being any realistic projection of future forest dynamics. Although the simulations may (debatably) represent the best we can presently accomplish in modeling this problem, I suggest that the absence of an internally-consistent ecological theory results in simulations which simply do not adequately represent any future outcomes of current environmental trends.

SIMULATIONS OF FUTURE ENVIRONMENTS

The examination of forest response to concurrent changes in climate and carbon fertilization begins with a simulation of climate change in the absence of fertilization effects. Then the analysis is broadened to include the few carbon fertilization effects measured in greenhouses, and fertilization effects greater than those measured. Forest dynamics were simulated in three places: a boreal forest in west central Ontario, a coniferous-deciduous transition forest in northwest Michigan, and a deciduous forest in east central Tennessee. These were the sites at which Solomon and West (1987) assessed potential reactions to CO_2-induced climate changes by the forest industry. The sites were also among twenty-one at which climate effects of CO_2 increases were simulated (Solomon 1986).

The initial simulations assume that climate begins to change after year 400 as atmospheric CO_2 increases without enhanced growth because of carbon fertilization. During the first 100 years of simulated warming (years 400 to 500), summer and winter temperatures rise respectively 2.5° and 5.0° C at the boreal site, 2.5° and 3.5° C at the transition site, and 3.0° and 2.0° C at the deciduous forest site, based on the climate simulations of Mitchell (1983) and Mitchell and Lupton (1984). At the same time, annual precipitation is not changed. During the next 200 years of warming (years 500 to 700), summer and winter temperatures rise respectively 2.5° and 4.0° C at the boreal and transition sites, and 2.0° and 3.0° C at the deciduous forest site. Simulated annual precipitation of the 200-year period declines about 25% at the transition and deciduous forest sites, but is unchanged at the boreal forest site.

The stand simulations of CO_2-induced climate changes resulted in temporally ordered sequences of forest stand destruction and regrowth (See Figures 2, 3, 4 of Solomon and West 1987). These sequences are likely to differ if carbon fertilization also affects forest growth, particularly because carbon fertilization should postpone mortality. This idea was tested at the three sites where forest response to climate change was simulated.

The model was modified to allow as much as a 20% increase in deciduous tree growth and an 11% increase in coniferous tree growth, with as much as a doubling of CO_2 [from 350 to 650 uL L^{-1} (Sionit et al. 1985)]. Moisture effects on growth were unchanged in coniferous trees (P.J. Kramer, personal communication 1984) and were decreased by as much as 18% in deciduous trees (Sionit et al. 1985), with as much as a doubling of CO_2. Figures 2 through 4 illustrate stand biomass response in the following ways: to climate change alone ("no CO_2 effect") and to climate change combined with carbon fertilization effects, as measured by Sionit et al., 1985 ("CO_2 effect as measured"); at twice those measured (40% increase in deciduous tree growth, 22% increase in coniferous tree growth, 36% decline in deciduous tree water use; "CO_2 effect doubled"); and at three times those measured (60% increase in deciduous tree growth, 33% increase in coniferous tree growth, 54% decline in deciduous tree water use; "CO_2 effect tripled").

Simulation runs generated several important results. For the current paper, the above-ground biomass of all trees on the simulated forest stands was used in diagrams as a simple statistic which could illustrate stand dynamics, although much more complex variables can be examined. Not surprisingly, continuously increasing carbon fertilization effects decreased the time required to repopulate the plot with new trees after the climate change had induced tree-dieback at the boreal site (Figure 2). Although the dieback began at about

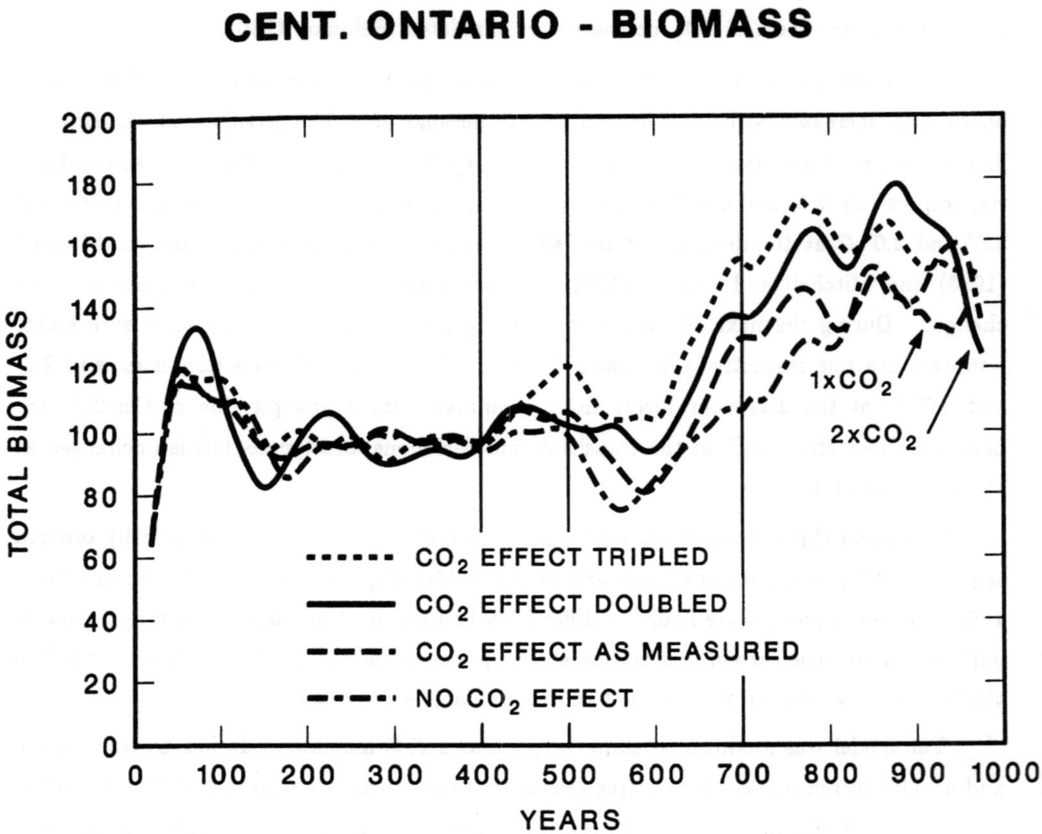

Figure 2. Simulated dynamics of stand biomass at the boreal forest site in west central Ontario under experimental conditions of changing climate, four different effects of CO_2 fertilization. Simulation climate conditions during years 0-400 include modern climate and climate variance; during years 400-500, climate gradually changing to that determined by doubled CO_2 at year 500; during years 500-700, climate gradually changing to that determined by quadrupled CO_2 at year 700; and years 700-1000, stable climate and climate variance determined by quadrupled CO_2. Simulation CO_2 fertilization conditions superimposed on the climate change include no fertilization ("no CO_2 effect"), the fertilization effect measured in long-term (3-year) glasshouse experiments with seedlings of deciduous and coniferous species ("CO_2 effect as measured"), twice the measured fertilization effect ("CO_2 effect doubled"), and three times the measured fertilization effect ("CO_2 effect tripled"). See text for details of experimental model treatments.

the same time, both with and without measured fertilization, recovery began 40 years earlier and was completed about 100 years later in the absence of carbon fertilization. The simulated dieback feature of forest response to climate change was almost absent when the carbon fertilization effect was doubled, and it disappeared entirely under a carbon fertilization effect three times that measured in growth chamber experiments (Figure 2). After climate stabilized at year 700, total stand biomass reached slightly greater values with carbon fertilization than without.

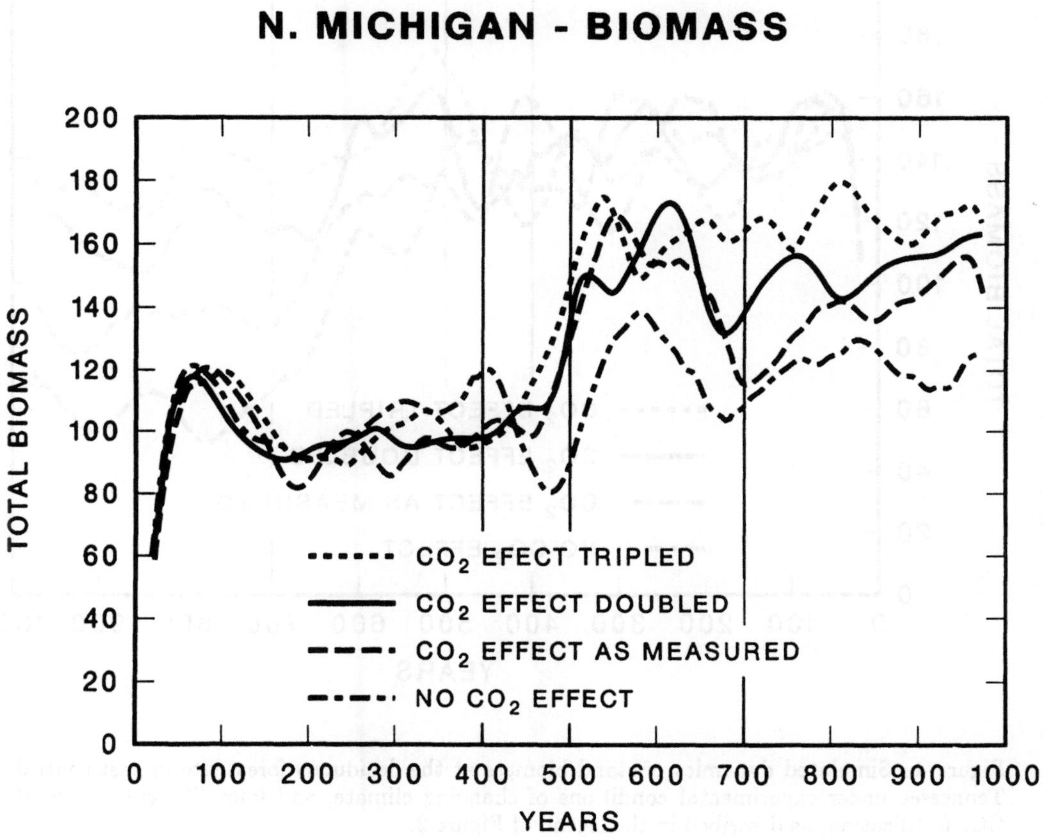

Figure 3. Simulated dynamics of stand biomass at the coniferous-deciduous transition forest site in northwest Michigan under experimental conditions of changing climate, and four different effects of CO_2 fertilization as described in the legend of Figure 2.

In contrast to results at the boreal site, the simulated forest at the coniferous-deciduous transition site (Figure 3) required only the measured carbon fertilization effect to balance, and thus eliminate, the dieback before a doubling of atmospheric CO_2 occurred at year 500. Deciduous tree communities eventually controlled the transition forests (Figure 2 of Solomon and West, 1987). They began to dominate much earlier with than without simulated carbon fertilization. In addition, increasing fertilization produced a rapid increase in stand biomass. The immediacy of changes is suggested by the fact that final rank order of stand biomass at year 1000 was first established less than 50 years after CO_2-

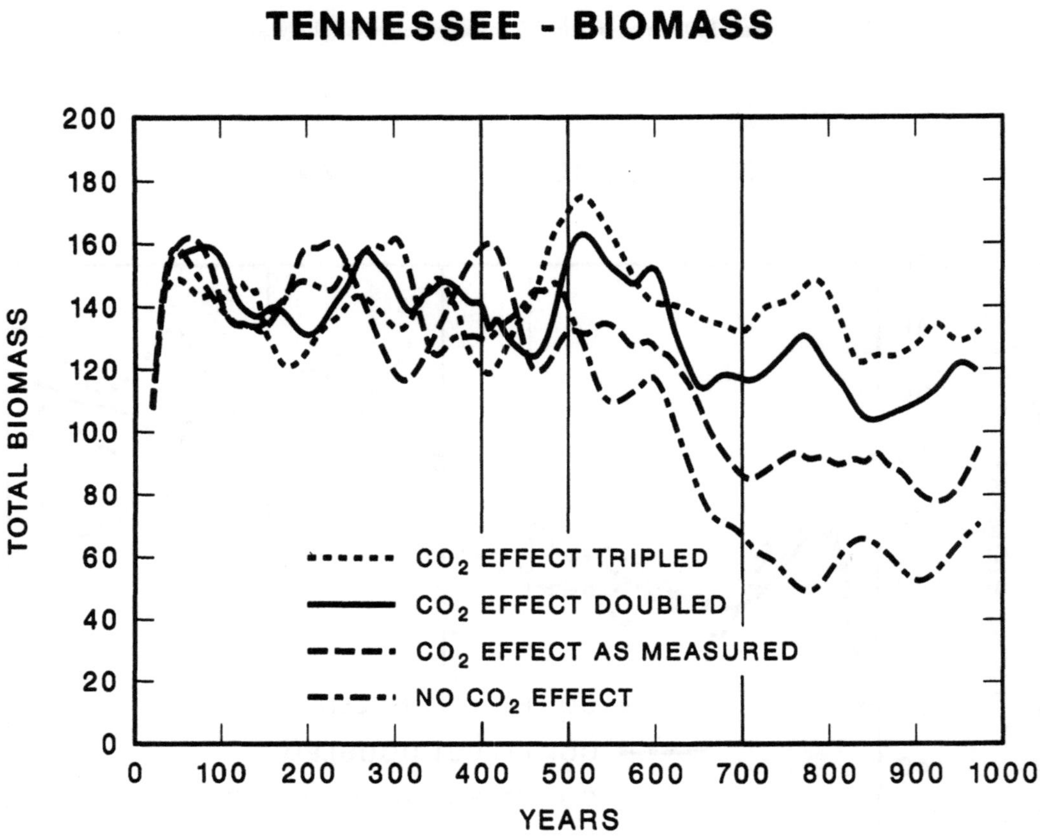

Figure 4. Simulated dynamics of stand biomass at the deciduous forest site in east central Tennessee under experimental conditions of changing climate, and four different effects of CO_2 fertilization, as described in the legend of Figure 2.

induced climate and growth effects began (about year 440 in Figure 3). The sensitivity of these simulated forests to changes in mortality rates was apparently great enough to generate an almost instant response to these subtle environmental changes.

The simulated deciduous forest most clearly responded to the successively greater effects of carbon fertilization (Figure 4). No response was evident before CO_2 doubled. Then, each succeeding increase in CO_2 treatment reduced the effect of climate-induced dieback and further postponed the time at which dieback began. Ultimately, the dieback phenomenon completely disappeared with the application of triple the measured CO_2 effect. As at the transition site, increasing fertilization effects were paralleled by increasing stand biomass, although the final sequence is not apparent until well after the time at which atmospheric CO_2 doubled. Once environmental stability was established after year 700, carbon fertilization effects produced large increases in stand biomass: the measured fertilization effect yielded a 50% increase in biomass (60 to 90 Mt/ha); double the fertilization effect, an additional 28% increase in biomass (90 to 115 Mt/ha); and triple the fertilization effect, an additional 17% increase in biomass (115 to 135 Mt/ha).

DISCUSSION

The foregoing simulations are based on a model of forested ecosystems which possesses little theoretical basis. Rather, the model is built upon the assumption that natural history traits and climatic limits unique to each species will control the outcomes of competition among individual trees for sunlight on a plot, and that this dynamic is all the theory required to generate the synthetic processes we detect at the ecosystem level.

Based upon this "theoryless" concept, the model simulated the kinds of forest ecosystem dynamics that might be expected to occur, if increasing atmospheric CO_2 concentrations induce opposing responses by forest trees to direct carbon fertilization and to the concomitant CO_2-induced climate shifts. Overall, forest biomass was affected more in the simulations by the chronic and continuous climate change than by the parallel change induced by carbon fertilization. The exception was the transition forests, in which many tree species grew at the edges of their geographic ranges and thus, were most sensitive to subtle environmental changes.

The dominance of climate changes is logical, considering the scenarios of change. For example, the 2° to 4° C increase in temperature which climatologists suggest could accompany a doubling of CO_2 represents the entire current range of mean July temperatures from south to north in the boreal forest of central Canada (Hare and Thomas 1979). In contrast, the parallel fertilization effect of a CO_2 doubling, as measured in long-term CO_2 fumigation experiments, represents only a 20% increase in growth of deciduous trees and an 11% increase in the growth of coniferous trees. If future direct CO_2 effects are to rival or exceed climatic effects, then other, unmodeled CO_2 fertilization features will have to be extremely important.

The simulation results also indicated other forest responses that may occur in the future. For example, significant changes in boreal or deciduous forest biomass may not be detectable during the first few decades of environmental changes that lead to doubled CO_2 and concomitant increased temperatures. However, the oncoming shifts in forest biomass may be presaged by early losses in seedlings and saplings relative to mature trees. Also, early detection of forest responses may be ordered geographically. The sensitivity of simulated transition forests to environmental change implies that coniferous-deciduous transition forests and other forests near tree growth limits may be the first to respond to changing CO_2 and climate. This implication is consistent with the suggestion by LaMarche et al. (1984) that they may have detected CO_2-derived increases in tree growth only at high altitude range edges.

These ideas are worthwhile working hypotheses only as long as the forcing and response functions simulated in the model are similar to the forcing and response functions that affect forests in the future. Accurate simulations are currently restricted by the lack of the specific growth chamber data required to characterize the alternative (dashed) lines in each of the response functions illustrated in Figure 1.

In addition, the scenarios of future climate change are subject to large errors, particularly in the effects of feedbacks among components of the climate system. For example, in the most recent projections of CO_2-induced climate change by Manabe and Wetherald (1986), temperature increases are twice as great as those used in the simulations discussed here, and soil moisture is 40% less than that used here. The larger climate effects occur because of moisture feedbacks that were not considered in earlier climate model experiments.

Even if vegetation models and climate scenarios embodied accurate reflections of current process and future change, respectively, impending vegetation responses to environmental changes could not be accurately predicted. The major deficiency revolves about the substitution in the model of empirically-derived values and functions for missing theoretically based processes. For example, the model is based upon species distributions which are defined by the presence of mountains, oceans, and other nonclimatic restrictions upon the geographic ranges of tree species. Lacking any coherent theory which can predict environmental limits to tree species, the current boundaries must be assumed to coincide with climatic barriers, although this is clearly not the case for some species.

This restriction is particularly striking if one considers our definition of western tree species boundaries as limited by soil moisture deficiency. Pre-Columbian tree ranges were limited on the west by high frequencies of prairie fires which have been suppressed for the past century. In place of suppressed fires, land-use (agriculture) is now the chief determinant of western tree migration. As continental summers warm and dry (Manabe and Wetherald 1986), our predictions of tree decline due to drought will become less and less accurate, because there is no means to determine how tree ranges will actually respond to drought.

Certain aspects of reality are neither simulated in the stand models, nor predictable from theory. Most important is the interacting effects of chronic diseases and atmospheric pollutants. Insects, disease, and their vectors (e.g., other insects, fungi, and bacteria) have their own, often complicated, life cycles which depend on weather and climatic events in a manner different from that of the host trees. No model has yet been applied to the complex ramifications of pathogen, insect, and tree-life-cycle interactions under CO_2-induced climate change and other environmental perturbations. A large-scale regional research program is under way at several cooperating institutions to determine the chronic effects of acidic precipitation on forests, based on field studies and the forest-stand model (for example, see McLaughlin et al. 1983). This effort might be extended to include insects, pathogens, and other air pollutants, as well as climate change. Yet, in the absence of adequate theoretical constructs, the result will remain empirical and, therefore, will be incapable of valid extrapolation.

Obviously, theory must be preceded by appropriate data sets. More data from many species on the responses of mature (as well as seedling) trees to increased atmospheric CO_2 concentrations are required to characterize potential CO_2 fertilization and increased water-use efficiency. Indeed, there is a critical need for evidence that any tree life stage besides seedlings will benefit from CO_2 fertilization. At present, we can expect such benefits only in plants growing in noncompetitive, nonlimiting agricultural systems. Thus, data on the presence and effects of CO_2 fertilization and water-use efficiency phenomena must be obtained from trees growing in unmanaged stands, in order to theorize, hypothesize, and then to reliably predict the effects of carbon fertilization on forests.

ACKNOWLEDGEMENTS

Much of the material in this paper was presented in 1986 in an unpublished and unnumbered report printed by the U.S. Environmental Protection Agency: "Effects of changes in stratospheric ozone and global climate, Volume 3: Climate Change," J.G. Titus, Editor, under the title of "Atmospheric carbon dioxide change: Agent of future forest growth or decline?" by A.M. Solomon and D.C. West. I greatly appreciate the help of M.L. Tharp, who programmed the simulations, and of D.C. West, who aided in the writing of the paper. I further appreciate the help of W.M. Post and V.H. Dale, who provided constructive reviews, and M. Brandl, who typed and edited the paper. Research was undertaken while the author was employed at Oak Ridge National Laboratory, Oak Ridge TN, and was sponsored jointly by the U.S. National Science Foundation under Interagency Agreement No. BSR84-17923, AO3, and the U.S. Department of Energy, under Contract No. DE-ACO5-

840R21400 with Martin Marietta Energy Systems, Inc. This is publication No. 69 of IIASA's Project "Ecologically Sustainable Development of the Biosphere."

REFERENCES

Bartlein, P.J., I.C. Prentice, and T. Webb III. 1986. Climatic response surfaces from pollen data for some eastern North American taxa. *J. Biogeogr.* 13:35-57.

Botkin, D.B., J.F. Janak, and J.R. Wallis. 1972. Some ecological consequences of a computer model of forest growth. *J. Ecol.* 60:849-72.

Dale, V.H., M. Hemstrom, and J. Franklin. 1986. Modeling the long-term effects of disturbances on forest succession, Olympic Peninsula, Washington. *Can. J. For. Res.* 16:56-67.

Emanuel, W.R., H.H. Shugart, and M.L. Stevenson. 1985. Climate change and the broad-scale distribution of terrestrial ecosystem complexes. *Climate Change* 7:29-43.

Goudriaan, J. 1986. Simulation of ecosystem response to rising CO_2, with special attention to interfacing with the atmosphere. Pages 49-53 in *Climate-Vegetation Interactions.* UCAR Report OIES-2, ed. C. Rosenzweig and R. Dickinson. Boulder CO: NASA, Office for Interdisciplinary Earth Studies, P.O. Box 3000.

Hare, F.K., and M.K. Thomas. 1979. *Climate Canada, 2nd ed.* Toronto: Wiley and Sons.

Johnson, A.H. 1983. Red spruce decline in the northeastern U.S.: Hypotheses regarding the role of acid rain. *J. Air Pollut. Contr. Assoc.* 33:1049-54.

Kramer, P.J., and N. Sionit. 1987. Effects of increasing carbon dioxide concentration on the physiology and growth of forest trees. Pages 219-246 in *The Greenhouse Effect, Climate Change, and U.S. Forests*, ed. W.E. Shands and J.S. Hoffman. Washington D.C.: Conservation Foundation.

LaMarche, V.C., D.A. Graybill, H.C. Fritts, and M.R. Rose. 1984. Increasing atmospheric carbon dioxide: Tree ring evidence for growth enhancement in natural vegetation. *Science* 225:1019-121.

Lemon, E.R. 1983. *CO_2 and plants: The response of plants to rising levels of carbon dioxide.* Boulder, Colorado: Westview Press.

Manabe, S., and R.T. Wetherald. 1986. Reduction in summer soil wetness induced by an increase in atmospheric carbon dioxide. *Science* 232:626-28.

McLaughlin, S.B., T.J. Blasing, L.K. Mann, and D.N. Duvick. 1983. Effects of acid rain and gaseous pollutants on forest productivity: A region scale approach. *J. Air Pollut. Contr. Assoc.* 33:1042-49.

Mitchell, J.F.B. 1983. The seasonal response of a general circulation model to changes in CO_2 and sea temperatures. *Quart. J. Roy. Meteorol. Soc.* 109:113-52.

Mitchell, J.F.B., and G. Lupton. 1984. A 4 x CO_2 integration with prescribed changes in sea surface temperatures. *Prog. Biometeorol.* 3:353-74.

Nilsson, S., and P.N. Duinker. 1987. Extent of forest decline in Europe: A synthesis of survey results. *Environment* 29:4-9, 30-31.

Oechel, W.C., and B.R. Strain. 1985. Native species responses to increased atmospheric carbon dioxide concentration. Pages 118-154 in *Direct effects of increasing carbon dioxide on vegetation*, eds. B.R. Strain and J.D. Cure. DOE/ER-0238. Washington D.C.: U.S. Dept. of Energy.

Olson, J.S., J.A. Watts, and L.J. Allison. 1983. *Carbon in live vegetation of major world ecosystems.* ORNL/TM-5862. Oak Ridge, Tennessee: Oak Ridge National Laboratory.

Plochmann, R. 1984. Air pollution and the dying forests of Europe. *Am. For.* 90:17-21, 56.

Regehr, D.L., F.A. Bazzaz, and W.R. Boggess. 1975. Photosynthesis, transpiration and leaf conductance of *Populus deltoides* in relation to flooding and drought. *Photosynthetica* 9:52-61.

Reichle, D.E., J.R. Trabalka, and A.M. Solomon. 1985. Approaches to studying the global carbon cycle. Pages 15-24 in *Atmospheric carbon dioxide and the global carbon cycle*, ed. J.R. Trabalka. DOE/ER-0239. Washington D.C.: U.S. Department of Energy.

Shugart, H.H. 1984. *A theory of forest dynamics.* New York: Springer-Verlag.

Shugart, H.H., and D.C. West. 1977. Development of an Appalachian deciduous forest succession model and its application to assessment of the impact of the chestnut blight. *J. Environ. Management* 5:161-179.

Shugart, H.H., and D.C. West. 1980. Forest succession models. *BioScience* 30:308-13.

Sionit, N., B.R. Strain, H. Hellmers, G.H. Riechers, and C.H. Jaeger. 1985. Atmospheric CO_2 enrichment affects the growth and development of *Liquidambar styraciflua* L. and *Pinus taeda* L. seedlings. *Can. J. For. Res.* 15:468-71.

Solomon, A.M. 1986. Transient response of forests to CO_2-induced climate change: Simulation modeling experiments in eastern North America. *Oecologia* 68:567-79.

Solomon, A.M., and H.H. Shugart. 1984. Integrating forest-stand simulations with paleoecological records to examine long-term forest dynamics. Pages 333-56 in *State and change of forest ecosystems -- Indicators in current research*, ed. G.I. Agren. Report 13. Uppsala, Sweden: Swed. Univ. Agric. Sci.

Solomon, A.M., and M.L. Tharp. 1985. Simulation experiments with late Quaternary carbon storage in mid-latitude forests. Pages 235-50 in *The carbon cycle and atmospheric CO_2: Natural variations Archean to present*, eds. E.T. Sundquist and W.S. Broecker. Washington D.C.: American Geophysical Union.

Solomon, A.M., and T. Webb III. 1985. Computer-aided reconstruction of late Quaternary landscape dynamics. *Ann. Rev. Ecol. and Syst.* 16:63-84.

Solomon, A.M., and D.C. West. 1985. Potential responses of forests to CO_2-induced climate change. Pages 145-70 in *Characterization of information requirements for studies of CO_2 effects: Water resources, agriculture, fisheries, forests and human health*, ed. M.R.

White. DOE/ER-0236. Washington D.C.: U.S. Department of Energy.

Solomon, A.M., and D.C. West. 1987. Simulating forest responses to expected climate change in eastern North America: Applications to decision-making in the forest industry. Pages 189-217 in *The Greenhouse Effect, Climate Change, and U.S. Forests*, ed. W.E. Shands and J.S. Hoffman. Washington D.C.: Conservation Foundation.

Solomon, A.M., H.R. Delcourt, D.C. West, and T.J. Blasing. 1980. Testing a simulation model for reconstruction of prehistoric forest-stand dynamics. *Quat. Res.* 14:275-93.

Solomon, A.M., J.R. Trabalka, D.E. Reichle, and L.D. Voorhees. 1985. The global cycle of carbon. Pages 15-24 in *Atmospheric carbon dioxide and the global carbon cycle*, ed. J.R. Trabalka. DOE/ER-0239.

Solomon, A.M., D.C. West, and J.A. Solomon. 1981. Simulating the role of climate change and species immigration in forest succession. Pages 154-177 in *Forest succession: Concepts and application*, eds. D.C. West, H.H. Shugart, and D.B. Botkin. New York: Springer-Verlag.

Solomon, A.M., M.L. Tharp, D.C. West, G.E. Taylor, J.M. Webb, and J.C. Trimble. 1984. *Response of unmanaged forests to CO_2-induced climate change: Available information, initial tests, and data requirements.* TR009. Washington D.C.: U.S. Department of Energy.

Strain, B.R., and J.D. Cure, eds. 1985. *Direct effects of increasing carbon dioxide on vegetation.* DOE/ER-0238. Washington D.C.: U.S. Department of Energy.

Trabalka, J.R., J.A. Edmonds, J.M. Reilly, R.H. Gardner, and L.D. Voorhees. 1985. Human alterations of the global carbon cycle and the projected future. Pages 247-87 in *Atmospheric carbon dioxide and the global carbon cycle*, ed. J.R. Trabalka. DOE/ER-0239. Washington D.C.: U.S. Department of Energy.

Part V

Modelling Approaches

The Role of Theoretical Research in the Design of Programmes for the Control of Infectious Disease Agents

G.F. Medley

Parasite Epidemiology Research Group,
Department of Pure and Applied Biology, Imperial College, Prince Consort Road, London SW7 2BB, UK

INTRODUCTION

The work which forms the subject of this paper directly addresses serious, medical problems for which there is a urgent need for solutions. The purpose of modelling the population dynamics of infectious disease agents is to gain greater conceptual understanding of their population dynamics. This knowledge can then be used to design control programmes with the associated alleviation of suffering and economic benefits. The greatest progress in development of control programme design has been achieved by proposing solutions to problems and monitoring their effects when implemented. As is generally the case in ecology, we can learn most about the processes operating by monitoring a population or ecosystem during a deliberate or accidental perturbation. Thus models have two roles in epidemiology: firstly to mimic population dynamics, and secondly to interact with observational and experimental data. It is the impact that theoretical ecology has on the collection and analysis of data that we wish to emphasise here.

Obviously models have a use as a means of exploring possible dynamics in a particular system. More or less complex models can be constructed and parameter space explored for different behaviours, or new processes introduced which may alter the dynamics. However, in order that theory be translated into application, eventually models must interact with data. Indeed, models (and hence their underlying assumptions and hypotheses) should be testable by experiment and field study, otherwise they do nothing toward advancing our knowledge.

Appropriately constructed models are can aid in the collection of data and design of experiments by providing some indication of which parameters and processes are the most influential on the dynamics of the system. Data collection can then be concentrated on those areas, the underlying theoretical framework used to analyse the data, and the results of the analysis used to modify the theory. Hopefully this continual interaction between data and theory increases our understanding of the system, which in turn increases our ability to control the disease.

In this paper we give two examples of considerable medical importance of the use of theoretical models in the gathering, analysis and interpretation of data. The first is a stochastic simulation model of helminth parasites of humans, and the second is the estimation of the incubation period of the Acquired Immunodeficiency Syndrome (AIDS).

POPULATION DYNAMICS OF HELMINTHS.

Descriptions of the basic biology and population dynamics of the common helminth parasites of man can be found elsewhere /1,2/. In summary, helminths are relatively large parasites which live within the body of their hosts. This work concentrates on those which live within the gut, and have a life cycle which involves the release of eggs into the environment which undergo development as free-living stages before infecting another host (gastro-intestinal geo-helminths). Examples are the roundworm (*Ascaris lumbricoides*), the whipworm (*Trichuris trichiura*) and the hookworms (*Ancylostoma duodenale* and *Necator americanus*).

Surprisingly little is known about the population parameters of human helminths due to the extreme difficulty in gathering data. For example there are no accurate estimates of the life expectancy of any of the species. The primary problem is that simply counting parasite populations involves killing the worms using available drugs, and collecting them when expelled. This destructive sampling prevents simple monitoring of the populations without intervention. The investigator also encounters all the ethical problems associated with studying a disease of humans.

There exist safe, cheap, effective drugs for all the gastro-intestinal helminths of man, and yet the number of people estimated to be infected with *Ascaris lumbricoides*, for example, is of the order of 1,000,000,000 /3/. It is our inability to design and implement community-based control programmes, despite our ability to cure individuals, that prevents eradication of disease /4/.

It is generally agreed that control of parasite numbers is a more attainable goal than complete eradication. The main obstacle to eradication is the ability of the worm population to recover rapidly from depressive perturbations. In some cases, the parasite population has reattained its precontrol level after only 1-2yr of reinfection /4,5/.

An almost universally observed feature of helminth populations is the overdispersed distribution of parasites among their hosts that ensures that most hosts (=people) harbour no/few worms and that a few hosts harbour most of the worm population. The negative binomial distribution has proved to be a good empirical description of the observed pattern. This heterogeneity is important in two respects. Firstly, the most influential regulatory processes governing the parasite population are thought to occur within the definitive (human) host. Obviously the majority of worms in the heavily infected hosts will experience greater density dependent effects than the minority in the lightly infected hosts. Secondly, this heterogeneity greatly alters the design of disease control and chemotherapy programmes. It is the minority of heavily infected hosts that experience the greatest disease and health risk. Presuming that the hosts at greatest risk may be distinguished in some way, drugs targetted at them will prevent disease and kill parasites most efficiently. This approach has been termed selective chemotherapy when applied to individuals, and targetted chemotherapy when applied to subgroups within the community (eg: particular age groups) /4/.

In order to investigate the interaction of this heterogeneity and chemotherapy application, I have developed a simulation model in collaboration with Professor R.M.Anderson FRS, Imperial College. A full description of the model will appear elsewhere /6/. The simulation is based on a human community of about 500 members, and uses the Monte Carlo technique to generate stochasticity in the events of host infection and parasite death and maturation. Previous models of helminth population dynamics have for the most part either assumed a Poisson (random) distribution of parasites among hosts, or have included the negative binomial distribution in a hybrid structure /7/. In the simulation the distribution of parasites is a dynamic entity that is maintained by some combination of processes. Application of chemotherapy (removing worms) within the simulation alters the distribution of worm numbers between hosts, and its recovery can be observed

Two contrasting assumptions were made about the processes operating in order to generate the observed distribution of parasites among hosts. The first, termed *Predisposition* or *Susceptibility*, assumes that all hosts have a different infection rate which is assigned to the host at birth and remains constant throughout the host's life. No biological basis is specifically assigned to this susceptibility factor, although it may have genetic, behavioural, social and nutritional components. The best evidence for the Susceptibility hypothesis comes from field studies which show that individual hosts tend to reacquire a similar worm burden (relative to the remainder of the host population) following chemotherapy /5,8/.

The second hypothesis, termed *Environmental*, assumes that all hosts are identical, and that heterogeneity is developed by a patchy environment. All hosts experience the same average infection rate, but at any one point in time, a few hosts are experiencing heavy infection. Quantitative evidence for the patchy distribution of infective stages is limited /9,10/ but the presence of defaecation sites and foci of contamination are undisputed.

The population dynamics of worms numbers within an individual host, i, may be modelled by a simple immigration-death process:

$$dM_i/dt = \Lambda_i - \mu M_i$$

where Λ_i is the immigration rate, μ is the death rate and $M_i(t)$ the number of adult worms at time t. In the case of the Predisposition hypothesis, $\Lambda_i = \beta L(t) s_i$, where β is a transmission coefficient, L(t) is the number of infective larvae in the environment at time t and s_i is the susceptibility factor assigned to host i. The s_i are distributed as a gamma distribution with mean unity and variance 1/h where h is a parameter.

Host i has contact with infective stages as a Poisson processes with rate Λ_i, and a single worm is picked up on each contact. In the case of the Environmental hypothesis, the contact rate is the same for each host: $\Lambda_i = \Lambda = \beta L(t) u$. On contact, the number of worms entering the host is drawn from a logarithmic distribution with probability density function:

$$f(x; p) = -(1-p)^x / (x \ln(p))$$

where $u = -p \ln(p) / (1-p)$, and p is a parameter.

The consequences of these two hypothesis may be seen in *Figure 1*, which shows the simulated worm burdens from 25 individual hosts. Note that at any particular point in time (a vertical line drawn through the graph) the observed distribution of worms per host is very similar for each hypothesis, despite the evident dissimilarity between the patterns of individual burdens through time. Obviously, neither of these hypotheses is wholely correct, but comparison of the patterns they generate may reveal something about the population

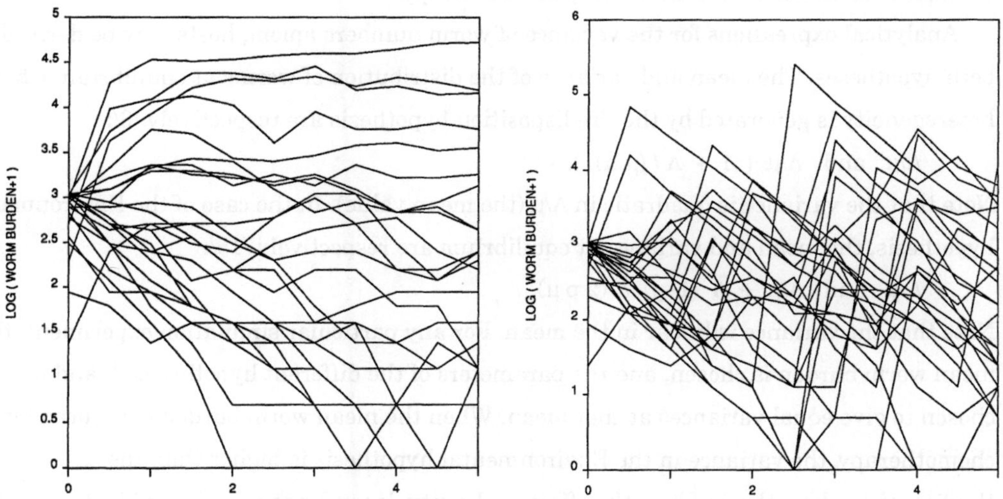

FIGURE 1. Simulated worm burdens of 25 individual hosts followed over 5 years shown on a logarithmic scale. Although the longitudinal patterns generated by the susceptibility (left) and environmental (right) hypotheses are strikingly different, the distribution of worms at any single point in time is very similar /6/.

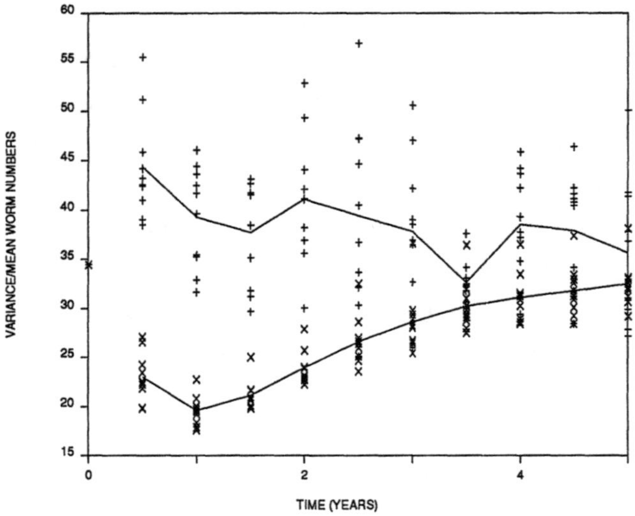

Figure 2. This figure shows the variability (as measured by the variance / mean ratio) in worm numbers distributed between hosts following chemotherapy for both susceptibility (×) *and environmental* (+) *hypotheses.* *Each point is a result from a single simulation, and the lines are the mean values. The variability is greater when the environmental hypothesis is assumed.*

dynamics of helminths.

Of prime importance in the evaluation of control programmes is the recovery rate of the parasite population to precontrol levels following drug application. Results using the simulation model consistently show that the susceptibility hypothesis displays much faster recovery than the environmental hypothesis. The reason for this is the difference in variability in worm numbers following chemotherapy.

Analytical expressions for the variance of worm numbers among hosts may be derived for both hypotheses. The mean and variance of the distribution of worms at equilibrium when heterogeneity is generated by the Predisposition hypothesis are respectively /6/:

$$\Lambda/\mu \quad \text{and} \quad \Lambda/\mu \; (1 + \Lambda / (\mu\, h)).$$

Note that the variance is quadratic in Λ/μ (the mean value). In the case of the Environmental hypothesis, the mean and variance at equilibrium are respectively /15/:

$$\Lambda/\mu \quad \text{and} \quad \Lambda/\mu + \Lambda \; (1-p) \; / \; (2\, p\, \mu).$$

Note that the variance is linear in the mean. For any particular simulation experiment, the mean worm burden is chosen, and the parameters of the different hypotheses (h and p) chosen to give equal variances at that mean. When the mean worm burden is reduced by chemotherapy, the variance in the Environmental hypothesis is higher than the Predisposition hypothesis. Thus the effective density dependence experienced by the worms in the Environmental hypothesis in greater, and their population growth is consequentially slower.

Chemotherapy also alters the variance of worm numbers. *Figure 2* shows the change in simulated variance in worm numbers following application of a 100% effective drug to 80% of the host population chosen at random (10 simulations for each hypothesis). The variance returns to its precontrol level much faster assuming the Environmental hypothesis. This is partly due to the contrast described above, and partly as a result of the speed of approach to equilibrium of each hypothesis /6/. The degree to which variability is altered by chemotherapy can be measured in the field. Results show that variability (as measured by the variance/mean ratio) is significantly reduced by chemotherapy /5,11,12,13,14/ which accords better with the Susceptibility hypothesis than the environmental hypothesis.

Thus, in this case, comparison between theoretical results and observations reveals something about the processes underlying the population dynamics: namely that intrinsic differences between individual hosts can create an appreciable proportion of the heterogeneity in worm numbers between hosts which mimics patterns observed during recovery from perturbation. This adds evidence for the existence of "wormy" people (those who are predisposed to a high worm burden), and consequentially affirms the likely usefulness of selective and targetted chemotherapy /4/.

THE INCUBATION PERIOD OF THE ACQUIRED IMMUNODEFICIENCY SYNDROME (AIDS)

Preliminary results of theoretical studies of the population dynamics of the Human Immunodeficiency Virus (HIV) — the etiological agent of AIDS — emphasised the importance of the disease incubation period to the overall pattern of the epidemic /16,17/. The incubation period is defined as the time between infection with HIV and diagnosis of AIDS, which is usually equated with the onset of severe illness. It is of crucial importance to gain accurate information regarding the form and magnitude of the incubation period if we are going to use models to predict the progress of the epidemic.

The difficulty in obtaining quantities of data with accurate times of infection and diagnosis are obvious. However, one such set was kindly given to us by Dr. T.A.Peterman, Centers for Disease Control, USA of patients who were infected by a single HIV-infected blood transfusion, so that the time of infection is known fairly accurately /18/. There are several complications in the data which make conventional analysis impossible. Firstly, the actual number of infective transfusions given is unknown - we only know about those that have been diagnosed. Secondly, the number of infective transfusions given per unit time was growing with increase in infected blood due to the growing epidemic. Thirdly, the portion of the incubation period observed is dependent on the time of transfusion.

We collaborated with Professor Sir David Cox FRS and Professor Lynne Billard in the analysis of this data /19,20/. The approach we adopted was to take models directly from the theoretical descriptions of the population dynamics. We proposed two functions for both the increasing incidence of the disease, and the incubation period. Different forms were used for these functions and were fitted using standard maximum likelihood techniques. The mean incubation period for all patients was 6.42yr (assuming an exponential increase in infective transfusions and a Weibull distributed incubation period). The results indicated that age and perhaps sex were influential factors.

The original models have been modified to enable explicit treatment of age and sex /20/. The introduction of age is statistically significant, and indicates that children have a much faster rate of disease development. The introduction of sex is not significant.

These results are the first to attempt to estimate explicitly different incubation periods for different age groups. This approach represents an attempt to recreate a population process from a collection of observations on individuals. The results are pertinent to both the population dynamics and the course of infection within individual patients. Thus theoretical models provide a means whereby a quantitative and qualitative perception of the form and magnitude of the incubation period can be gained, the results will aid in the further development of theoretical models of the epidemic, which, in turn, should contribute to further data collection and analysis.

CONCLUSIONS

Populations are collections of individuals, and hence population dynamics are some combination of processes occurring at an individual level. If we know what the individual processes are and how they combine together, we can predict the population dynamics. A simple example is the AIDS incubation period: measuring and analysing the incubation rates from individual patients assists in formulation of a simple, population level description of the process (the incubation period distribution).

However we rarely have such information, and must try to elucidate the constituent processes from observations and experiments on whole populations. With appropriate models we can learn from contrasts between prediction, observation and experiment. For example, measuring heterogeneity in worm numbers before and after a chemotherapeutic perturbation reveals something about the processes operating at an individual host level which generates the heterogeneity. Attempts to control or modify populations without understanding of the underlying processes can have disastrous results. It is only by making models applicable by their interaction with data that the gap between understanding and practical achievements is narrowed.

REFERENCES

/1/ Cox, F.E.G.: *Modern Parasitology*. Blackwell Scientific Publications, Oxford 1982.

/2/ Anderson, R.M.: Population dynamics and control of hookworm and roundworm infections. In *Population Dynamics of Infectious Diseases: Theory and Applications* (ed. R.M.Anderson). Chapman & Hall, London 1982.

/3/ Peters, W.: Medical aspects: comments and discussion II. in *The Relevance of Parasitology to Human Welfare Today.* (eds. Taylor, A.E. & Muller, R.). Blackwell Scientific Publications, Oxford 1978.

/4/ Anderson, R.M. & Medley, G.F.: Community control of helminth infections of man by mass and selective chemotherapy. *Parasitology*, **90** (1985), 629-660.

/5/ Bundy, D.A.P., Cooper, E.S., Thompson, D.E., Anderson, R.M. & Didier, J.M.: Age-related prevalence and intensity in a St. Lucian Community. *Trans. Roy. Soc. Trop. Med. Hyg.*, **81** (1987), 85-94.

/6/ Medley, G.F. & Anderson, R.M.: manuscript in preparation.

/7/ Anderson, R.M. & May, R.M.: Helminth infections of humans: mathematical models, population dynamics and control. *Adv. Parasit.*, **24** (1985), 1-101.

/8/ Haswell-Elkins, M.R., Elkins, D.B. & Anderson, R.M.: Evidence of predisposition in humans to infection with *Ascaris*, hookworm, *Enterobius* and *Trichuris* in a South Indian fishing community. *Parasitology*, **95** (1987), 323-337.

/9/ Wong, M., Project Coordinator, Parasite Control Programme, Montserrat, West Indies: personal communication.

/10/ Hominick, W.H., Dean, C.G., Schad, G.A.: Population biology of hookworms in West Bengal: analysis of numbers of infective larvae recovered from damp pads applied to the soil surface at defaecation sites. *Trans. Roy. Soc. Trop. Med. Hyg.* (in the press).

/11/ Bundy, D.A.P., Cooper, E.S., Thompson, D.E., Didier, J.M. & Simmons, I.: Epidemiology and population dynamics of *Ascaris lumbricoides* and *Trichuris trichiura* infection in the same community. *Trans. Roy. Soc. Trop. Med. Hyg.*, (in the press).

/12/ Bundy, D.A.P., Lecturer, Parasite Epidemiology Research Group, Imperial College: personal communication.

/13/ Elkins, D.B.: The epidemiology and control of *Ascaris lumbricoides* in a Indian fishing community. *Ph.D. Thesis, University of London* (1987).

/14/ Elkins, D.B., Haswell-Elkins, M. & Anderson, R.M.: The epidemiology and control of intestinal helminths in the Pulicat Lake region of Southern India. I. Study design and pre- and post-treatment observations on *Ascaris lumbricoides* infection. *Trans. Roy. Soc. Trop. Med. Hyg.*, **80** (1986), 774-792.

/15/ Tallis, G.M. & Leyton M.K.: Stochastic models of populations of helminth parasites in the definitive host. *Math. Biosci.*, **4** (1969), 39-48.

/16/ Anderson, R.M., Medley, G.F., May, R.M. & Johnson, A.M.: A preliminary study of the transmission dynamics of the human immunodeficiency virus (HIV), the causative agent of AIDS. *IMA J. Math. App. Med. Biol.*, **3** (1986), 229-263.

/17/ Anderson, R.M., Medley, G.F., Blythe, S.P. & Johnson, A.M.: Is it possible to predict the minimum size of the acquired immunodeficiency syndrome (AIDS) epidemic in the United Kingdom? *The Lancet* **i** 1987; 1073-1075.

/18/ Peterman, T.A., Drotman, D.P. & Curran, J.W.: Epidemiology of the acquired immunodeficiency syndrome (AIDS), *Epidemiol. Rev.*, **7** (1985), 1-21.

/19/ Medley, G.F., Anderson, R.M., Cox, D.R. & Billard, L.: Incubation period of AIDS in patients infected via blood transfusion. *Nature*, **328** (1987), 719-721.

/20/ Billard, L., Cox, D.R., Medley, G.F. & Anderson, R.M.: The distribution of the incubation period for the acquired immune deficiency syndrome (AIDS). (manuscript in preparation).

Microinteractive Predator-Prey Simulations

W.F. Wolff

IFF der KFA, P.O. Box 1913, D-5170 Jülich, Fed. Rep. of Germany

1. Introduction

Theoretical population dynamics often entails the use of differential equations to describe the growth and decline of populations. Such mathematical models are designed either for predictive purposes to make accurate short–term forecasts, or to identify generic characteristics and underlying principles. The former require detailed information about the system being modelled and are usually too complicated to extract information concerning basic underlying mechanisms. The latter, called 'strategic models' by Holling (1966), are specifically designed to elucidate these underlying mechanisms, so accuracy in forecasting and prediction is usually traded for mathematical simplicity. Despite this fact, strategic models still quickly reach the limit of being analytically tractable. Furthermore, they are hampered by hidden assumptions, which while being justified and well–founded in some cases, might change the qualitative outcome and give misleading clues. Just to name a few, most approaches assume local equilibrium and large numbers within the populations involved. In the terminology of statistical physics these models could be classified as mean–field type of models.

Populations consist of individuals and therefore the (time) evolution of these populations is determined by the actions of individuals, such as giving birth to one or more offspring or dying. Since these events are largely uncorrelated and independent from each other the most basic equation describing the time evolution is the master equation for the probability distribution $P(N,t)$ of having exactly N individuals at time t (comp. e.g. Ch. 6 in Nisbet and Gurney, 1982).

Though it may be possible to formulate such an equation it usually cannot be handled by analytical techniques, let alone solved, except in the most simple cases (van Kampen, 1983). Due to this lack of tractability one has to resort to mathematically and/or biologically motivated approximations such as finding equations for average variables neglecting correlations or looking for quasistationary states (Nisbet and Gurney, 1982). These approximations and their underlying assumptions are usually not even mentioned and one often starts immediately with the averaged equations. Fluctuations, which are to a large extent neglected, are then 'artificially' reintroduced by using stochastic models instead of deterministic ones.

Most of the methods above describe populations by specifying the number of individuals present in the population as a function of time (and space in inhomogeneous systems). While this has the advantage of encorporating well-developed mathematical tools for differential equations and also for stochastic 'birth-and death' processes (comp. e.g. van Kampen, 1983; Goel and Richter-Dyn, 1974), the price paid is that the population is described by aggregate parameters and variables. Furthermore, evolutionary processes, i.e. natural selection, put constraints on the properties of individuals and thus on the population. Determination of the distribution of phenotypic characteristics at the genetic level is largely impossible and the incorporation of these properties into models using global variables like population numbers and averaged parameters is often difficult, if at all possible. One of the most promising attempts to solve this problem has been the concept of an 'evolutionary stable strategy' (Price and Maynard-Smith, 1973; Maynard-Smith, 1982) which yields criteria on the behaviour of the individual. On the other hand, it is well-known that there are large differences in resource competition within a species or a population (see e.g. Ch. 10 in Krebs and Davis, 1987). The incorporation of these properties, not to mention the differences among individuals into population dynamical models, is to a large extent unclear.

One possibility is to accept the fact that the *individual* is the fundamental unit of a population. In doing so one has to specify what the individual does and how it behaves in a given situation. It is clearly not feasible to formulate a mathematical equation describing as complicated an individual as, say, a fox (compare, however, the contribution by N. van der Hoeven in these proceedings) and try to capture its behaviour in every detail. One possibility is to describe the individual and its basic behavioural patterns by way of rules or strategies, e.g., one might specify the rule *"when hungry, then hunt"*. Generic rules of this type might capture the basic features underlying behavioural actions, namely what a given individual is most likely to do in a given situation. Naturally, simplifications have to be made, but trying to keep such rules as simple as possible also helps to clarify their importance for the behaviour of the whole system or the population. Although these rules might be only a mere caricature of the real world, they should nevertheless capture the basic features. What these features are is, however, a question of debate and one can always be confronted with the valid remark that 'Nature is more complicated than that'. Yet one should undertake such endeavours in order to test the findings with results which can be obtained by more traditional methods. The next step would be to build models with more complicated rules and operating in more complex systems.

In the following an approach is presented which attempts to simulate the dynamics of populations, their interaction and time evolution in a predator-prey type of system. To be concrete and to present the methods and ideas behind this approach some well-known and rather simple textbook examples are considered. The ultimate object of this study is to suggest that in quite a number of cases where existing methods cannot be successfully applied or extended to explain phenomena in ecological systems, this approach using simulations might give useful insights.

This work concentrates *exclusively* on a two-species predator-prey system and is structured as follows: In Sect. 2 a description of the rules governing the behaviour and the action of the individuals is given. One of these rules, namely the one for the migration of individuals, is described in some detail. The third section is divided into several subsections. In the first, Sect. 3.1, some basic results are recovered which can also be obtained with a simple Lotka-Volterra type of model or generalizations thereof. In Sect. 3.2 the influence and the importance of various parameters are assessed, in particular the environmental carrying capacity and also the success rate with which predators are capable of capturing prey individuals. In the remaining two sections, 3.3 and 3.4, several choices for rules are compared with each other. Specifically, and a comparison is made between the simplest possibile rule governing migration , namely a random walk, and rules which depend on the nearby configuration of other individuals. The last section contains a discussion.

2. Explanation of Method and Algorithm

In this chapter the rules used in the simulations are described. One of these rules, namely that which determines migration or the movement of individuals within the system, is considered in some detail. For simplicity it will be assumed that individuals move on a square array of $L \times L$ cells (this choice of a lattice structure does not change the general results obtained in the simulations, but simplifies the discussion).

At any one interval of time an individual may move from its cell to one of its four closest neighbours. The rule describing this motion first requires a decision of the type *"either stay or move"*. The first choice, i.e. to stay within the current cell, is assumed to occur with a probability p_{stay}, the second, i.e. to move, with $p_{move} = 1 - p_{stay}$. The probability p_{stay} may depend on the configuration of other individuals around this cell and also on the current internal state of the specific individual. The second choice, *'move'*, immediately requires a subsequent decision, namely which one of the four neighbouring cells the individual is going to enter. Two possible decision rules are considered here.

The simplest possibility is to select at random any one of the neighbouring cells with probability $p_{random} = 1/4$ thereby neglecting the configuration of other individuals or the internal status of the individual. In other words, each individual performs a random walk over the array, and in what follows this particular rule will be called *'random'*.

The second possibility accounts in a rather simple way for the distribution of other individuals when choosing into which cell to move. Predators are more likely to choose cells where they can find and capture prey over those with few or no prey at all. This predeliction can be incorporated in the rule by assigning the cells a probability proportional to the number of prey within that cell. The probability of moving to cell i containing n_i prey individuals is thus given by

$$p_i^{hunt} = n_i/N \ , \tag{2.1}$$

where N is the total number of prey individuals in all four neighbouring cells, whence $p_1 + p_2 + p_3 + p_4 = 1$. This rule will be called the hunting rule or *'hunt'* for short.

Similarly, prey prefer cells with as few predators as possible. For example, the probability of choosing a cell may be inversely proportional to the number of predators within that cell, i.e.

$$p_i^{escape} \sim 1/n_i \ , \tag{2.2}$$

where n_i is the number of predators in cell i. This particular choice will be referred to as the escape rule or *'escape'*.

Both rules presented above are probabilistic and allow for the possibility of mistakes, e.g. a predator may move to a cell with only one prey when there is another adjacent cell with, say, five prey. This flexibility is to be contrasted with deterministic rules which rigidly specify the transition. An example of a deterministic rule is one where the predator always chooses the cell with largest number of prey (and vice versa for the prey).

Although other migration rules are not considered in this work, it should be noted that by slightly modifying of, say, the hunting rule it becomes possible to interpolate smoothly between *'hunt'* and *'random'*. Thus, if one chooses the probability p_i not according to Eq. (2.2) but as

$$p_i = \frac{n_i + \delta}{N + 4\delta} \ , \tag{2.3}$$

then one recovers *'hunt'* for $\delta \to 0$, while for $\delta \to \infty$ one gets $p_i = 1/4$, i.e. *'random'*. This reformulation may be seen either as an extension of the hunting rule or as a mixed strategy, where *'random'* is chosen with a probability p_{random}, and *'hunt'* (according to Eq. 2.1) with probability $p_{hunt} = 1 - p_{random}$, but where neither p_{random} nor p_{hunt} depend on the particular configuration of other individuals. A similar observation holds for the escaping rule.

The foregoing specifications apply only to cells inside the array. At the boundaries the probabilities are calculated as before but with the cell inside the array counted twice to make up for the missing one. At the corners of the array a similar procedure was used.[1]

[1]At the boundaries this has the effect of creating a bias towards the center. It has been checked that the overall behaviour is insensitive to various possible choices of boundary rules.

All other rules used in the present simulations pertain to the individual and thus are exactly the same within the array as on the boundaries. These rules describe the birth and death processes of predator and prey individuals and the predation process where predators attack and try to capture a prey in the same cell.

The prey population is assigned a finite carrying capacity K, which is assumed to be constant throughout the system. The probability of a prey having *one* offspring is thus given by

$$p_{prey}^{birth} = b_{prey}(1 - \frac{n_i}{K}) \ , \quad n_i \leq K; \quad \ldots = 0, \quad n_i \geq K, \tag{2.4}$$

where n_i is the number of prey within this cell (newly born individuals are not included in this number), K is the carrying capacity, and b_{prey} is the overall birth rate of the prey population.

The probability of predator reproduction depends on their internal feeding state, namely whether or not the predator had been able to capture a prey individual in the previous time step. In other words, the probability for reproduction of a predator is smaller for a hungry predator than for a satisfied one, $b_{hungry} < b_{satisfied}$.

Similar rules describe the mortality of both predator and prey. Prey individuals are assumed to have a constant mortality rate. Their probability of dying is constant at every time step, m_{prey}, while the corresponding mortality for predators again depends on whether or not they had been able to capture prey recently. As with birth rates, the probability of a predator mortality depends on their state of hunger, and hungry ones die more often than those that are satisfied , $m_{hungry} > m_{satisfied}$.

The final rule concerns the predation process. The following simple scenario is followed: A predator picks out a particular prey individual in the same cell (if available!), attacks and tries to capture it. The probability of success is assumed to be the same for all predators, p_{pred}, independent of the internal states of the individual involved or of its spatial position. If a predator has successfully attacked a prey, its internal state is changed from *'hungry'* to *'satisfied'*, the captured prey is deleted from the population, and then the next predator is considered.

One time step in the simulation is divided into three processes:

1.) A *migration* process in which each individual may move once or several times according to the rules *'random'*, *'hunt'* or *'escape'* given above.

2.) A *predation* process where each predator attacks available prey and tries to capture it.

3.) A *birth-death* process in which each individual may produce *one* offspring according to the probabilities given above and, finally, each individual is subjected to the appropriate probability of dying.

In each of these processes an individual is chosen, a rule is applied an executed, and then the focus shifts to the next individual until all members of both populations have been considered.

During each run, every individual within the system was kept track of, including the spatial position and internal states (e.g. 'hunger' or 'satisfaction' for predators). The output generally consisted of the number of individuals of each population, general information, such as average values of internal variables, and (at selected times), the spatial distribution of the individuals.

All simulations presented below were executed on a 30×30 square array. Simulations on smaller arrays, e.g. 5×5, were also carried out but resembled more the situation in a single cell than that which transpires on an extended array. In the single-cell case dispersal does not occur, and in certain cases the system can be modelled with differential delay equations. In some cases this limiting behaviour was used as a check on the general behaviour of the simulation. Very large systems (larger than roughly 200×200) behave as an aggregate of smaller subsystems. Oscillations are greatly attenuated and,for the most part, behaviour resembles fluctuations about a steady state.

Simulations were usually run for 2000 time steps (fewer, if one of the species shpuld become extinct). The initial configuration consists of a total of 1000 individuals, randomly distributed across the array. The ratio of prey to predators was either 3 to 1 or 5 to 1[2]. The mobility p_{move} was 0.45 for the prey and 1.0 for the predators. The reproduction probabilities were $b_{prey} = b_{satisfied} = 0.4$, $b_{hungry} = 0.01$ and the mortalities, $m_{prey} = m_{satisfied} = 0.05$ and $m_{hungry} = 0.08$. With these parameters the average life expectancy of a prey is 20 time steps (in the absence of predators). A satisfied predator lasts about as long, while a hungry predator remains for only 12.5 time steps. During its life time an average satisfied predator produces 8 offspring and a prey at most 8 offspring (but only for a large enough carrying capacity, K, and a small prey density). A hungry predator has an accumulated probability of only 1/8 to produce one offspring. The average area covered by a randomly migrating individual varies as the square–root of its age. A prey individual living its full life expectancy of 20 time steps covers an area of approximately 4×4 lattice spacings centered around its place of birth. The carrying capacity K and predator success rate p_{pred} were changed during several of the runs and are given separately in the simulations shown below.

[2]The results were largely independent of the initial configuration.

3. Results

3.1. Lotka–Volterra Predator–Prey System

The simplest equations describing predator–prey relationships are of the standard Lotka–Volterra type (Lotka, 1925; Volterra, 1926)

$$\frac{dN_1}{dt} = b_1 N_1 (1 - \frac{N_1}{K}) - c_{12} N_1 N_2 \quad , \tag{3.1a}$$

$$\frac{dN_2}{dt} = -b_2 N_2 + c_{21} N_1 N_2 \quad . \tag{3.1b}$$

Here N_1 denotes the prey population with an intrinsic birth rate b_1. K is the environmental carrying capacity which specifies how many prey can be sustained by the environment in the absence of predators[3]. The effect of predators on the prey is measured by the 'functional response' $c_{12}N_1N_2$. The predator population N_2 has an intrinsic death rate $-b_2$ and a growth rate given by the 'numerical response' $c_{21}N_1N_2$, which depends on the prey abundance.

For an infinite carrying capacity, $K = \infty$, the prey density, i.e. the N_1^2 term, in (3.1a) is missing and the solutions of (3.1) are always periodic, i.e. a limit cycle with the well–known cycles of overexploitation–underexploitation (May, 1973). In this case the system (3.1) is structurally unstable rendering it hard to simulate numerically. It has been said of such a system that 'it has no place in biology' (R. M. May in Hallam and Levin, 1982).

For finite carrying capacity K the system exhibits a stationary point, which is approached by both populations in either exponential or oscillatory fashion for $b_1 < 4Kc_{21}(Kc_{21}/b_2 - 1)$ (for a review see Goel et. al., 1971; Nisbet and Gurney, 1982). Furthermore, systems with a finite carrying capacity are easier to simulate because K gives an approximate upper bound for the *local* prey abundance.

One of the basic assumptions underlying the differential equations (3.1) is that the system should be homogeneous and in global equilibrium like, for instance, a chemical reaction. To achieve this in the simulation one may choose two approaches. Either one considers a small array with only a single (or very few) cell(s), or one considers a larger array with many cells, where the movement of individuals is chosen appropriately. The second approach is one used here. Individuals have an infinite mobility and each of them chooses at random *any* cell of the array which amounts to a complete mixing of the system during each migration step.

A typical result of such a simulation is shown in Fig. 1. Instead of damped oscillations eventually leading to a fixed point, cyclic behaviour (Fig. 1a) and a limit

---

[3]Note that in the simulation K denotes the carrying capacity per cell while in Eq. (3.1) the carrying capacity refers to the whole system. In order to prevent confusion the same symbol is used for both.

1a)

1b)

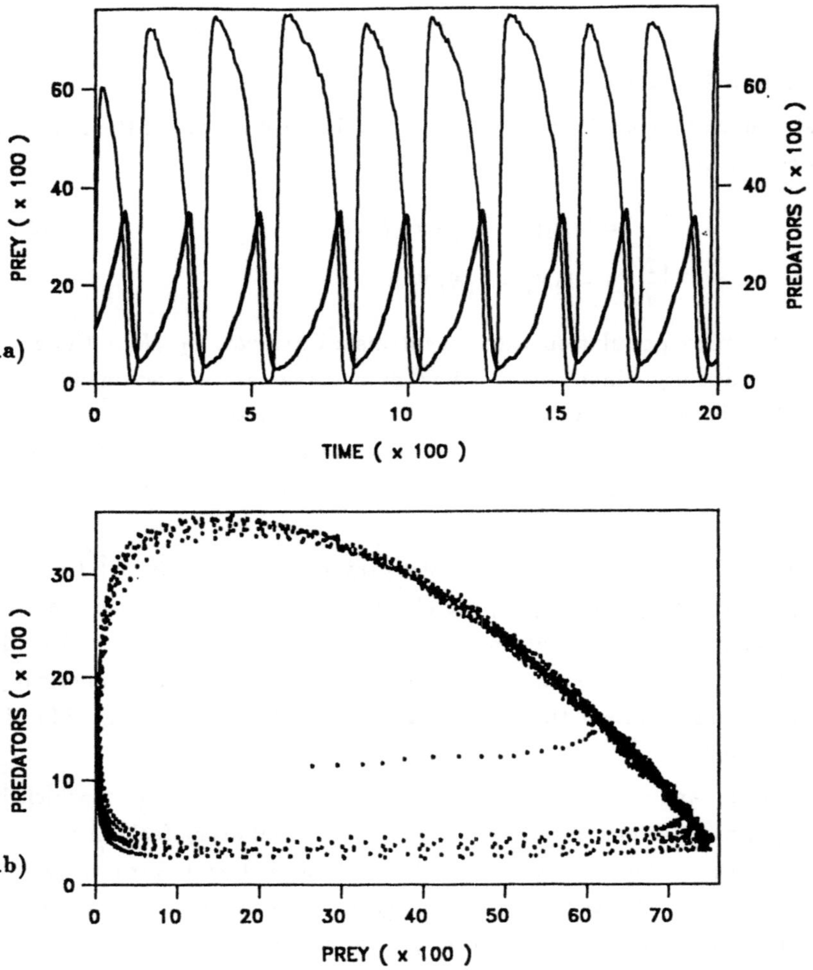

Fig. 1: (a): Time evolution of a predator–prey system for carrying capacity $K = 10$ *and success rate* $p_{pred} = 0.2$. *Individuals may migrate to any cell during each migration step. (Predators are shown by the thick curve.) (b): Limit cycle corresponding to (a).*

cycle in the predator–prey plane (Fig. 1b) are observed. The cycles are undamped, which can be clearly seen when the system is run much longer. This qualitative difference between the solution of the system (3.1) and the simulation is mainly due to the fact that in the simulations each predator has a limited capacity of attacking and eventually capturing prey, namely once every time step, whereas in Eq. (3.1a) the binary collision term $\sim N_1N_2$ does not saturate, and the feeding capacity of predators increases linearly with prey numbers. This, loosely speaking, *finite-belly* effect can be incorporated into the Lotka–Volterra system (3.1), e.g. by using a Holling-type II functional response (Holling, 1965).[4] The binary term in (3.1a) is changed to $N_1N_2/(1+\beta N_1)$ which saturates for large prey numbers $N_1 \gg 1$ at N_2/β, i.e. at $1/\beta$ per predator. Eqs. (3.1) then become

$$\frac{dN_1}{dt} = b_1 N_1 (1 - \frac{N_1}{K}) - c_{12} N_2 \frac{N_1}{(1+\beta N_1)} \quad , \tag{3.2a}$$

$$\frac{dN_2}{dt} = -b_2 N_2 + c_{21} N_1 N_2 \quad . \tag{3.2b}$$

This system of equations exhibits a limit cycle already for finite but high values of the carrying capacity K (comp. Section 3.2). It does not, however, adequately describe the results of the simulation. Nevertheless, it is possible to describe the curves in Fig. 1a (or, at least, parts of them) quite accurately by analytical expressions derived from a set of differential equations where the numerical response must be chosen appropriately (Wolff, Kerszberg, Ruján, 1988).

This simple example indicates that the results obtained by more traditional approaches can be at least qualitatively recovered if the assumptions inherent in these approaches are explicitly taken into account.

3.2. Sensitivity to Carrying Capacity and Success Rate of Capture

To test the sensitivity of the results of the simulations to parameter changes, two cases will be considered. The first concerns the effect of varying the carrying capacity K. This gives rise to a phenomenon called the *'paradox of enrichment'* by Rosenzweig (1971). Secondly, the general behaviour of the system is studied when the success rate with which predators are able to capture prey is increased. In both cases only the *'random'* migration rule is used here for both predator and prey.

3.2.1. Sensitivity to Carrying Capacity

The carrying capacity K determines the density of the prey population. For a given carrying capacity K (per cell) and n_i prey individuals in cell i the probability p that each prey will have *one* offspring is given by Eq. (2.4), and thus the probability to have n_i+m prey individuals in the next time step is given by the Bernoulli distribution

$$P(n_i + m | n_i) = \binom{n_i}{m} p^m (1-p)^{n_i - m}, \quad n_i < K \; . \tag{3.2}$$

The equilibrium number of prey per cell in the absence of predators is then given by (ignoring migration)

$$\overline{n_i} = \frac{K}{b_{prey}} (b_{prey} - m_{prey}) \; , \tag{3.4}$$

which increases linearly with K.

[4]This and other possibilities are discussed in (May 1973, 1981).

In general, a finite environmental carrying capacity K has a stabilizing influence on the behaviour of the populations. In the Lotka–Volterra model (3.1) *any* finite K produces a stable fixed point as compared to a limit cycle that arises when the carrying capacity is infinite, $K = \infty$. In more general systems like (3.2) limit cycles may already occur for finite K. In loose terms, if the carrying capacity K is much larger than the equilibrium prey density in the presence of predators (for the system (3.2) this is approximately given by $B^* = b_2/c_{21}$), then the stabilizing influence contributed to the dynamics by the prey density dependence should be relatively weak. Increasing K lowers the stability, and cyclic behaviour of the solution becomes possible. In particular, the system (3.2) exhibits such a transition. When K is increased, with all other parameters fixed, one first finds a stable fixed point solution, then damped oscillations decaying to a fixed point, and eventually a limit cycle when $K > 2b_2/c_{21} + 1/\beta$. This can be seen by considering the predator and prey isoclines (e.g., using the graphical analysis by Rosenzweig and MacArthur, 1963), or by performing a stability analysis around the stationary points. (See also Theorem 2.2 from the contribution by T. G. Hallam in Hallam and Levin, 1982).

Although the populations in the simulation are not adequately described by the Eqs. (3.2), a similar transition, namely from fixed point to (approximate) cyclic behaviour, nonetheless occurs. In Figs. 2a–d the carrying capacity K is increased. At first, fiexed point behaviour for the populations is apparent. Small carrying capacity results in a small prey density, and in many cells there are no prey at all. Hence, predators do not encounter prey individuals at every time step, and searching time becomes important. This leads to a very weak interaction between the two populations which stabilize at fixed levels.[5] At higher carrying capacities, more prey can be sustained, and more cells of the array become populated, which reduces searching time. Then predators can proliferate and are more capable of controlling the prey population. Oscillations gradually emerge and become more prominent as one moves from the K–limited regime (K small) to a predator–controlled regime (K large) (Figs. 2b, c).

However, unlike the system (3.2) where a further increase of the carrying capacity did not lead to qualitatively new behaviour, the latter simulations show that the system as a whole changes its character. The amplitude of the oscillations increases faster than the mean values of the populations. Therefore, the prey population decreases to lower numbers at the bottom of the cycles, while predators become more abundant at the peaks. This opens up the possibility for the prey to become extinct, which happens when the predator population is large enough to cover (and thus to

[5]When K is too small not enough prey can be sustained to ensure the survival of the predator population. For the chosen initial conditions this happens e.g. at a carrying capacity $K \leq 2$.

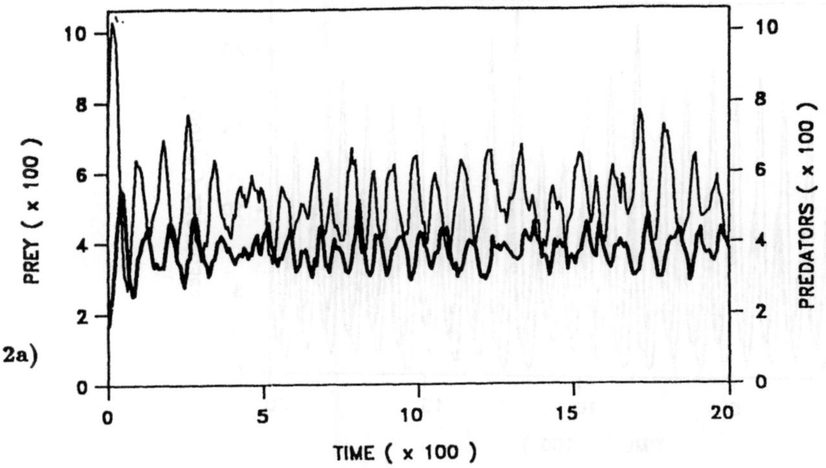

2a)

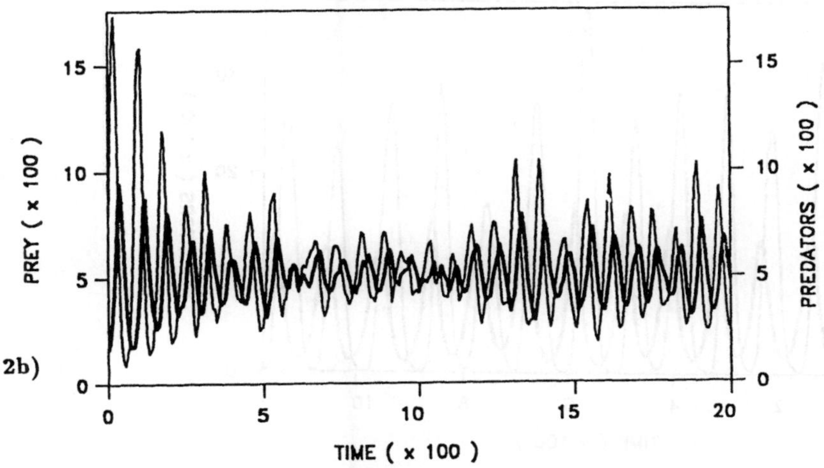

2b)

Fig. 2: Time evolution of a predator - prey system for several values of the carrying capacity $K = 3$ (a), $K = 4$ (b), and success rate $p_{pred} = 0.4$. (Predators are shown by the thick curve.)

control) the whole system. If this is not the case the prey still might have a chance to survive in a small part of the system.

In Fig. 2c the predator population increases up to 1000 individuals at the peaks which is indeed large enough to occupy all cells of the array. However, because of local fluctuations, predators are not distributed uniformly across the system. In some areas they are more abundant, while in other araes only a few exist. Although predators will deplete the prey in most parts of the array and then decline because of lack of food, a small number of prey have a chance to survive in those areas with few predators.

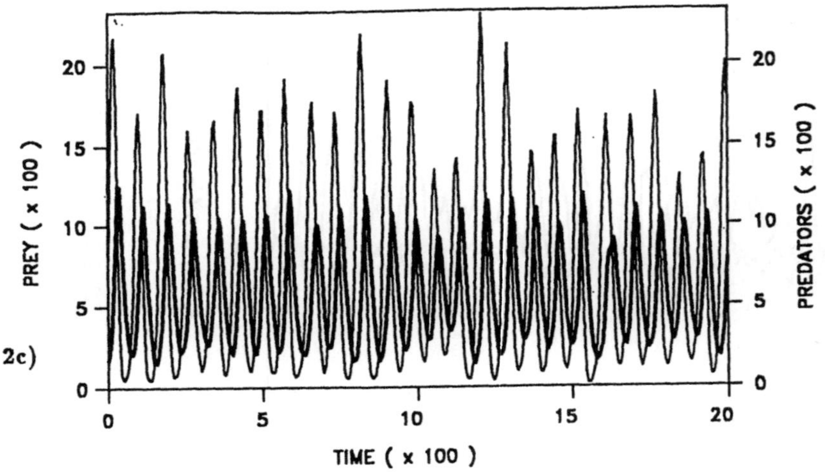

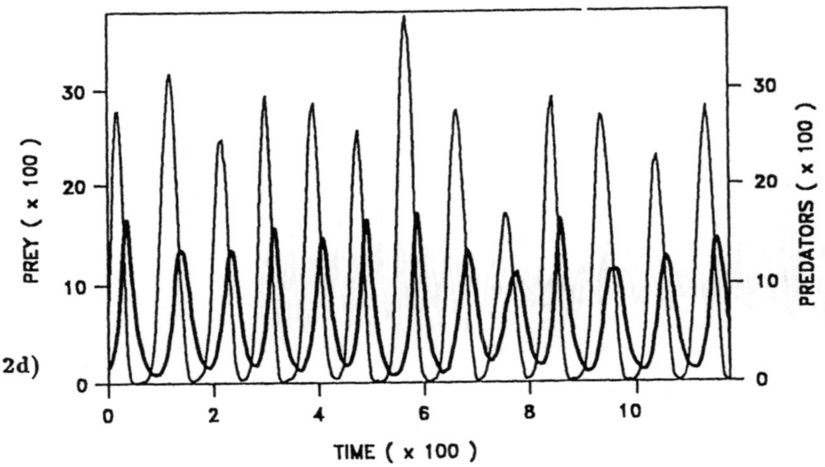

Fig. 2(ctd.): Same as before: carrying capacity $K = 5$ (c), $K = 6.15$ (d). Note that in (d) a different horizontal scale has been used. (Predators are shown by the thick curve.)

These surviving prey proliferate and repopulate most of the system. Then predators again find enough prey, and their population increases leading to another cycle.

However, if the predators have proliferated too much and can cover most cells of the array, these predator–rare patches are too rare or disappear. Then the prey have little chance to survive anywhere in the system. Although they might succeed for a while, eventually there will appear a spatial arrangement of predators such that the prey will be unable to escape and their population becomes extinct. An example is shown in Fig. 2d, where the prey was able to survive for twelve cycles, even though during one cycle their population had declined to a small patch of six individuals. From this it seems evident that extinction is caused by demographic stochasticity and a particular spatial arrangement of the predators.

Thus, persistence of both populations can only be achieved if the prey are able to find areas in the systems which are more or less free of predators and in which new prey subpopulations can be built up before being discovered by pursuing predators. Broadly speaking, the picture that emerges is almost the same as the one from the laboratory experiments by Huffacker and co–workers on predator–prey interactions (Huffacker, 1958; Huffacker et. al., 1963; see also Hilborn, 1975).

3.2.2. Sensitivity to Success Rate of Capture

The predation process during which predators feed on prey is the only direct interaction between the two populations. The strength of this interaction is measured by the success rate p_{pred} with which predators can capture the prey.

Inefficient predators have a small rate of success, resulting in a rather weak interaction. Hence variations in the predator abundance have little effect on the prey population. Furthermore, prey will not be controlled as much by the predators but rather be limited by the carrying capacity. As in the example in the previous section, one will see essentially fixed point behaviour (Fig. 3a).

To sustain their population at its current level, predators must capture κN_2 prey individuals, $1/\kappa = 1 - (b_{satisfied} - m_{satisfied})/(b_{hungry} - m_{hungry})$; $\kappa = 1/6$ for the parameters used in the simulation. If prey are sufficiently available, at most $p_{pred} N_2$ prey individuals can actually be captured. Therefore predators will surely become extinct whenever $p_{pred} < \kappa$, otherwise they might proliferate. (Because of this, the rate of success was always chosen to be larger than κ.) In Fig. 3a $p_{pred} = 0.2$, and predators proliferate rapidly initially when prey are sufficiently abundant. Vice versa, the rapid decline of the prey population is caused solely by the growth of the predator population. After this initial transient both populations settle at fixed values.

From this and the above considerations it can be inferred that at every time step roughly the same number of prey individuals is captured, namely just as many as are needed to sustain the predators at their current level. Prey are thus harvested with a constant rate, and the time evolution of its population can be understood in quantitative terms using a simple logistic model with harvesting, similar to 'general production models' first used in the management of marine fisheries where such models (Schaefer, 1954, 1957; Pella and Tomlinson, 1969; for reviews see e.g. Clark, 1976, 1981; Conrad, 1982).

An increase in the success rate drives the system away from the previous situation where the predators managed merely to sustain their population at a fixed level. Predators begin to overexploit their resource, and a destabilization of the system and

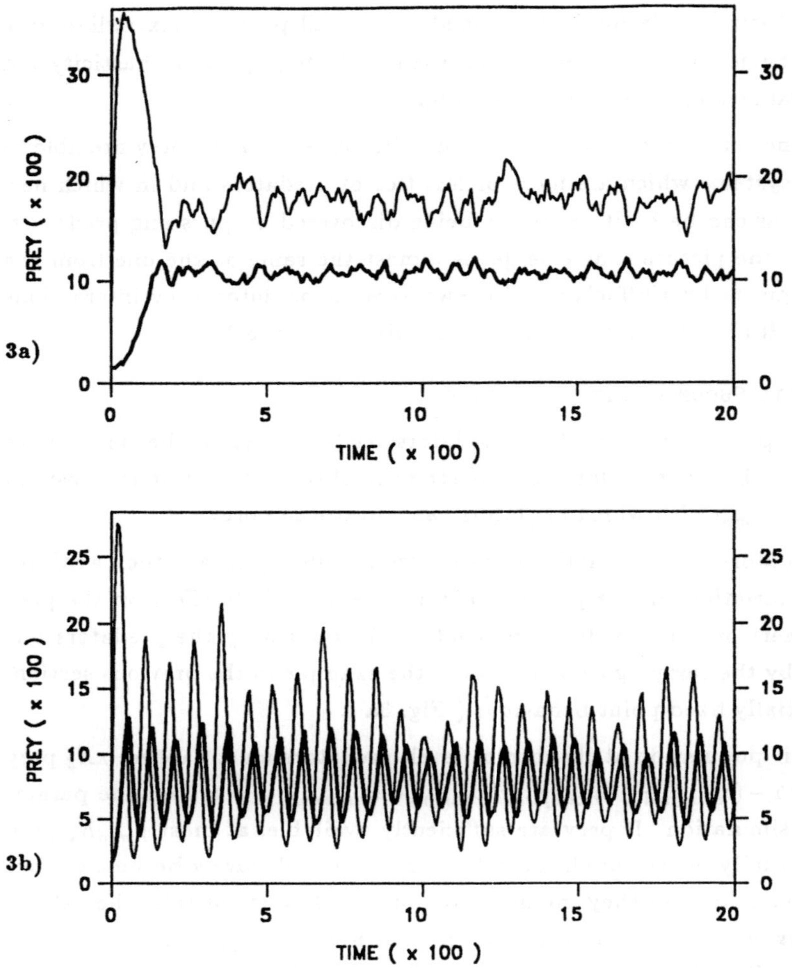

Fig. 3: Time evolution of a predator - prey system for several values of the success rate $p_{pred} = 0.2$ *(a),* 0.3 *(b), and carrying capacity* $K = 5$. *(Predators are shown by the thick curve.)*

thus cyclic behaviour ensues. Examples with typical results are shown in Figs. 3b–e.

Oscillations are rather regular at lower predator success rates (Fig. 3b), but turn out to vary strongly in magnitude for larger success rates (Fig. 3c) with occasional outbursts of prey and predator populations. These outbursts are more the rule than the exception and could be attributed to a very patchy distribution of the populations. Patches poor in prey caused by an overexploitation of the prey and lead to extinction of both populations in large parts of the array.

Increasing the success rate even further enables the predators to deplete the prey

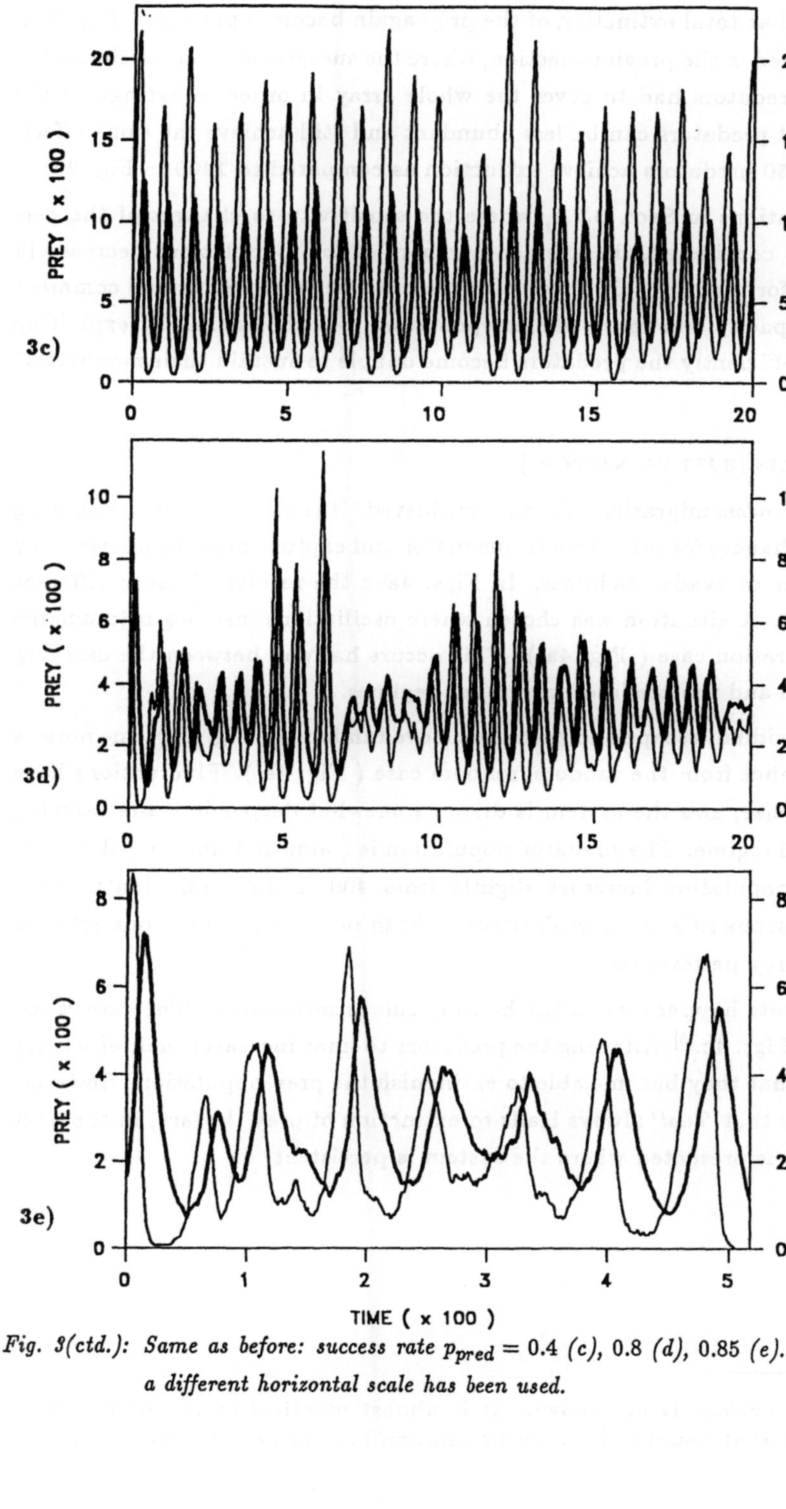

Fig. 3(ctd.): Same as before: success rate $p_{pred} = 0.4$ *(c),* 0.8 *(d),* 0.85 *(e). Note that in (e) a different horizontal scale has been used.*

to such an extent that total extinction of the prey again becomes possible (Fig. 3e). In the case considered in the previous section, where the success rate was much smaller (Fig. 2d), the predators had to cover the whole array in order to extinguish the prey. More efficient predators can be less abundant and still achieve the same effect. In Fig. 3e barely 550 predators achieve extinction as compared to 1400 in Fig. 2d.

Unlike the situations in Sect. 3.2.1, where the sensitivity to changes of the carrying capacity was considered, the average numbers of *both* populations decrease in magnitude. While for the prey this dimunition seems to be an almost trivial comment (fixed carrying capacity, but more efficient predators), it shows that by exploiting their resource too efficiently the predators become unable to sustain their numbers at the previous levels.

3.3. Change of rules (hunt vs. random)

So far only the *random* migration rule was considered. It is obvious that the hunting rule increases the chances for predators to encounter and capture prey. Similarly, prey can *'escape'* in order to evade predators. In Figs. 4a–c the results of using different rules are compared. A situation was chosen where oscillations just begin to emerge in the random migration case (Fig. 4a). This occurs halfway between the carrying capacity–controlled and the predator–controlled regimes.

When prey individuals escape and predators move randomly (Fig. 4b), one notices only a small difference from the random/random case (Fig. 4a). Fluctuations have become smaller in size, and the system is driven somewhat deeper into the carrying capacity–controlled regime. The predator population is (almost) unchanged in size, the average prey population increases slightly from 400 to 460 individuals. This suggests that the escape rule has a weak effect on both populations and does not lead to a much better prey performance.

Quite the opposite happens when the hunting rule is introduced. The case *'hunt-escape'* is shown in Figs. 4c.[6] Allowing the predators to hunt increases their efficiency to such an extent that they become able to extinguish the prey population. However, this does not imply that *'hunt'* always leads to extinction of prey. In fact, in the next section an example is presented where the system is persistent.

[6]The case *'hunt–random'* is not shown. It is almost identical to the first part of Fig. 4c, except that usually the prey population becomes extinct after just one cycle.

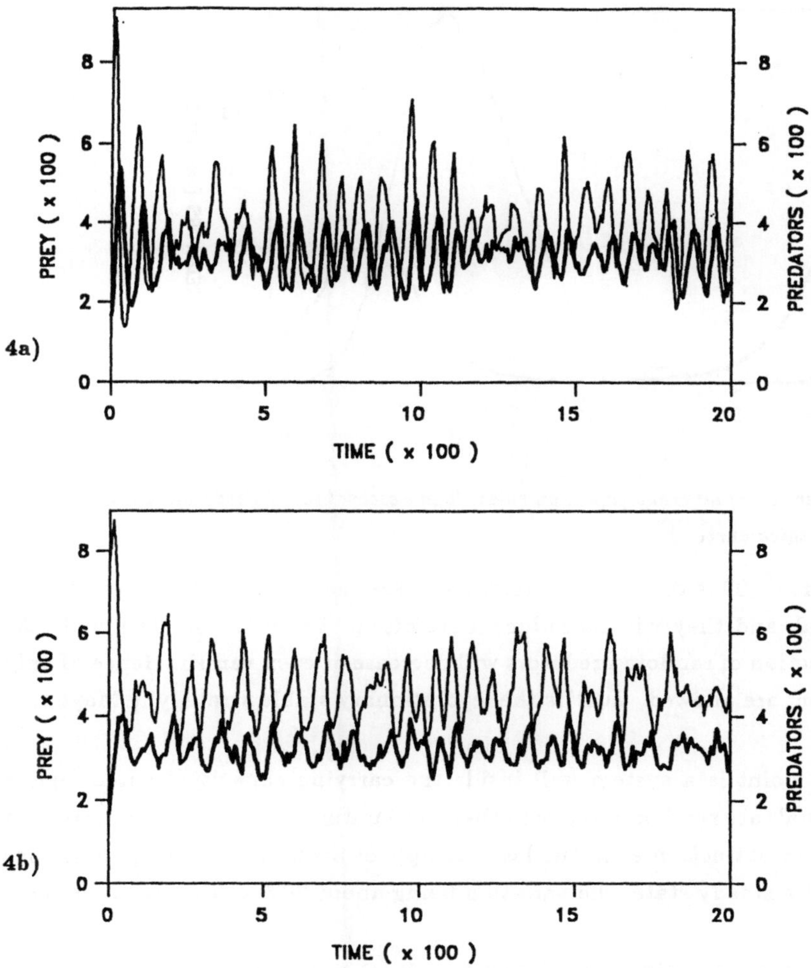

Fig. 4: Comparison of different migration rules: 'random–random' (a), 'random–escape' (b). Carrying capacity $K = 3$, success rate $p_{pred} = 0.5$. (Predators are shown by the thick curve.)

3.4. Selection of Rules

In this last example the three migration rules, *'hunt'*, *'escape'*, *'random'*, are left to compete against each other. A more efficient strategy will lead to a greater fitness of the individual and to a larger number of offspring. In the long run the fittest strategy should then be adopted by all (or almost all) of the surviving members of the population. One cannot specify *a priori* a pay–off matrix to determine the relative merits of the three strategies. However, it is clear that if some predators use *'hunt'* and some use *'random'*, the 'hunters' have a better chance to encounter prey than do the

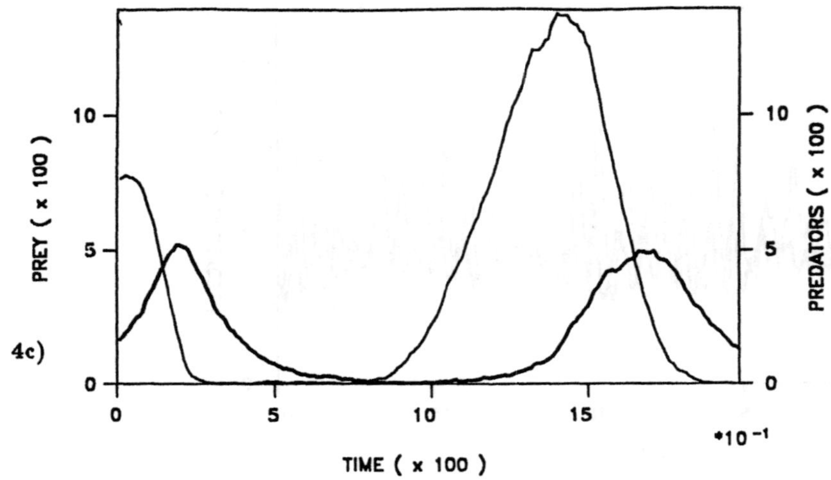

Fig. 4(ctd.): Comparison of different migration rules: 'hunt–escape' (c). (Predators are shown by the thick curve.)

random predators.[7] This difference in turn increases the hunters' chance to capture a prey individual, and they will reproduce more often. Therefore, hunters will take over, and the fraction of random predators will decrease or even vanish. Hence, if only *'hunt'* and *'random'* are allowed, *'hunt'* is the 'evolutionary stable strategy' (Maynard-Smith, 1982).

The starting point is a system well inside the carrying capacity–limited regime when all individuals are random walkers (otherwise hunting predators would drive the prey population to extinction as in the last example of Sect. 3.3). Both populations fluctuate around a steady state, the variation being about 20% of the mean numbers (Fig. 5a).

In order to compare the different rules with each other, individuals are constrained to use one strategy during their whole life. For example, predators use either *'random'* or *'hunt'*. This strategy is passed unchanged to their offspring, i.e. there is *no mutation.* Initially 2% of the populations are 'hunting' or 'escaping', the remaining 98% use *'random'*.

In Fig. 5b the fraction of 'hunters' and 'escapers' is shown as it evolves in time. After about 250 time steps the predator population is completely dominated by 'hunters'; hunting predators quickly outperform the random walkers. On the other hand, selection pressure on the prey is much weaker than on the predators, since it takes much longer until all prey become 'escapers'. At first, when there are still random preda-

[7]Both do equally well, if *all* neighbouring cells are occupied by enough prey. This possibility is disregarded for the moment.

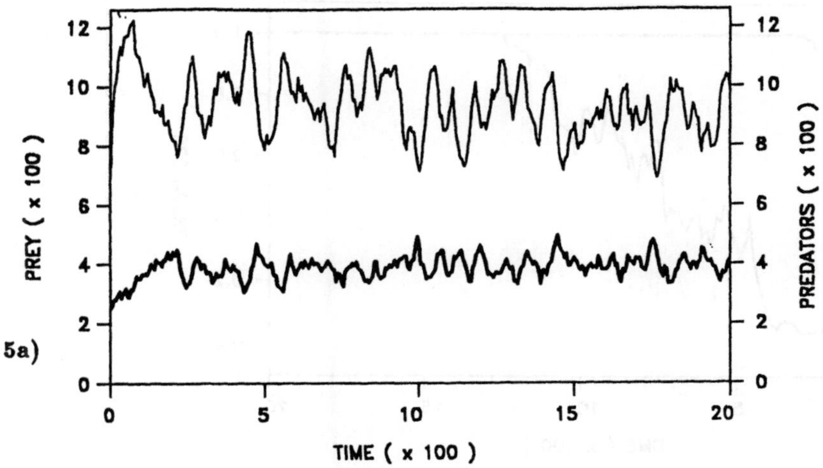

Fig. 5a: Both prey and predators move randomly. Carrying capacity $K = 3$; success rate of predators $p_{pred} = 0.3$. For further details see text. (Predators are shown by the thick curve.)

tors, there seems to be little advantage for prey individuals to use *'escape'*. This is indicated by the small increase in the fraction of 'escapers'. Hunting predators exert a much greater pressure on the prey. After predators become dominated by 'hunters', it then pays more for the prey to escape, and the fraction of escapers increases sharply followed by a steady rise until all prey attempt to escape (Fig. 5b). This was a feature observed in most simulations and corroborates the findings of the last example (Figs. 4a,b), where in the presence of random predators, escaping produced only a weak change in the size of the prey population.

Fig. 5c shows the time evolution of both populations. Because hunters have completely taken over after about 250 time steps, the remaining part of the curve what would have occured if hunt had been chosen right from the beginning. In this region one observes large, almost regular oscillations, in particular after the prey population is dominated by 'escapers'. The average number of prey individuals decreases by almost half during the development of 'escapers' (compare Figs. 5a and 5c). The introduction of different rules (in particular *'hunt'*), has driven the system out of the relatively stable K–limited into the predator–controlled regime which is less stable. As the escaping rule seems to have only a weak effect on the overall behaviour of the system, this change from fixed point behaviour to oscillations can be attributed mostly to the domination of hunting predators.

The situation where both prey and predators move randomly (Fig. 5a) resembles the real world situation where zooplankton feed on phytoplankton in highly turbid waters (Hart, 1986). In such silt–laden waters visual predation does not exist. The op-

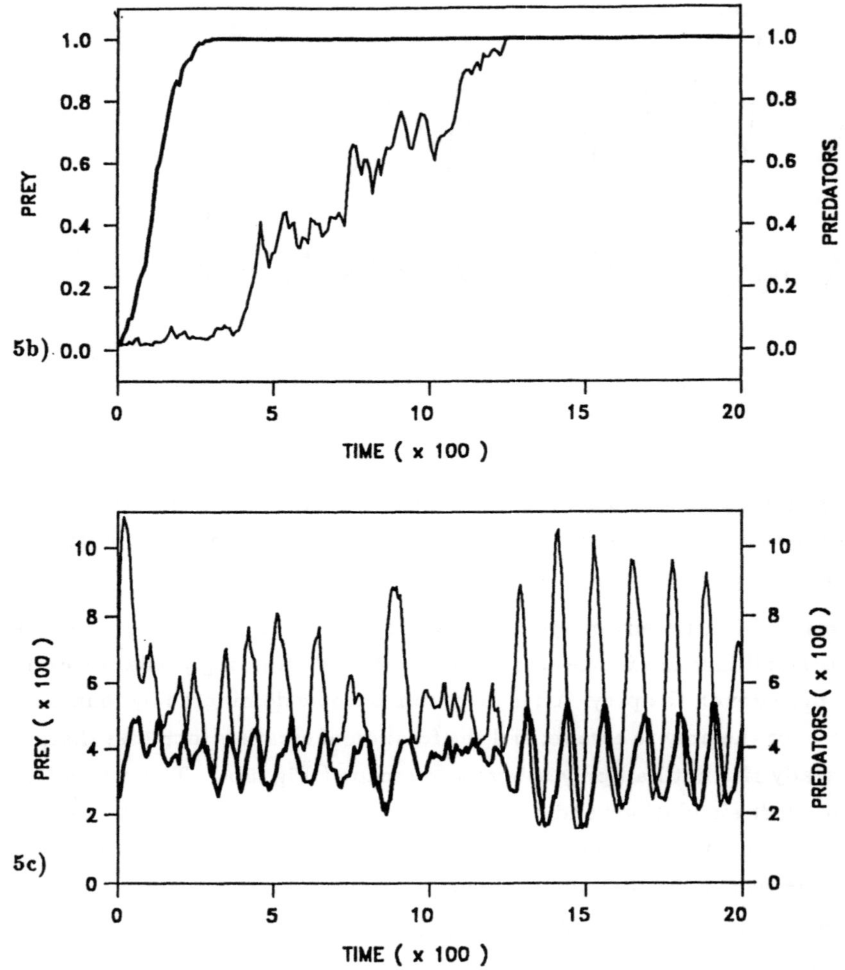

Fig. 5 (ctd): Selection of migration rules. Fig. 5b shows the fraction of escapers and hunters. Fig. 5c shows the population numbers. For further details see text. (Predators are shown by the thick curve.)

posite occurs in clear waters where visual predation takes place, and the hunt/escape situation as prescribed in Fig. 5c prevails (e.g. Likens, 1985).

4. Discussion and Conclusion

It has already been remarked in Section 2 that the size of the array used in a simulation plays an important role in the overall behaviour. The appropriate choice of array size depends on the problem considered. In general, no clear cut answer can be given, but (at least for the cases considered in this work) rules of thumb are possible.

Small arrays tend to resemble the single–cell case. Simulations have shown that deviations from single–cell behaviour start to develop when the array size becomes appreciably larger than the area covered by a typical, randomly migrating individual. This area is about 4×4 cells for a prey with a life expectancy of 20 time steps, and arrays of roughly this size behave like the single–cell case.

Medium–sized arrays, ranging from about 15×15 up to approximately 60×60, share the same behaviour. The mean number of individuals increases linearly with the number of cells such that densities remain constant and oscillations are of comparable magnitude. For larger arrays the size of the oscillations decreases rapidly. For example, the oscillations in the prey abundance shown in Fig. 3c cover almost two orders of magnitude. The same simulation on a 200×200 array shows fluctuations of less than 20% about the mean value. Nevertheless, in smaller subarrays, e.g. 30×30, one still finds rather well–developed oscillations slightly smaller than the ones in Fig. 3c. However, these oscillations in different parts of large arrays are generally out of phase. In the large system they will thus cancel each other, and only small fluctuations about the mean values result. Increasing the size of the array thus increases the stability and persistence of the system.

The simulations in this work have been done on a 30×30 array. This size turns out to be an appropriate choice for the situations considered. On the one hand, it is considerably larger than the single–cell case; on the other, it is still small enough to prevent subarrays from oscillating (too much) out of phase. This particular choice was also partly motivated by the desire to save CPU time without qualitatively changing the results. It should be remarked that the algorithm has been implemented in such a way that the execution time depends only on the average number of individuals within the system, but not on the size of the array (which, however, influences population size). E.g., the simulation shown in Fig. 1 required about 140 sec, and the one in Fig. 3d 40 sec on a CRAY X-MP/22.

A comparison of the simulations presented in Sects. 3.3 and 3.4 reveals that different strategies may have distinctly different effects. *'Hunt'* and *'escape'* are more efficient strategies than random migration. Both are capable of displacing a population in which almost all individuals use *'random'* (Fig. 5b) and hence represent the evolutionary stable strategies (Maynard–Smith, 1982) among the various migration rules.[8] *'Escape'*, however, offers only a small advantage when playing against *'random'* (compare Figs. 4a,b, and also the short time behaviour in Fig. 5b). The advantage to *escape* becomes much bigger when it plays against *'hunt'*, since the fraction of escapers is forced to grow rapidly as soon as the predators become dominated by 'hunters' (Fig. 5b).

Contrary to this, *'hunt'* shows a comparable increase in efficiency when playing against either prey strategy. It always leads to a large decrease in prey abundance (Figs. 5a,b) accompanied by a destabilization of the system, i.e. by the emergence of oscillations. Even extinction of prey is possible (Fig. 4c). One might have thought that *'hunt'* for predators could be counterbalanced by an appropriate change in the migration rule for the prey. The simulations suggest that this does not seem to be the case.

A change of strategy is also not equivalent to a change of one (or a few) parameter(s). For example, *'hunt'* increases the efficiency of predators in capturing prey. Yet, the introduction of *'hunt'* produces different results than a corresponding increase in the success rate p_{pred} with which predators kill prey. Although both lead to destabilization and oscillations, the frequency (and magnitude) of these oscillations is much larger in the latter case (compare Figs. 3b,c and Fig. 5b). The average prey abundance decreases in both cases. But while the average number of predators remains almost unchanged when *'hunt'* is introduced (Figs. 5a,c)[9], a decrease in the predator abundance eventually occurs as the predator success rate becomes larger. In fact, both populations decrease approximately as the inverse of the success rate.[10]

The above observations appear to be common to a number of other simulations in which different strategies were compared to each other. They suggest that, in general, elaborate escape strategies do not lead to a marked increase in the performance of prey. On the other hand, simple hunting rules, like the ones used in this work, already change the qualitative behaviour of the system, in most cases for the worse.

There are, of course, many possible generalizations and extensions of rules and situations which might be amenable to computer simulations. One such generalization (besides looking at more than just two species) is to consider the effects of spatial inhomogeneity, e.g. by allowing the carrying capacity to change in space. Another is to consider age–structured populations and/or sexually reproducing populations.

---

[8]Note that only *pure* strategies are considered.

[9]In Fig. 5a the average numbers are 943 prey individuals and 381 predators; the corresponding numbers for Fig. 5c are 553 and 366, resp.

[10]At least for the prey this can also be obtained from Eqs. (3.2). The success rate should be roughly equivalent to the strength of the numerical response, c_{21}, which is inversely proportional the average prey population $B^* \sim b_2/c_{21}$.

Furthermore, the rules could be made more realistic, taking into account, e.g., the cost of migration, cost of reproduction, several (not just one) states of hunger, and mating strategies, just to name a few.

Acknowledgement

I would like to thank S. Cousins, M. Kerszberg, P. Rujan, for helpful discussions and, in particular, J. Grobbelaar, B. Huberman, J. McGlade, and R. Ulanowicz for a critical reading of the manuscript.

References

Clark C. W. (1976) Mathematical Bioeconomics: The Optimal Management of Renewable Resources. *Wiley-Interscience, New York*

Clark C. W. (1981) Bioeconomics. In May R. M. (1981) *op. loc.*

Conrad J. M. (1982) Bioeconomics and the Management of Renewable Resources. In Hallam T. G., Levin S. A. (1982), *op. loc.*

Goel S. N., Maitra S. C., Montroll E. W. (1971) On the Volterra and Other Nonlinear Models of Interacting Populations. *Rev. Mod. Phys.* **43**, *231*

Goel S.N., Richter–Dyn N. (1974) Stochastic Models in Biology. *Academic Press, London*

Hallam T. G., Levin S. A. (1982) Mathematical Ecology. *Biomathematics* **17**, *Springer, Berlin*

Hart R. C. (1986) Aspects of the Feeding Ecology of Turbid Water Zooplankton. *In situ* Studies of Community Filtration Rates in Silt–Laden Lake le Roux, Orange River, South Africa. *J. Plankt. Res.* **8**, *401*

Hilborn R. (1975) The Effect of Spatial Heterogeneity on the Persistence of Predator–Prey Interactions. *J. Popul. Biol.* **8**, *346*

Holling C. S. (1965) The Functional Response of Predators to Prey Density and its Role in Mimicry and Population Regulation. *Mem. Ent. Soc. Can.* **45**, *43*

Holling C. S. (1966) The Strategy of Building Models of Complex Ecological Systems. In Watt K. E. F. (ed.) (1966) Systems Analysis in Ecology. *Academic Press, London*

Huffacker C. B. (1958) Experimental Studies on Predation: Dispersion Factors and Predator–Prey Oscillations. *Hilgardia* **27**, *343*

Huffacker C. B., Shea K. P., Herman S. G. (1963) Experimental Studies on Predation. *Hilgardia* **34**, *305*

Krebs J. R., Davis N. B. (1987) An Introduction into Behavioural Ecology. *Blackwell Scientific Publishing, Oxford*

Likens G. E. (ed.) (1985) An Ecosystem Approach to Aquatic Ecology: Mirror Lake and its Environment. *Springer Verlag, New York*

Lotka A. J. (1925) Elements of Physical Biology. *Williams and Wilkins, Baltimore* (Reissued as: Elements of Mathematical Biology. *Dover, 1956*)

May R. M. (1973) Stability and Complexity in Model Ecosystems. *Princeton University Press, Princeton N. J.*

May R. M. (ed.) (1981) Theoretical Ecology: Principles and Applications. *Blackwell Scientific Publishing, Oxford*

Maynard–Smith J. (1982) Evolution and the Theory of Games. *Cambridge University Press, Cambridge*

Nisbet R. M., Gurney W. S. C. (1982) Modelling Fluctuating Populations. *John Wiley & Sons, Chichester*

Pella J. J., Tomlinson P. K. (1969) A Generalized Stock Production Model. *Bull. Inter–Amer. Trop. Tuna Comm.* **13**, *421*

Price G., Maynard–Smith J. (1973) The Logic of Animal Conflict.*Nature* **246**, *15*

Rosenzweig M.L. (1971) Paradox of Enrichment: Destabilization of Exploitation Ecosystems in Ecological Time. *Science* **171**, *385*

Rosenzweig M. L., MacArthur R. H. (1963) Graphical Representation and Stability Conditions for Predator–Prey Systems. *American Naturalist* **97**, *209*

Schaefer M. B. (1954) Some Aspects of the Dynamics of Populations Important to the Management of Commercial Marine Fisheries. *Bull. Inter–Amer. Trop. Tuna Comm.* **1**, *25*

Schaefer M. B. (1957) Some Considerations of Population Dynamics and Economics in Relation to the Management of Marine Fisheries. *J. Fish. Res. Board Can.* **36**, *859*

van Kampen N. G. (1983) Stochastic Processes in Physics and Chemistry. *North Holland, Amsterdam*

Volterra V. (1926) Variazioni e Fluttuazioni del Numero d'Individui in Specie Animali Conviventi. *Mem. Acad. Lincei.* **2**, *31*

Wolff W. F., Kerszberg M., Ruján P. (1988) unpublished notes

Biological Attractors, Transients, and Evolution

S.P. Blythe[1] *and T.K. Stokes*[2]

[1]Department of Pure and Applied Biology,
Imperial College London, Prince Consort Road, London, SW7 2BB, UK
[2]Ministry of Agriculture, Fisheries and Food,
Directorate of Fisheries Research, Pakefield Road, Lowestoft,
Suffolk, NR33 0HT, UK

ABSTRACT

Biological systems are usually undergoing transients, or are evolving, or both. Estimates of attractor dimension depend on trajectories lying "near" the attractor, so they may seem inappropriate for population biological data. We discuss this problem, and offer a few simple tests of the correlation dimension applied to transient and evolving systems.

INTRODUCTION

The existence of chaotic behaviour, and fractal attractors, has been demonstrated in a wide variety of physical and chemical systems, and is suspected in some ecological systems both on the basis of appearance, and because some of the mathematical models used to describe population dynamics can themselves produce chaos. There are profound consequences for the analysis of population data (eg seeking density-dependence) if real populations fluctuate chaotically /7/, so it is necessary that data analysis techniques be available which can identify chaos. The calculation of exponents and dimensions from time-series data in principle provides such a means of identification.

Some difficulties met with in using exponents and dimensions, associated with small sample size and the distorting effect of environmental noise, are well recognised (eg /1/), but there are others. We mention two here, and raise some open questions on the use of dimension calculation in population biology.

First there are what we may call transient effects, some aspects of which have been considered. Exponent and dimension calculations rely upon the observed trajectory being in the neighbourhood of the attractor (in the appropriate phase space), but in most experiments we follow the system from some set of initial conditions, for a finite period of time, with no guarantee that the attractor will be approached closely. This of course is not confined to biological systems: in any chaotic system the time taken for initial value effects to die away will be very large, and it has been shown (/3/, /7/) that the calculation of Lyaponov exponents will give erroneous results unless extremely long time-scales are used. It has also been shown (/13/) that the estimate of the Mandelbrot dimension of an ensemble of classical mechanical particles will approach its limiting (Brownian) value as $(1\text{-}at^{-1/2})$, ie will be severly underestimated unless time t is large. Less well recognised is the interaction between transient and noise, whereby a system nominally undergoing a damped oscillatory approach to equilibrium can, in the presence of noise of the appropriate spectrum, exhibit quasi-cycles (runs of noisy cycles separated by bursts of incoherent noise) /11/. Can the method of dimension calculation distinguish between quasi-cycles, noisy limit cycles, and random fluctuations? How much distortion does a non-chaotic transient impose on the dimension calculation?

The second potential source of difficulty for exponent or dimension calculation is one peculiar to, and characteristic of, biological systems: biological evolution. The response to selection pressures of individuals in a population may cause some of the demographic parameters of the population to change with time. The greater the response, the faster and/or greater will be the change in parameter values. Certainly in many laboratory experiments, where conditions are markedly different from those pertaining for the initial "wild stock," we could expect strong selective pressure, so that (if a population persists) we are likely to observe non-trival parameter changes /12/. Disease dynamics, where control measures continually apply new selective pressure (and mortality) on the disease organism, also provide a good example of an occasion where changing demographic parameters are to be expected (in fact the concept of a "moving equilibrium" where the parameter variation due to changing control measures causes the expected equilibrium value of the system to change with time is well known in epidemiology /5/). Even without strong selective pressure, in an isolated population

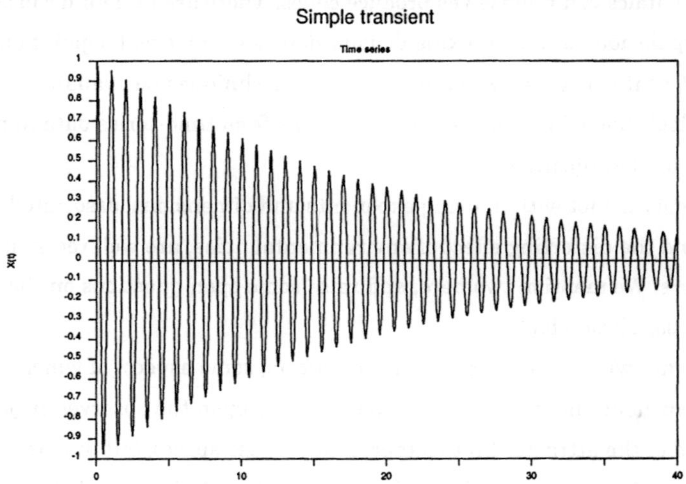

Fig 1. Simple oscillatory transient, solution to Eq (1) with ω=2 and b=0.051,Δt=0.05. There are 800 points shown.

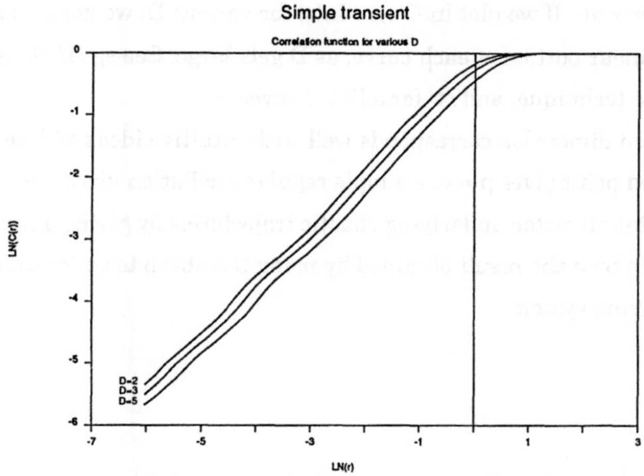

Fig 2. Behaviour of correlation function for various embedding dimensions D, calculated from the data set of Fig 1 with lag k=10. linear sections of curves are parallel.

genetic drift, the founder effect etc will probably give rise to some parameter change, and of course we must view viable mutations as the incorporation of certain random variations directly into the system. If parameters are time-varying, how are we to view the attractor of the system? Is the state of the system continuously moving through parameter space, so that the nature of the attractor (and hence the kind of behaviour observed) is continually changing, or are we observing an indefinite transient in a system where the parameter values might themselves be thought of as solutions to dynamic equations?

We regard the above as open questions, which we feel must be addressed before the methods of exponents and dimensions are to be used uncritically in population biology. In what follows, we shall make a very modest start at addressing the problem of transients and evolution. We shall calculate the correlation dimension for various time-series, and comment upon the results in the light of the above observations. The correlation dimension may be defined as

$$\rho = \lim_{r \downarrow 0} \lim_{D \uparrow \infty} \left\{ \frac{d \ln(C(r))}{\ln(r)} \right\}$$

where D is the embedding dimension, and C(r) is the correlation function for interpoint distance r between N points $\{Y_i\}$ in D-dimensional space, each Y_i being a vector of data points displaced by multiples of a fixed lag $k\Delta t$ (Δt the sample interval). We have

$$C(r) = N^{-2} \sum_{i=0}^{N-1} \sum_{j=0,\, i \neq j}^{N-1} H(r - |Y_i - Y_j|)$$

where $|Y_i - Y_j|$ is the distance between points Y_i and Y_j in D-dimensional space, and $H(x) = 0$ for x< 0, H(x) = 1, for x >0. If we plot ln(C(r)) v ln(r) for various D, we get ρ as the limit of the gradient of the linear portion of each curve, as D gets large. See eg /1/,/8/ for a more complete description of the technique, and /4/ for a list of caveats.

The correlation dimension corresponds well with intuitive ideas of dimensionality for simple attractors (a fixed point gives $\rho = 0$, a simple regular oscillation gives $\rho = 1$, etc), but can also identify the fractal attractor underlying chaotic trajectories by giving a non-integer value of ρ.

We will denote by σ the result obtained by using the above technique on data from a transient or evolving system.

THREE TESTS

We shall use both real and artificial data to illustrate the difficulties involved in calculating σ for transient or evolving systems.

Simple Deterministic Transient

For our first example, we shall use a thoroughly unbiological model to illustrate the effects of simple (rather than chaotic) transients on the calculation of σ. The second order ODE

$$x''(t) + 2b\,x'(t) + \omega^2 x(t) = 0 \;,\quad x(0)=1, x'(0)=0, \qquad (1)$$

has damped oscillatory solutions tending to x=0, for $\omega^2 > b^2$. Fig 1 shows an example with $\omega = 2$ and $b = 0.051$, giving 40 cycles of decreasing amplitude for t up to 40 units.

Calculating C(r) for this transient, with a sample interval of $\Delta t = 0.05$, and $k = 10$, gives the

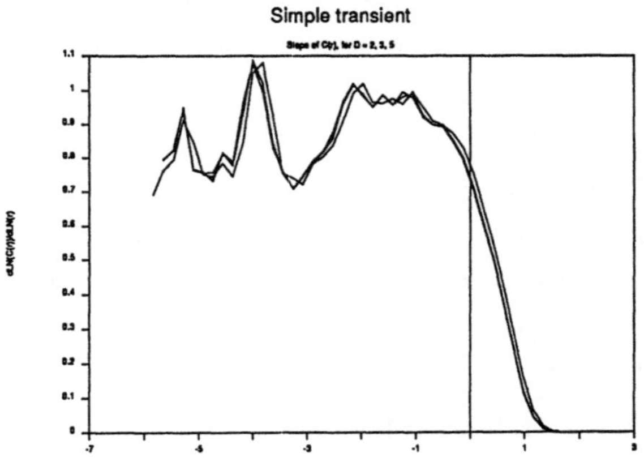

Fig 3. Gradients of correlation function curves shown in Fig 2. The horizontal scale has been adjusted so that the curves superimpose.

results shown in Fig 2, for D = 2, 3, 5. It is evident that the curves of ln(C(r)) v ln(r) are approximately parallel for much of the ln(r) range, and that the gradient should be a good estimate of σ.

Fig 3 shows the gradient of the curves in Fig 2 at each point, with the Fig 3 horizontal scale adjusted so that the results are superimposed. It is clear that the linear region of Fig 2 has a gradient of $\sigma \approx 0.9$, indicating that the correlation function C(r) cannot distinguish between a closed orbit (cycle) and a spiral (oscillatory transient).

Quasi-Cyclic Transient

Remaining with an artificial example for the sake of clarity, we use the non-autonomous extension of Eq (1)

$$x''(t) + 2b\, x'(t) + \omega^2 x(t) = \gamma(t)\,,\ x(0)=1,\ x'(0)=0, \tag{2}$$

with b and ω as before, and $\gamma(t)$ a source of Gaussian random noise with zero mean and given variance. Solutions x(t) can exhibit quite coherent quasi-cyclic behaviour during the transient. Fig 4 shows an example with $\gamma(t)$ having a variance of 20.0 (b and ω same as in Fig 1).

When we calculate C(r) for embedding dimension D = 2, 3, 5, 7, 9, 11, we see (Fig 5) no evidence for a limiting value to the linear sections of ln(C(r)) v ln(r). In fact, there is no sign (see Fig 6) of a linear section on the curve for high D at all, possibly due to "compression" of

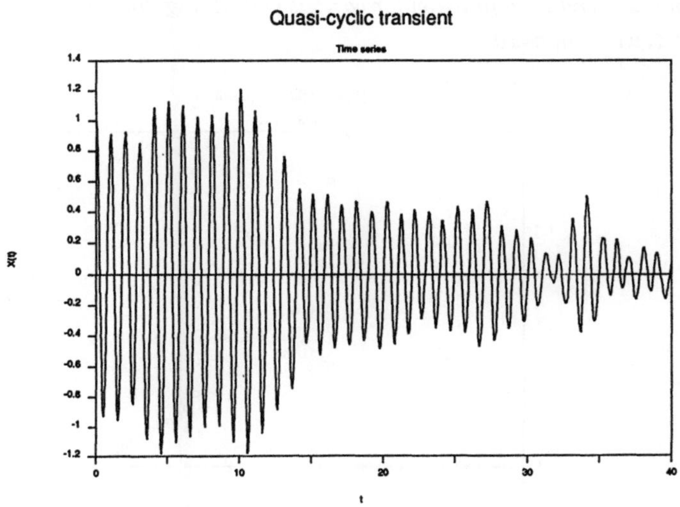

Fig 4. Solution to Eq (2), with ω=2, b=0.051, var(γ)=20.0,Δt=0.05. There are 800 points shown.

mid-scale r values as D gets large /1/. We could not say anything useful about σ from these results without knowing the true dimensionality ρ, and would probably conclude that we cannot distinguish the observed behaviour from random fluctuations on the basis of the correlation test (although spectral analysis of the x(t) signal would certainly pick out the dominant frequency).

An Evolving Population?

One of A.J. Nichnolson's experiments /9/ on the Australian sheep blowfly Lucilia cuprina (Wied.) produced a striking pattern of variation in adult fly numbers (see Fig 7), starting with

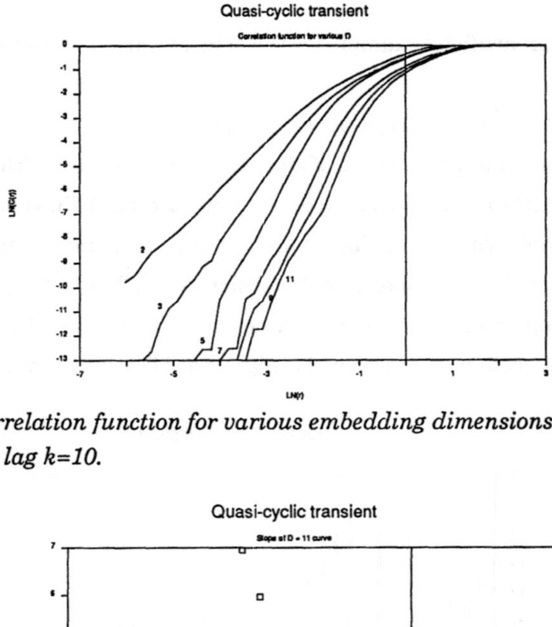

Fig 5. Behaviour of correlation function for various embedding dimensions D, calculated from the data of Fig 4, with lag k=10.

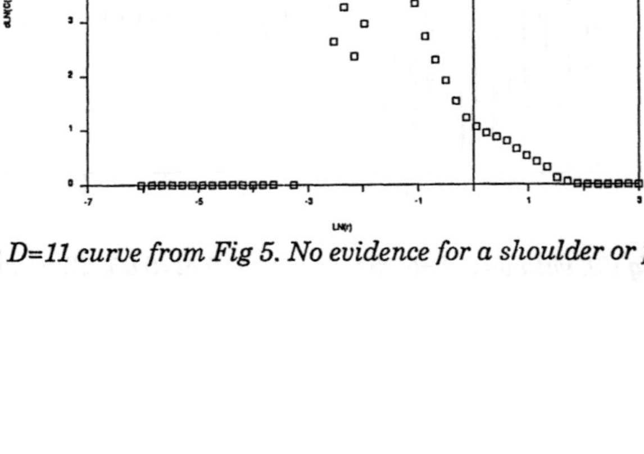

Fig 6. Gradient of the D=11 curve from Fig 5. No evidence for a shoulder or plateau.

irregular cyclic behaviour, and ending (after 2 years) with large amplitude, seemingly erratic fluctuations.

This behaviour has been claimed as chaotic, on the basis of appearance, and because a simple difference equation model of the population dynamics of one of Nicholson's populations (a larval food limited case) exhibits chaotic behaviour for the appropriate parameter values /4/. Nicholson himself provided a plausible alternative explanation, observing that female blowflies adapted to the harsh experimental conditions, so that individuals at the end of the experiment required much less protein to produce eggs than did the wild stock introduced at the beginning /10/. A detailed simulation of the experiment, with three demographic parameters varying with time, indicates that such adaptation could in fact produce the progressively changing pattern of behaviour shown in Fig 7 (/12/).

There is thus reasonably good evidence that the population is an evolving one, with (probably rather strong) selective pressure acting on females. What does C(r) look like? Fig 8 shows ln(C(r)) v ln(r) for D = 2, 3, 4, 5, 6, 7, 8, 9, 10. There is a region on each curve that is approximately linear, so we can estimate the gradient p there, and see how it varies with D. Fig 9 shows p against D, estimated by two different techniques.

The results in Fig 9 are tantalising: although no value of σ (a limit to the gradient p as D gets large) can be extracted directly, it is arguable that p is bounded by a value in the region of 5.0 - intriguingly /2/, also the number of independent variables in the simulation!

DISCUSSION AND CONCLUSION

The tests described above are in no way intended to be exhaustive or definitive. They are a

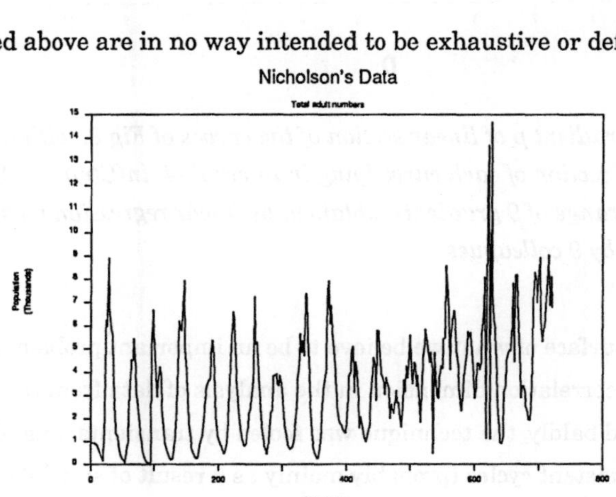

Fig 7. Adult blowfly numbers in Nicholson's experiment /8/. There are 360 points.

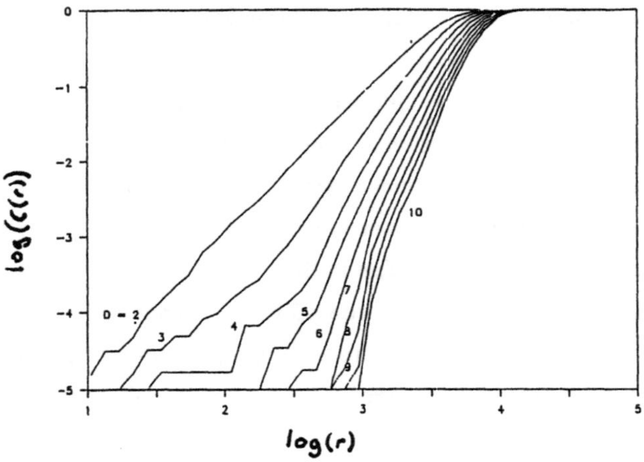

Fig 8. Behaviour of correlation function for various emdedding dimensions D, calculated from the data of Fig 7, with lag k=2.

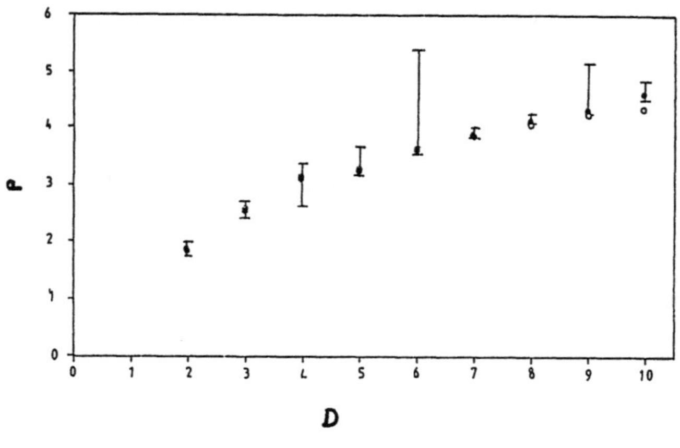

Fig 9. Change in gradient p of linear section of the curves of Fig 8. with increasing D. Open circles: gradient of section of each curve lying in interval -4 ln(C(r)) -1 . Solid circles with error bars: median and range of 9 gradients, obtained by linear regression on segments of each curve in Fig 8 identified by 9 colleagues.

scratching at the surface of what we believe to be an important problem in the uses of measures like the correlation dimension in the analysis of data from real biological populations. Stated baldly, the technique was fooled by transients, mistaking damped oscillations for persistent cycles (probably mainly as a result of small data-run N), and not distinguishing between a quasi-cyclic transient of high coherence and random noise; the technique did however produce encouraging, but anbiguous, results when applied to a putative evolving system.

We suspect that in the study of quasi-cycles, the most valuable use of a dimension calculation of the sort we consider is to provide *no* support for an hypothesis of chaos, thus freeing us to use more traditional techniques such as spectral analysis. However, the diagnostic power of σ is not yet fully developed.

The question of evolution seems to be a fruitful source of ideas. For example, if we consider an evolving system as moving from one region of parameter space to another, then it may be possible to calculate σ for various sections of the time-series, and observe changes in the attractor dimensions /1/ (although how to subdivide the time-series, and the overlap of time-lagged variables, are unresolved problems). Alternatively, if we assume that the parameter changes arise as a result of system behaviour ie they are really variables not parameters, then subdivision of the time-series may well be wrong. More detailed work is obviously required on these topics.

We hope that the simple tests and examples given here will stimulate some interest in the use of C(r) and the correlation dimension where problems of transients and evolution exist, particularly with a view to biological systems. They may also serve as a warning to the uncritical user of exponent and dimension methods on population data.

REFERENCES

/1/ Abraham, N.B., Albano, A.M., Das, B., Guzman, G. de, Yong, S., Gioggia, R.S., Pucciona, G.P. and Tredicce, J.R.: Calculating the dimension of attractors from small data sets. Phys. Lett. 114A (1986) 217-221.

/2/ Essex, C., Lookman, T. and Nerenberg, M.A.H.: The climate attractor over short time scales. Nature 326 (1987) 64-66

/3/ Gaumbado, J.M., and Tresser, C.: Some difficulties generated by small sinks in the numerical study of dynamical systems: two examples. Phys. Lett. 94A (1983) 412-414.

/4/ Grassberger, P.: Reply to Nicolis and Nicolis. Nature 326 (1987) 524

/5/ Hassell, M.P., Lawton, J.H. and May, R.M.: Patterns of dynamical behaviour in single species populations. J. Anim. Ecol. 45 (1976) 471-486

/6/ Hethcote, H.W. and Yorke, J.A.: Gonorrhea transmission dynamics and Control (Lect. Notes Biomath. 56). Springer-Verlag, Berlin 1984.

/7/ May, R.M.: Chaos and the dynamics of biological populations. to be published.

/8/ Nicolis, C. and Nicolis, G.: Is there a climatic attractor? Nature 311 (1986) 529-532.

/9/ Nicholson, A.J.: The self adjustment of populations to change. Cold Spring Harbour Symp. Quant. Biol. 22 (1957) 153-173.

/10/ Nicholson, A.J.: The role of population dynamics in natural selection. In: Evolution after Darwin Vol 1 - The evolution of life. Tax, S. (Ed.) University of Chicago Press, Chicago 1960, p.477-521.

/11/ Nisbet, R.M. and Gurney, W.S.C.: Modelling fluctuating populations. Wiley, Chichester 1982

/12/ Stokes, T.K., Gurney, W.S.C., Nisbet, R.M. and Blythe, S.P.: Parameter evolution in a laboratory insect population. Theor. Pop. Biol. (in the press)

/13/ Toxvaerd, S.: Fractal dimension of classical mechanical trajectories. Phys. Lett. 114A (1986) 159-160

Unstable Determinism in the Information Production Profile of an Epidemiological Time Series

F. Drepper

Programmgruppe Systemforschung und technologische Entwicklung, Kernforschungsanlage Jülich, P.O. Box 1913, D-5170 Jülich, Fed. Rep. of Germany

Keywords: time series analysis/ nonlinear dynamics/ generalized Kolmogorov-Sinai entropy/ local divergence rate/ unstable determinism/ chaotic attractors/ epidemiological time series/ measles

Abstract

The information production profile of a time series reveals the short term predictability as a function of certain dynamical patterns in the most recent past. It offers the possibility to distinguish deterministic phases of development from stochastic ones even in cases where the determinism is locally unstable or chaotic.
This new instrument for time series analysis of nonlinear dynamical systems has been applied to epidemiological data on measles cases in England and Wales. It is shown that there are phases of development within the two year cycle, where the deviations from periodicity are dominated by local instabilities of the underlying determinism.

Introduction

Like many other natural and social sciences ecology deals with open multicomponent systems whose individual components have autocatalytic or cooperative interactions resulting e.g. in reproduction or imitation. The cooperative behaviour of the individuals may lead to positive feedback and collective effects on the level of the whole system. Typical examples are "lock in" phenomena on the one hand and selfgenerated fluctuations on the other. The latter ones are at the center of interest of the present study. Such systems are characterized by the intuitive impression that the whole is more than the sum of its parts. There are simple cases, where the "more" can either be guessed by intuition or revealed by Fourier analysis. But possibly there are many other cases, where it is not evident a priori what the "more" is and how it can be made explicit and objective.

A preliminary definition of the "more" could be: Structure and functional dynamics which can be described by much less information than the one necessary to describe the dynamics of all individual components. The low information flow characterizing the functional motion of the whole system is a manifestation of low dimensional deterministic dynamical systems which govern the macroscopic behaviour. The deterministic laws of motion may differ for different phases of the behaviour. In contrast to the high dimensional (micro)dynamics on the individual level, the low dimensional one is called macrodynamics.

Typical examples, where neither the intuitive approach nore the one by Fourier analysis is able to distinguish the low dimensional dynamics from (high dimensional) noise processes are the so called chaotic systems, which show sensitive dependence on initial conditions. Particular cases are the phenomenon of turbulence in hydrodynamics (Brandstätter et al. 1983) and the phenomenon of speculation in economics (Drepper 1988). As has been shown by Schaffer (1987) this applies also to some childhood diseases in some large North American cities. Many natural and social sciences are in a phase of paradigmatic change concerning their view of the origin of fluctuations in such systems. According to the old point of view, these fluctuations are mainly due to external noise sources, whereas the new view considers these fluctuations to be endogenous, resulting from instabilities of the underlying macrodynamics.

The idea that a low information flow contained in an empirical time series implies low dimensional deterministic macrodynamics is a concept, which can be made operational by new methods of time series analysis originating from statistical physics and information theory, in particular from turbulence research. The central point is that certain characteristic features of appropriate information quantities allow to distinguish between random (stochastic) motion on the one hand and low dimensional deterministic motion on the other. In the deterministic case the methods allow the additional discrimination between stable and unstable dynamics. In stable deterministic systems nearby trajectories stay close together during the dynamics whereas they diverge in unstable ones. The special feature of the new instrument of time series analysis to be presented here is that both distinctions can be made separately for different regions of the state space.

Apart from the new inductive methods of time series analysis, statistical physics has also provided a general deductive statement about the long time behaviour of the macrodynamics of open multicomponent systems. For reasons similar to those leading to the second law of thermodynamics for closed systems, there are also restrictions to the long time behaviour of open systems, whose system size remains limited. If such systems have a deterministic macrodynamics (due to the large number of their individual components) this dynamics has to be dissipative. That means: the subset of states visited during the dynamics in the long time limit is much smaller than the set of all possible initial states. The asymptotic set, to which the system trajectories retreat after an initial transitory phase, is called limit set or attractor.

Equilibrium systems are special dissipative systems with (fixed) point attractors. In contrast to closed systems, open ones may have limit sets which are not isolated points. The best known examples of nonequilibrium attractors are limit cycles, i.e. periodically closed orbits. Since the pathbreaking work of Lorenz (1963) on the one hand and Ruelle and Takens (1971) on the other, the scientific interest has also been caught by aperiodic so called chaotic attractors, which are characterized by an exponential divergence of nearby trajectories (on time average). Whereas Lorenz was the first to discover a strange (or chaotic) attractor in a simplified hydrodynamical system, Ruelle and Takens showed under relatively general assumptions that for open systems, which are operating sufficiently far away from stable equilibria and whose macrodynamics has more than 3 degrees of freedom, chaotic attractors are the normal case.

The analysis of model systems with noise sources shows that it may be relatively difficult to distinguish between chaotic attractors on the one hand and periodic or quasi-periodic ones on the other. The characteristic feature of chaotic systems is that the long time avarage of the divergence (Lyapunov exponent) is positive. The determination of the Lyapunov exponent is known to be difficult for systems with noise sources whose dynamics is only known from a finite number of observations. For such systems it turns out to be easier to discriminate between attractors, which are stable everywhere and those, which have local instabilities (divergence of trajectories) somewhere on the attractor.

For the epidemiological example to be presented here, the last distinction is of special interest. Childhood diseases like measles tend to occur in epidemics, which often appear at highly regular intervals. On the other hand it has been noticed that the macroscopic dynamical pattern may substantially differ from period to period. The first feature has led to an interpretation of the epidemiological dynamics as noisy limit cycles (Anderson et al. 1984), whereas the second feature has led to the suggestion that the macrodynamics of childhood diseases might be dominated by chaotic attractors (Schaffer 1987).

The purpose of the present study is twofold. In the first part the information production profile is introduced as a new instrument of time series analysis. In the second part this new instrument is applied to the analysis of the epidemiological dynamics of the measles incidence in England and Wales. In this case the main problem is not to decide, whether there is a deterministic nonequilibrium attractor underlying the dynamics or not. However, the information production profile will turn out to be a useful tool to analyse the subtle interplay between unstable nonlinear determinism and noise sources in generating uncertainty of prognosis. In particular it can help to discriminate regions in state space, where the "attracting force" gets weak so that the noise dominates the motion, from situations, where the nonequilibrium attractor plays an active role in the creation of uncertainty due to local instabilities of the deterministic macrodynmamics. The main result of the analysis will be that the deviations from periodicity are mainly due to instabilities of the underlying deterministic dynamical system.

The Information Production Profile

For the identification and characterization of nonequilibrium attractors in nonlinear dynamical systems, one can use two different properties of such dynamics. The first one is the order of the attracting set of states, which carries the dynamics in the long time limit (Grassberger and Procaccia 1983a). This (possibly fractal) order leads to a characteristic scaling behaviour of the average information obtained by a single observation of the state of the system. The scaling exponent describing the increase of information, when the resolution of observation is increased is called the information dimension.

A second group of analytical methods aims at the disclosure of a characteristic order in the sequence, with which the points are visited in state space (Grassberger and Proccaccia 1983b). The starting point of the second type of analysis are information quantities to characterize whole sections of trajectories in state space, more specifically the increase of information, when the length of these sections is increased. For dynamic processes with autocorrelation (e.g. due to a deterministic law of motion), the information of a trajectory does not increase proportional to the trajectory length. This is equivalent to the effect of redundancies in symbol sequences. The more the knowledge about a dynamic process extends into the past, the less is the uncertainty of a one step prognosis. This leads to the fact that the increase of information for a one step extension of a trajectory is a monotonously decreasing (non increasing) function of the trajectory lenght. One of the characteristic features of deterministic dynamics will be shown to be the fact that this increase of information (production of information) is independent of the resolution of observation. For noise processes on the other hand, there is a monotonous increase of the information (uncertainty) production with increased resolution.

It is a common feature of all these methods of time series analysis that they rely upon the observation of a complete set of state variables of the dynamical system. This would be a major drawback in studies of empirical systems whose dynamics is not known a priori. In this situation, one can make use of the fact, that in dynamical systems with a finite number of state variables the information on the momentary value of all state variables can be replaced by information on the recent history of a part of the variables - in the extreme case even by information on the history of a single variable. More precisely: Taking f equidistantly delayed values of a single variable of a d-dimensional dynamical system as coordinates of an f-dimensional space, one generates a trajectory in an f-dimensional space (Drepper 1987). If the dimension of this auxillary state space (embedding space) is high enough (and the delay time not too small), the trajectory will not intersect itself like the one in the original space. That means that both trajectories are topologically equivalent. (Wittney's embedding theorem garanties $f \leq 2d+1$.) Therefore topological invariants like the information dimension, the avarage information production rate (Kolmogorov-Sinai entropy) and the Lyapunov exponent have the same value in both spaces (Takens 1981, Packard et al. 1980). In particular it is possible to identify an attractor in the auxillary state space as the limit set of a deterministic dynamical system, (or one can decide whether an attractor is chaotic or not). The appropriate dimension of the embedding space does not have to be known prior to the time series analysis, but can be determined via the invariance property of one of the topological invariants. The minimal embedding dimension f is an upper bound to the number of state variables and the information dimension a lower bound.

Specializing to the case of a single observable, the time series analysis to be described in the following will refer to a trajectory $\{\vec{X}_i \mid i = (f-1)\tau + 1, \ldots, N\}$ in an f-dimensional embedding space, which is spanned by vectors

$$\vec{X}_i = \{X_i^1, X_i^2, \ldots, X_i^f\}, \tag{1a}$$

the coordinates of which are defined as

$$X_i^\alpha = Y_{i-(f-\alpha)\tau} \qquad for \qquad \alpha = 1, 2, \ldots, f. \tag{1b}$$

$\{Y_n \mid n = 1, \ldots, N\}$ is the original scalar time series (equidistant in time) and τ the delay time given as an integer multiple of the elementary time intervall.

There have been numerous approaches to estimate the average information production rate of empirical systems. For systems with continuous state variables the information obtained from an observation of a state is dependent on the resolution of observation. To define information quantities for such systems, one needs something like an abstract measuring apparatus, which defines the accuracy of observation. The usual way to define the resolution is via a fixed cell partitioning of the state space. (The abstract measuring device can only distinguish states belonging to different cells.) For practical reasons a different type of abstract device will be used here. It is assumed to allow for all pairs of points in state space the discrimination whether their distance is smaller than a given radius r or not. This leads to an easily implementable estimate of the avarage information production rate H for empirical systems, which is described in more detail by Termonia (1984), Cohen und Procaccia (1985) and Grassberger (1986a). We expect H to be a function of the resolution, measured e.g. as 1/r, and of the length s of the time intervall for which the history is known at each time step. For a finite length N of the sample time series we further expect a dependency on N.

$$H(r,s,N) = \frac{1}{N-(f-1)\tau-s} \sum_{i=(f-1)\tau+s}^{N-1} h(r,i,s,N) \tag{2a}$$

with

$$h(r,i,s,N) = -\log\frac{f(r,i+1,s+1,N)}{f(r,i,s,N)}, \tag{2b}$$

$$f(r,i,s,N) = \sum_{\substack{j=(f-1)\tau+s \\ j\neq i}}^{N} \Theta(r - Max\{|\vec{X}_{j-s+1}-\vec{X}_{i-s+1}|, \dots, |\vec{X}_j-\vec{X}_i|\}). \tag{2c}$$

$$\Theta(x) = \begin{Bmatrix} 0 \\ 1 \end{Bmatrix} \text{ für } \begin{Bmatrix} x \le 0 \\ x > 0 \end{Bmatrix}.$$

$\Theta(x)$ represents the so called Heaviside step function. In combination with the summation in (2c) it serves to count the frequency, with which its argument is greater than zero. f(r,i,s,N) therefore indicates the frequency with which the distances of all s consecutive points of a trajectory section $\{\vec{X}_{j-s+1}, \dots, \vec{X}_j\}$ to the corresponding s points of a given reference section, are smaller than r. The index i refers to the end point of the reference section $\{\vec{X}_{i-s+1}, \dots, \vec{X}_i\}$.

If the average spatial distance of points adjacent in time is small compared to the overall extension of the attractor, then the condition j ≠ i in (2c) must be replaced by $j \neq \{i-v, \dots, i+v\}$ (Grassberger 1986b, Theiler). In this case, $v > 0$ must be selected in such a way that the average of $|\vec{X}_j - \vec{X}_{j-v-1}|$ is of the order of magnitude of the extension of the attractor. The limit $s \to \infty$, $r \to 0$, $N \to \infty$ of H(r,s,N) represents the so-called Kolmogorov-Sinai entropy. The quotient in (2b) indicates the factor by which the number of adjacent trajectories decreases, when their length is extended by one step.

Equation (2a) represents an average over time. h(r,i,s,n) therefore may be interpreted as the information production rate at time i. (2a) can also be regarded as a spatial average with a weight function corresponding to the relative density of trajectories on the attractor. It is a typical feature of dynamical systems to have an uneven spacial distribution of the information production due to local instabilities (hyperbolic fixed points) of the deterministic laws of motion. Analogously to equation (2a) local information production rates for given regions of the state space can be defined. A relatively simple way to parameterize different sections of the attractor is to use the time relative to the begin of a typical cycle as a parameter. The begin of the cycle can e.g. be defined by a Poincaré plane cutting through the center of the attractor. The relationship between the relative time and the corresponding region in the state space is given by the motion on the attractor. To get narrower sections one may use a Poincaré window (Drepper 1986,1988). For a visualization of this relationship graphical methods are helpful, for which it may be necessary to reduce the dimension of the embedding space. (This restriction however, does not apply to the estimation of information production rates.) In the special case of periodic dynamics, the relative time t corresponds to the phase within a period of lenght T. Dropping the possibility to use a Poincaré window, the different regions in state space can be defined by index sets

$$I_t = \{ i \mid i \bmod T = t \} \qquad \textit{for} \qquad t = 0, 1, \dots, T-1 . \tag{3a}$$

Restricting the average in (2a) to the index set I_t (which is assumed to have $N(I_t)$ elements), we obtain the local information production rate H(r,t,s,N) of the section of the attractor represented by the point set $S_t = \{ \vec{X}_i \mid i \in I_t \}$ as

$$H(r, t, s, N) = \frac{1}{N(I_t)} \sum_{i \in I_t} h(r, i, s, N) \tag{3b}$$

When discussed as function of r and t, (3b) will be called an information production profile. To derive its characteristic behaviour in the case of deterministic dynamics, we consider the limiting case of a very long time series $(N \to \infty)$ and hight resolution of observation $(r \to 0)$. For large N the relative frequencies f in (2c) may be replaced by joined probabilities. Their quotient may then be represented as a conditional probability.

$$\begin{aligned} h(r, i, s) &\equiv \lim\ h(r, i, s, N) \\ &= -\log\ P(\vec{X}_{j+1} \in B(r, \vec{X}_{i+1}) \mid \vec{X}_j \in B(r, \vec{X}_i), \dots, \vec{X}_{j-s+1} \in B(r, \vec{X}_{i-s+1})) \end{aligned} \tag{4a}$$

Equation (4a) is the conditional probability for the end point $\vec{X}_{j+1}$ of some trajectory section to fall inside the sphere $B(r, \vec{X}_{i+1}) \equiv \{ \vec{X} \mid |\vec{X} - \vec{X}_{i+1}| \leq r \}$ with radius r about the center $\vec{X}_{i+1}$ if the s preceding points have already fulfilled corresponding conditions. For deterministic systems (and for Markow processes), the conditional probability can be simplified as

$$h(r, i) = -\log\ P(\vec{X}_{j+1} \in B(r, \vec{X}_{i+1}) \mid \vec{X}_j \in B(r, \vec{X}_i)). \tag{4b}$$

Without loss of generality, we may assume that the dynamics during each time interval can be described by a discrete map. For the sake of simplicity, we start with a one-dimensional dynamic system, which is assumed to be defined by the map

$$X_{i+1} = f(X_i)$$

and the initial state X_1 . In order to calculate the conditional probability (4b), it is useful to distinguish two cases at each point in time discribed by the index i.

1. The absolute value of the slope of the map at the point X_i is smaller than 1, i.e. $| f'(X_i) | \leq 1$. In this case all trajectories consistent with the initial condition $| X_j - X_i | \leq r$ also fulfil the condition $| X_{j+1} - X_{i+1} | \leq r$. The local information production rate thus disappears: h(r,i) = 0.

2. The absolute value of the slope of the map at X_i is greater than 1, i.e. $| f'(X_i) | > 1$. In this case the fraction of the trajectories fulfilling the condition $| X_{j+1} - X_{i+1} | \leq r$ is inversely proportional to the absolute value of the slope at the center of the initial interval.

We therefore obtain for the one-dimensional case:

$$h(r,i) \underset{r \to 0}{\simeq} Max(0, \lambda(X_i)) \qquad \text{with} \qquad \lambda(X_i) \equiv \log |f'(X_i)| \tag{5}$$

In transferring this to several dimensions we can assume that in the limit r → 0 there is always a complete set of spacial directions (k = 1,...,f) for which the information production takes place independently, i.e. additively. Further it is important that a potential gain in the accuracy of prediction for one direction cannot compensate for a positive information (uncertainty) production in another direction. If we denote the information production in the k-th direction as $\lambda_k(\vec{X}_i)$, we obtain in generalization of (5)

$$h(r,i) \underset{r \to 0}{\simeq} \sum_{\substack{k=1 \\ \lambda_k(\vec{X}_i)>0}}^{f} \lambda_k(\vec{X}_i). \tag{6}$$

We see that there is a relatively close connection between the local information production rate, the conditional probability of prognosis and the local divergence rate of adjacent trajectories.

For the interpretation of the information production profile the important point is that for deterministic dynamics the information production rate is independent of the resolution of observation 1/r. This is in contrast to noise processes, for which there is a monotonous increase of the information production. Under the simplifying assumption that all trajectories, which can be distinguished, occur at the same frequency, the information obtained by an observation of a trajectory is equal to the logarithm of the number of different possible trajectories (Shaw 1981). h(r,i) = 0 for deterministic systems means that the number of trajectories which can be resolved scales like the number of different initial states of the trajectories. For noise processes on the other hand it is evident that the number of different trajectories increases faster with increased resolution than the number of initial states (Drepper 1986). It can be shown formally that this leads to an increase of h(r,i) proportional to log 1/r and that the slope reflects properties of the autocorrelation of the noise.

Summarizing we obtain the following result for the interpretation of the information profile.

- For pure noise processes, the information production rate increases linearly with the logarithm of the resolution of observation, whereas

- for deterministic motion the information production rate becomes independent of the resolution, if the accuracy of observation is high enough.

  - a stationary information production at level 0 indicates a stable (laminar) or strong determinism.

  - a stationary information production at a finite level indicates a chaotic (turbulent) or weak determinism.

The question now arises whether a determinism can be identified and characterized if the motion to be analysed does not follow a strict determinism but rather proceeds as a random process in the basin of attraction of a nonequilibrium attractor. If we assume that the stochastic process does not frequently leave the attractor by more than the distance δ and that this distance δ is small compared to the overall extension of the attractor then the following behaviour of the information production profile may be expected:

- For resolutions corresponding to radii r greater than δ, the information production rate can still be expected to be stationary (independent of r).
- However, for radii smaller than r, we may expect a behaviour as for a random process. This means that the local information production will increase linearly with the logarithm of 1/r.

We thus obtain an upper and a lower cut-off radius for the scaling range in which the above-mentioned information quantities may be interpreted in the sense of a deterministic attractor. The upper cut-off radius is defined by the global geometry of the attractor and the lower radius by the amplitude of the random fluctuations δ.

Both the upper and the lower cut-off radius generally vary in the different regions of the attractor. Thus even for systems with relatively weak noise sources there is no longer any range of r in which the information production rate averaged over the entire attractor (Kolmogorov-Sinai entropy) is independent of the resolution of observation. This problem may be avoided by estimating the information production separately for different sections of the attractor. When choosing the size of these sections one has to compromise between uniform cut-off radii of the scale behaviour on the one hand and the statistical significance of the estimated information quantities on the other.

It has become apparent recently that apart from the usual (Shannon) definition of the information H_S of a probability distribution $\{P_i \mid i = 1, \dots, M; \sum P_i = 1\}$

$$H_S = -\sum_{i=1}^{M} P_i \log P_i, \qquad (9a)$$

a whole class of generalized (Renyi) informations

$$H^{(q)} = -\frac{1}{q-1} \log \sum_{i=1}^{M} P_i P_i^{q-1} \qquad (9b)$$

is also suitable for defining characteristic quantities of dynamic processes on fractally ordered sets useful for identifying and characterizing nonequilibrium attractors. In the limit $q \rightarrow 1$, the Renyi information (9b) is converted into the Shannon information (9a) according to L'Hospital's rule.

The case q = 2 has special advantages for estimating information quantities of empirical systems (Grassberger 1986a). Whereas the Shannon information can be interpreted as the average logarithm of probabilities, the information $H^{(2)}$ can be regarded as the logaritm of average probabilities. With finite samples the problem arises that the relative frequencies can only be considered to be good estimates of the corresponding probabilities, if the frequencies are significantly greater than 1. This restriction defines a lower limit to r, below which the empirically determined estimates may no longer be interpreted as informations. In the case q = 2, the information estimates have a good statistics, since they are based on arithmetic averages of the frequences. It can be shown both for the case of deterministic dynamics as well as for Brownian motion that the information production rates of the order q = 2 have the same asymptotic behaviour for $r \rightarrow 0$ as the corresponding informations for $q = 1$. The information profile for q = 2 can therefore be interpreted in the same way as the one for $q = 1$.

The information production profile of the Renyi order q = 2 can be estimated in analogy to equation (3b). It shows the locally averaged information production rates of order 2 as a function of both the relative time t within a typical cycle as well as of the logarithm of the observation accuracy 1/r. The time t = 0 refers either to the passage of the reference trajectories through a Poincaré window or - in the case of periodic dynamics - to the start of a new period. In order to get a further improvement of the statistics, the information production rates are averaged over a set $\{ s \mid s_1 \leq s \leq s_2 \}$ of different lengths of trajectory sections terminating all at the same end point. Here we use the second characteristic feature of deterministic dynamics that the uncertainty of prognosis is independent of the length of the known history (s.a. equation 4b). To allow also for cases where the embedding dimension is not high enough, the minimal length s_1 can be chosen greater than one.

$$H^{(2)}(r,t,N) = \frac{1}{s_2 - s_1 + 1} \sum_{s=s_1}^{s_2} \log \frac{\sum_{i \in I_0} f(r, i+t, s, N)}{\sum_{i \in I_0} f(r, i+t+1, s+1, N)} \tag{10}$$

The average of the local information production rates $H^{(2)}$ can be described as a generalized Kolmogorov-Sinai entropy and is a topological invariant too. This is approximately true also for an average along a typical cycle on the attractor. Although the local information production rate is not a topological invariant (since it is e.g. dependent on the time delay τ in (1c)), the information production profile of the motion in the auxiliary state space can nevertheless be interpreted qualitatively. Apart from the possibility of identifying and characterizing determinism, the information profile gives rise to a relatively sensitive method of validating dynamic models of unstable or chaotic attractors.

Analysis of epidemiological data on the incidence of measles

The input to the present analysis is the time series of the logarithm of the number of measles cases in England and Wales during 1092 weaks in the time interval from the first week in 1948 up to the last week in 1969 as shown in fig. 1. The end of the time interval corresponds roughly to the introduction of public vaccinations which changed the dynamical pattern considerably. The time series was first analysed by Anderson et al. (1984) by using spectral methods. One of their findings was a relatively strong Fourier component at a period of two years, which led them to the interpretation of the dynamics as a noisy limit cycle.

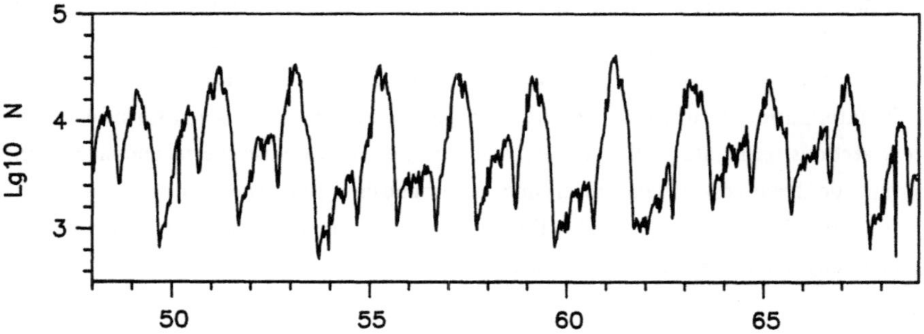

Fig. 1 shows the logarithm of the number of measles cases in England and Wales as obtained by Grenfell, Dep. of Biology, University of Sheffield.

There is no doubt that the system has a nonequilibrium attractor, which governs the fluctuations of the system, but the question remains, whether the deviations from periodicity are only due to noise or whether the determinism underlying the dynamics has instabilities. In a first step towards an answer to this question a 3 dimensional reconstruction of the nonequilibrium attractor in a state space generated by delayed coordinates is displayed graphically. As can be seen from fig. 1, there are two different time scales, which are relevant for the dynamics of this epidemiological system: a smaller one of about 10 - 15 weeks and a larger one of about 50 - 100 weeks. Fig. 2 shows a reconstruction with a delay time τ = 22, which turns out to be a minimal value for a resonable resolution of the full two year cycle. The trajectory displayed in fig. 2 corresponds to 9 full cycles starting at the 27th week in 1949. For better visibility a gentle spline smoothing has been applied (for this figure only).

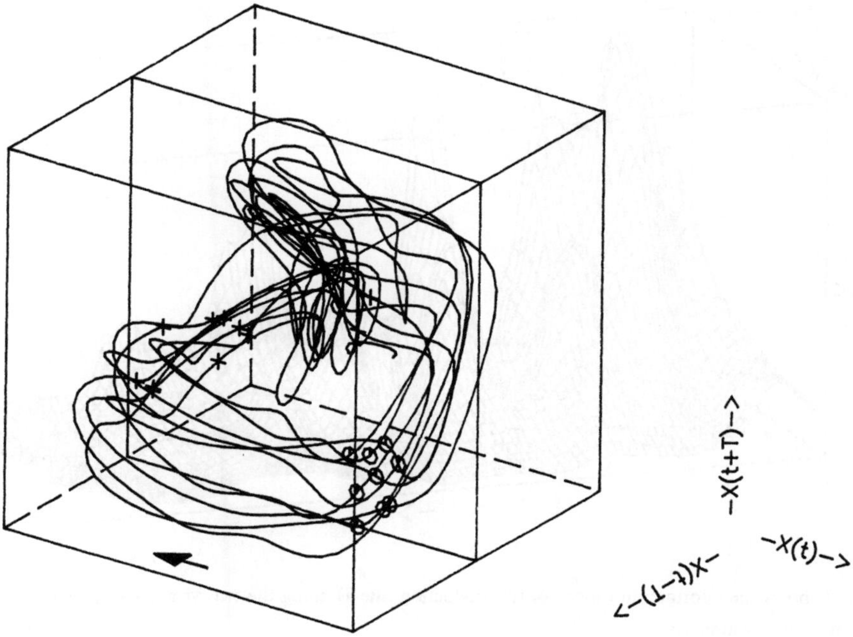

Fig. 2 shows the attractor in the auxiliary state space reconstructed from the time series in fig. 1 by using delayed coordinates with a delay of 22 weeks and a gentle spline smoothing. For better spacial imagination, the attractor has been cut by the indicated Poincaré plane

From figures (1) and (2) one can see that there are phases within the two year cycle, where the deviations from periodicity are especially strong, but it turns out to be difficult to decide, whether these deviations result from an exceptionally strong noise source or from instabilities of the underlying attractor.

This question, however, can be answered from figures 3 and 4, which give the information production profile for the time series in fig. 1. Fig. 3 shows the short time predictability as a function of the resolution of observation of points in state space and as a function of the week within the two year cycle. The begin (t = -26) and end of the cycle (t = 79) correspond to the 27th week of a year with an odd number. Fig 4 shows a cut through the surface in fig. 3 at an intermediate level of resolution (where the determinism becomes visible at most of the peaks). The two relevant time scales are accounted for by taking an embedding dimension f = 14 with a delay time τ = 1 on the one hand and a minimal trajectory length $s_1 = 52$ weeks. The maximal lenght s_2 was set to 3 years.

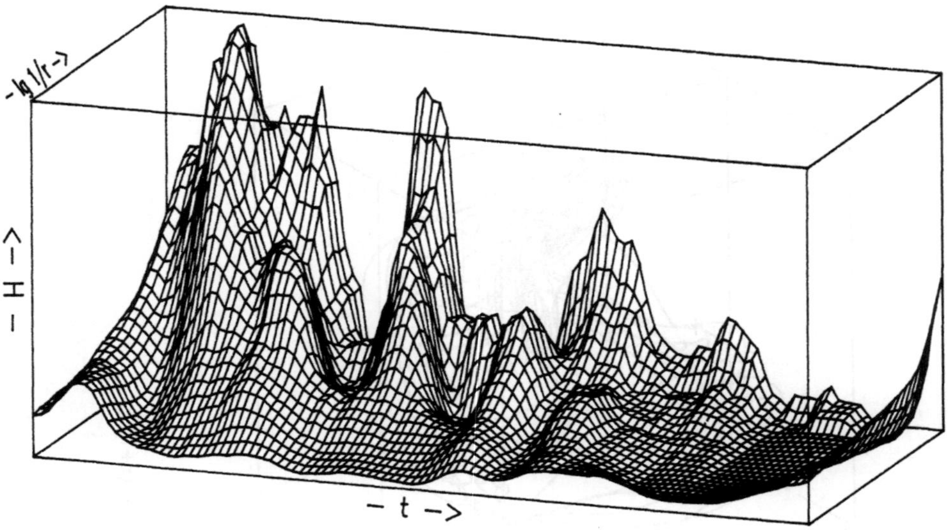

Fig. 3 shows the information (uncertainty) production rate H along the two year cycle as function of the resolution of observation 1/r.

To a nonspecialist in epidemiology two main features of the system under study can be revealed by an inspection and interpretation of figures 3 and 4.

- As was already visible from fig. 2, there is a very uneven distribution of the short term predictability of the dynamics. There are bursts in the short term unpredictability mostly in connection with school- and kindergarten holidays. In the periods between these bursts the uncertainty of prognosis is very low.

- Most of these information production bursts have a ridge with a horizontal slope in the direction of increased resolution and should therefore be interpreted as phases of development with a deterministic divergence of trajectories due to an instability of the dynamics. As far as the encounter of the instabilities coincides with school holidays, these instabilities may be seen as exogenously generated due to sudden changes in the contact rates between children of the relevant age classes. Since this external forcing of the epidemiological system has a regular rhythm, the temporary instabilities can be detected as part of the deterministic system. The instability in autumn may be due to the yearly variation of the climatic conditions, which is known to be able to produce chaos in the SEIR model (Schaffer 1987).

The additional question, whether in the case of the measles in England and Wales the instabilities are sufficiently strong to generate a chaotic attractor, if one could take away the noise from the system, can not be decided on the basis of a single information production profile alone.

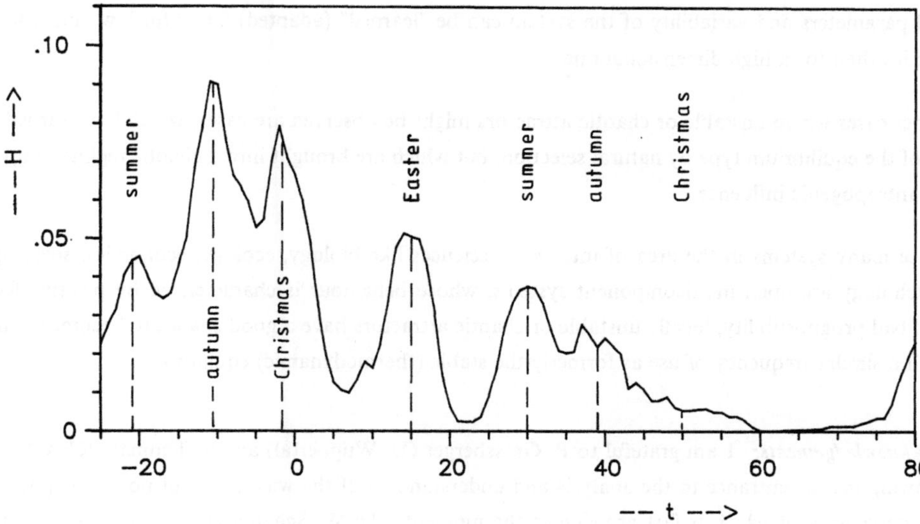

Fig. 4 shows a cut through the information production surface of fig. 3 at a level of resolution, where most of the deterministic instabilities are visible. t = 1 corresponds to the first week of a year with an even number.

A further analysis of the epidemiological system could be obtained by deductive methods in form of either macroscopic or microscopic simulation models. Since models with unstable determinism have a sensitive dependence of present states on the noise in the past, the simple comparison of model results with a single real history may not be a good means of verification of such models. In this situation the comparison of the information production profiles can be used as a more reliable instrument to check the validity of a model. An additional feature of the information production profile which could be checked in appropriate epidemiological models are the autocorrelation properties of the noise hidden in the asymptotic slope in the direction of increased resolution. (This deserves further theoretical studies).

Summarizing one can say: The information production profile seems to be a useful tool to identify and characterize low dimensional nonequilibrium attractors governing the macrodynamics of naturally generated multicomponent systems. This applies also to cases, where the determinism is unstable or even chaotic. In the last case the length of observation of the time interval during which the attractor has to keep its shape may turn out to be a severe limitation.

As far as it does not apply to the planetary system as a whole, natural selection can be seen to operate not only on the individual level but also on the system level. Without doubt there are many cases, where a stable equilibrium has the highest selective advantage on the system level. But there are probably other cases, where nonequilibrium attractors can be seen as means to generate a natural variability, which is optimally controlled by the system itself. (Note: Though chaotic systems are not accessible to long time prediction, they may well be accessible to closed (or open) loop control.) To establish the control function, low dimensional attractors have a selective advantage, because the relationship between control parameters and variability of the system can be "learned" (adapted) faster for low dimensional dynamics than for a high dimensional one.

Other cases where unstable or chaotic attractors might be observed are cases, which have turned out to be of the equilibrium type by natural selection, but which are brought into a chaotic regime of operation by antropogenic influence.

Since many systems in the area of interest of sciences like biology, ecology, economics, soziology and psychology are open multicomponent systems, whose behaviour is characterized by positive feedback and bad prognostibility, locally unstable or chaotic attractors have a good chance to become a paradigm with a similar frequency of use as formerly the stable (thermodynamic) equilibria.

Ackknowledgements: I am grateful to P. Grassberger (U. Wuppertal) and S. Thomae (KFA Jülich) for showing me the entrance to the analysis and understanding of the wast realm of nonlinear phenomena, the perception of which is fast growing at the moment. To W. Schaffer (U. of Arizona), G. Sugihara (Scripps, San Diego) and R. May (U. of Oxford) I owe the confidence that the new methods of time series analysis are also relevant for ecological systems. To B. Grenfell (U. of Sheffield) I am grateful for giving me access to the epidemiological time series and its interpretation.

Literatur

Anderson R. M., Grenfell B. T. and May R. M., J. Hyg., Camb. 93 (1984)
Brandstätter A. et al., Phys. Rev. Lett.51 (1983) 1442
Cohen A. and Procaccia I., Phys. Rev. A 31 (1985) 1872
Drepper F., Analyse Chaotischer Dynamik: Ein Makroskop für Vielkomponentensysteme, yearbook 1986 - 87 of the Kernforschungsanlage Jülich (GMBH), p.o. box 1913, D 5170 Jülich, FRG
Drepper F., Determinism in the Information Production Profile of a Stock Market Index, in Fritsch B. and Keller J.U. (ed.s), Dissipative Structures in Integrated Systems, series ed.: Forschungsstelle für gesellschaftliche Entwicklung , Mannheim, Nomos Verlag 1987
Grassberger P. (1986a), Estimating the Fractal Dimensions and Entropies of Strange Attractors, in Holden A.V. (ed.), Manchester Univ. Press 1986
Grassberger P. (1986b), Nature 323, (Okt. 1986) 609
Grassberger P. and Procaccia I., Physica 9 D (1983a) 189
Grassberger P. and Procaccia I., Phys. Rev A 28 (1983b) 2591
Kolmogorov A.N., Entropy per Unit of Time as a Metric Invariant of Automorphisms, Dokl. Akad. Nauk. SSSR 124 (1959) 754, Engl. summary in Math. Rev. 21, 2035
Lorenz E.N., Deterministic Nonperiodic Flow, J. Atmos. Sci. 20 (1963) 130
Packard N. et al., Phys. Rev. Lett. 45 (1980) 712
Ruelle D. and Takens F., On the Nature of Turbulence,
Commun. Math. Phys. 20 (1971) 167
Schaffer W. M., Olsen L. F., Truty G. L., Fulmer S. L. and Graser D. J., in Markus et al. (ed.s), From Chemical to Biological Organization, Springer, Berlin, 1987
Sinai Ya. G., On the Concept of Entropy of a Dynamical System,
Dokl. Akad. Nauk. SSSR 124 (1959) 768
Shaw R., Z. Naturforsch. 36a (1981) 80-112
Takens F., Lecture Notes in Math. 898, Springer, Heidelberg - New York, 1981
Termonia Y., Phys. Rev. A 29 (1984) 1612
Theiler J., Phys. Rev. A 34 (1986) 2427

Random Elements in a Population Model Based on Individual Development

N. van der Hoeven

Department of Biology, MT-TNO, P.O. Box 217, NL-2600 AE Delft, The Netherlands

INTRODUCTION

General models in population dynamics tend to neglect the fact that populations consist of individual animals (or plants). In these models populations are able to grow by an arbitrary amount and newly formed 'animals' are often immediately capable of reproduction. In reality new-born animals are smaller than adults, use less food and are not able to reproduce. Incorporating age and size structures in population models will make these models much more realistic, but will also severely limit their analysis. (See for a survey of recent work on the analysis of such models Metz & Diekmann, 1986). Other techniques have thus to be developed for the study of such models.

A first step in exploring the behaviour of populations with an age and size structure can be made using appropriate computer simulations. We have constructed a computer programme in which all individuals are followed separately in time. They can eat, grow, reproduce and die. The rate at which they do so will depend on their age and size.

Such a simulation method is very (computer) time consuming, but does not require difficult partial differential equations and complicated numerical procedures. An additional advantage of such an approach is that it is possible to incorporate all kinds of stochastic differences between the individuals in a natural way.

THE UNDERLYING INDIVIDUAL MODEL

The foundation for our models is the development of the waterflea Daphnia magna. In our laboratory, and in many others, the growth, reproduction, feeding behaviour, and other properties of this organism has been extensively studied. Kooijman (1986a) has developed a model based on physiological properties of the daphnids, fairly accuraty describing their development. We have extended this model to population level (Kooijman et al. in press). The parameter values used are in the same order of magnitude as our estimates from experiments with individual daphnids.

In making the extrapolation from an individual model to a population model we have assumed that the individuals only interact by depleting their common food supply.

The individual model on which the simulation was based makes the following assumptions:

- Individuals store the ingested food in a storage compartment.
- Individuals deplete their storage linearily per unit of weight (size-specific storage).
- Individuals need a certain amount of energy (food) for maintenance metabolism. This need is proportional to the body weight.
- For each unit of growth a certain amount of energy is necessary.
- The energy utilized is divided according to a fixed ratio between reproduction and differentiation on the one hand, and maintenance and growth on the other. Adults spend that fixed fraction on reproduction, juveniles use it in a juvenile specific way, for instance for differentiation.
- If the portion of energy allocated to maintenance and growth is not enough to cover the maintenance needs, part of the energy for reproduction is spent on maintenance.
- If the total energy utilization is not enough to cover the maintenance needs, the individual will die of starvation.
- The time of egg development and the energy investment per young can be calculated given the assumption that the young will have the same size-specific storage at birth as its mother had at egg formation, moreover the birth size of the young is fixed.
- The food uptake depends hyperbolically on the food concentration in the environment.
- The young are born in clutches.
- The eggs develop inside the mother.
- The animals reproduce parthenogenetically.
- The death rate as a consequence of senescence can be described by a Weibull distribution.

A mathematical derivation of this model has been given by Kooijman (1986a,b). The (scaled) model equations are given in the appendix.

METHODS USED IN THE SIMULATION

Computer simulations were based on the individual growth, reproduction and survival dynamics as described by Kooijman (1986a,b); each individual was followed separately. The individuals were assumed to produce their young in clutches. If the adult dies during pregnancy, the young also died. The death rate due to senescence was described by a Weibull distribution. At birth the time of death due to senescence was calculated. If the individual had not died of starvation before that time, it disappeared from the population at that moment.
The differential equations describing the behaviour of each individual were approximated using a fourth order Runga-Kutta method. The iteration step in most simula-

tions was 0.05 'days'. Since the food concentration seems to vary more severely than the individual properties, it was investigated whether the simulations improved if an iteration step ten times smaller was used for the approximation of the food concentration. Since this did not made a noticable difference, some of the simulation studies were performed with, and others without, a shorter step for the iteration of the food concentration.

In some of the simulations it was assumed that the ingestion rate of the individuals varied between individuals and with time. In cases where the ingestion rate varied with time, the new ingestion rate for the individual was calculated for each iteration step; the new value was the mean of the old ingestion rate of the individual and the individual specific ingestion rate. The individual specific ingestion rate was calculated for each individual at birth. If the ingestion rate varied, the coefficient of variation of the variance between and within the individuals was given. Sometimes the searching rate was varied instead of the maximum food uptake. Changes in the searching rate were expressed as changes in the saturation constant of the functional response (Holling, 1959).

The influence of a patchy environment was tested by dividing the environment into N patches. Several food regimes were explored. In the first place, the food inflow per patch was assumed to be equal. The patch in which an individual was located was recorded. Individuals could leave a patch at each iteration step. If they left their current patch, each patch (including the current patch) had an equal probability of being choosen. The leaving rate was identical for all patches.

The effect of an unequal food supply to the patches was examined for one specific distribution: 10 patches, one receiving 50% of the total food input, four each receiving 10%, and the remaining five patches each receiving 2%. In this case the effect of diffusion of the food, and of animals remaining longer in patches with a high food concentration were also investigated.

The equations describing growth, reproduction, death rate and spatial behaviour are given in the appendix.

THE EFFECT OF RANDOM ELEMENTS

Three different random elements in the stochastic behaviour of the individuals were investigated.

1. The effect of a temporal variance in the searching rate of the individuals.
2. The effect of random differences between the individuals in ingestion rate.
3. The effect of a simple patchy structure in the environment, and of random switching of individuals between these patches.

The effect of the temporal variance appeared to be small. The synchronization in sizes was slightly affected (Fig. 1). The oscillation cycle appeared to be fairly stable.

Random differences between the individuals broke down the size synchronization. If the variation coefficient became large (>0.2) the oscillation became less regular and the amplitude decreased considerably. Some examples are given in Fig. 2.
The effect of a patchy environment increased if the exchange rate decreased. A patchy environment always led to a less rigid size synchronization; it also led to a less regular oscillation if all patches get the same food supply and if the exchange probability per unit of time is of the order of 0.1 or less. Some examples are given in Fig. 3.

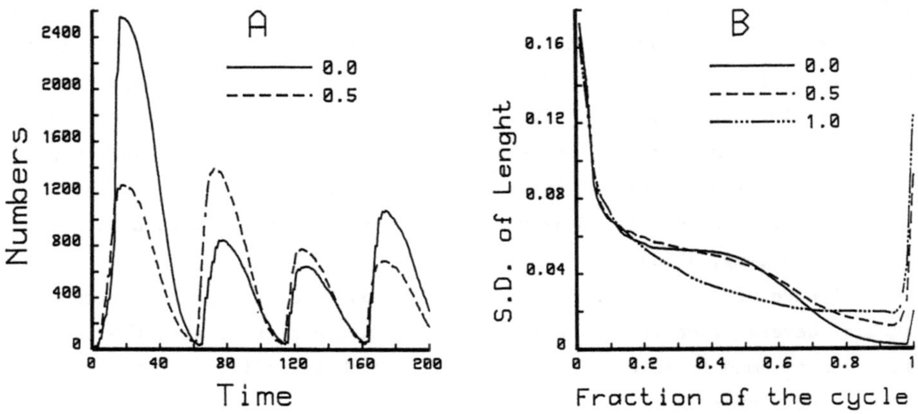

Fig. 1. a. The population numbers in time for no temporal variation and for a variation coefficient of the searching rate of 0.5.
b. The standard variation of the lengths of the simulated animals during an oscillation period. Coefficients of variation 0.0; 0.5; 1.0.

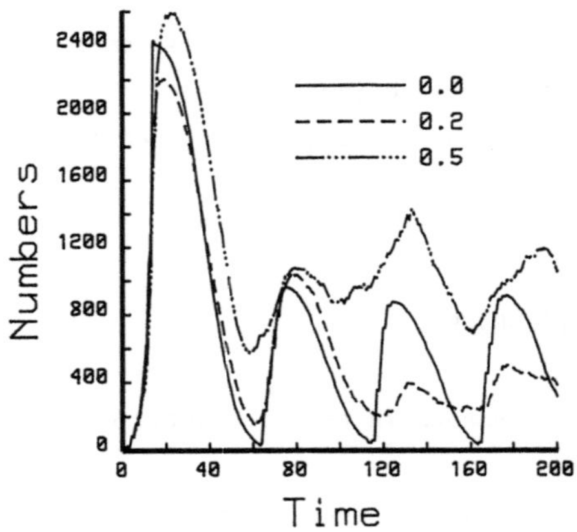

Fig. 2. The number of organisms for a variation coefficient between the individuals of 0.0, 0.2 and 0.5.

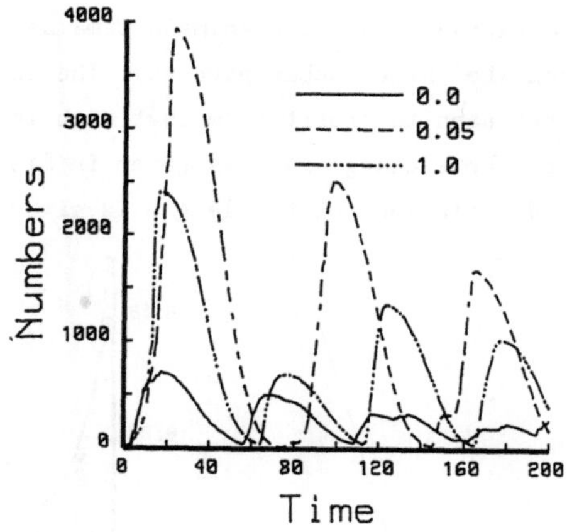

Fig. 3. The numbers of animals in a patchy environment (10 patches). The exchange rate between the patches is 0.00, 0.05 and 1.0.

The effect of a non-equal food supply to the patches was investigated for a limited number of cases. If the daphnids left their patches irrespective of the food concentration, and if no food was exchanged between the patches, the effect of the leaving rate of the daphnids became very strong. If that rate was small, the animals tended to remain in a patch for a long time. In the patch with the highest food input, they were able to grow faster and to produce more young, leading to a higher birth rate in that patch, and therefore to more animals in it. If the leaving rate was very high, animals spent on average one tenth of their life span in the patch with the highest food input. Since, in this case there is no relation between the patch in which the young are conceived and that in which they are born, and animals did not remain in a patch with a high food level, the average number of animals in each patch was the same. This led to an accumulation of food in the patch with the highest input, while the animals were not able to use that food for more then one tenth of their life span. The total number of animals remained smaller, while the total amount of food increased to a very high level.

Since it seems extremely unlikely that food would not diffuse between the patches in one way or another, one simulation in which food diffused, and a number of simulations in which the animals tended to remain longer in a patch if the food concentration in that patch is high, were run. If diffusion between the patches occured the effect of an unequal food supply to the patches disappeared to a large extent. The main effect was that far more animals died of hunger.

If the animals remained longer in a patch containing more food, on average more animals were in the patches with the higher food level, i.e. the patches with the higher food input.

The number of animals dying of hunger increased, especially if the leaving rate of a patch with a high food level was still relatively high. If the exchange rate increased, the mean food level also increased but the food was used less efficiently (beneath a certain minimum exchange rate the animals remained so long in a patch that food could also accumulate in the other patches). The animals left the patches with the high food level too soon to profit from that high food level. The mean food level as a function of the exchange rate is given in Fig. 4. Furthermore the oscillation of the standard deviations of the length is given for some exchange rates.

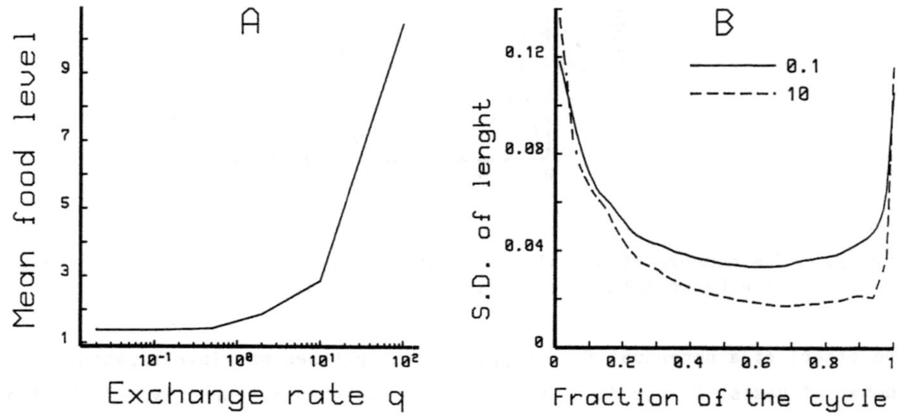

Fig. 4. a. The mean food level as a function of the exchange rate q between the 10 patches (unequal food distribution, the leaving rate for a patch with a high food level is less.
b. The standard variation of the lengths of the simulated animals during an oscillation period. Exchange rate 10 and 0.1.

EFFECTS OF THESE RANDOM ELEMENTS IN EXPERIMENTS

All these three random processes, and many more, will occur in a real Daphnid population. However, in a carefully executed experiment their effect may be kept quite small. The first random process has little effect anyhow. The coefficient of variance in the second process is related to the variance in the length of the daphnids. Experiments have shown that the variance in length in a daphnid clone is quite small (in the order of 0.02 to 0.05) (van Leeuwen et al., 1986). At that level of variance the effect on the population level is quite small.

If the food is administered homogeneously the third process only has a considerable effect if the daphnids tend to remain very long in one patch. However, in experimental environments which are not unusually large, daphnids will swim throughout the whole volume in a relatively short time.

Some slightly more realistic spatial models were also investigated: even in a carefully planned experiment it is inevitable that the algae will clot together. Furthermore, the algae can never be administered homogeneously, and the daphnids may be able to discriminate between better and worse searching sites.

A spatial structure in which algae are added at one point, diffuse slowly throughout the whole medium, attract each other slightly, and in which the daphnids can only move to the adjacent patches seems thus more realistic. Such a complicated spatial model was however, not investigated. A more simple spatial structure shows that the effect of diffusion of the food cells tend to counterbalance the effect of a nonhomogeneous food supply, although the food distribution remains extremely unequal.

DISCUSSION

An extensive computer simulation study on the effects on population level of a realistic model for individual development has been performed. It appears that a population of such individuals will strongly oscillate in numbers. The sizes of the individuals will synchronize.

The effect of random elements was investigated. It appears that all kinds of random effects (individual differences in the maximum attainable size, temporal variation in searching rate and random movement between different patches in the environment) mainly affect the extreme size synchronization of the individuals. In most cases the amplitude of the oscillation was slightly lessened by stochastic variation. The amplitude and regularity of the oscillation cycle were only severely affected in the case of considerable variation coefficients, or very slow movement of the individuals between the patches.

The results of the computer simulation have to be considered in the light of the species on which the individual model was based, namely the waterflea Daphnia magna. It is to be expected that at least in an laboratory experiment the variation coefficients of the modelled processes will be small for that species. Therefore random factors will not have an important effect. Only the size synchronization will not be as absolute as predicted by the purely deterministic model. Experiments should be performed to test whether a Daphnid population under conditions as in the simulation, behaves in a similar manner to the simulated population.

REFERENCES

Holling, C.S., 1959
Some characteristics of simple types of predation and parasitism.
Canad. Entomol. 91: 385-398.

Kooijman, S.A.L.M., 1986a
Population dynamic on basis of budgets.
In: J.A.J. Metz and O. Diekmann (eds.): The dynamics of physiologically structured populations. p. 266-297.
Springer Lecture Notes in Biomathematics.

Kooijman, S.A.L.M., 1986
What the hen can tell about her eggs: egg development on basis of budgets.
J. Math. Biol. 23: 163-185.

Kooijman, S.A.L.M., N. van der Hoeven and D.C. van der Werf (in press)
Population consequences of a physiological model for individuals.

Metz, J.A.J. and O. Diekmann, 1986
The dynamics of physiologically structured populations.
Springer Lecture Notes in Biomathematics.

Leeuwen, C.J. van, M. Rijkeboer and G. Niebeek, 1986
Population dynamics of Daphnia magna as modified by chronic bromide stress.
Hydrobiologia 133: 277-285.

APPENDIX
THE MATHEMATICAL EQUATIONS OF THE INDIVIDUAL MODEL

The derivation of the mathematical model used in this paper is the work of Kooijman (1986a,b). It is based on a combination of theoretical assumptions and observations on the waterflea Daphnia magna.
The lenght of the animals is denoted by $\tilde{L}$. Their size is assumed to be proportional to the cube of their length and their food intake to the square of their length. Animals can reach a maximum length, L_m, depending on their maximum assimilation rate $[\dot{A}_m]$, the costs of maintenance $\dot{\zeta}$, and the fraction of energy spend on growth and maintenance κ.

$$L_m = \kappa[\dot{A}_m]/\dot{\zeta}$$

The animals store their assimilated food. The amount of stored food per unit of body size is $[S]$. The maximum amount of stored food is $[S_m]$.
The equations will be given in the relative length $L = \tilde{L}/L_m$ and the relative storage per size $S = [S]/[S_m]$. The equation for S depends on the (scaled) food concentration in the environment, X.

$$dL/dt = (S-L)/3(b+aS) \text{ if } S>L \text{ and } 0 \text{ otherwise}$$

and

$$dS/dt = (f(X)-S)/aL$$

with $a = \kappa[S_m]/\dot{\zeta}$, and $b = \eta/\dot{\zeta}$, with η the cost for growth.
If $S<\kappa L$ the individual does not have enough energy for maintenance, and will die (Kooijman, 1986a).
Animals can only reproduce if their size L is larger than L_j, the size at which adulthood is reached. Their reproductive effort R is given by

$$dR/dt = (1-\kappa)L^2 \; S(L+b/a)/(b+aS) \text{ if } S>L$$

and

$$dR/dt = L^2 \; (S-\kappa L)/a \qquad \text{if } \kappa L<S\leq L$$

Young are born with length L_b. Their stored energy is identical to the stored energy of their mother at the moment of egg formation. The (scaled) energy per young S_0 can be calculated assuming that the young inside the egg follows the same growth equations as after hatching. An approximation of S_0 is

$$S_0 = L_b^{\,3} \; \{S_b \; (1-L_b/4S_b)^{-3} + 1.1b/a\}$$

The development time of the eggs T_b can be approximated by:

$$T_b = a \left[\begin{array}{l} 3/2\ S_0\ 1/4\ [1/2\ \ln\{(U^2 + U\sqrt{2} + 1)/(U^2 - U\sqrt{2} + 1)\} + \mathrm{arctg}\ \{U\sqrt{2}/1-U^2\}] \\ + L_b^{1/2}\ \ln\{1+b/a\} \end{array} \right]$$

with $U = (4S_b/L_b - 1)^{-1/4}$ (Kooijman, 1986b).
The death rate of the individuals is assumed to depend only on their age. It is described by a Weibull distribution.

$$F(t) = \exp(-(t/u)^c)$$

Each individual depletes the food supply with rate $f(X)L^2$, so the change in the food concentration is

$$dX/dt = I - f(X)\ L_i^2$$

with I the inflow rate of the food.
The functional response function $f(X) = X/(K + X)$. The food concentration X is a scaled value. The real food concentration $\tilde{X}$ is divided by the maximum food intake at the maximum length.
If the maximum food intake differs between the individuals $f_i(X) = \varepsilon_i\ X/(K + X)$, with ε_i a random (lognormal distributed) variable for each individual. If the searching rate of an individual has a temporal variation the saturation constant K_i of an individual at each moment is a random variable with as expectation the mean of the last searching rate and the (individual) mean searching rate. Sometimes the environment is divided into N patches. In the simple case all patches have the same food inflow and no spatial structure between the patches exists. All patches are quit with equal probability. The leaving rate is denoted by q. In the more complicated cases the inflow differs between the patches. Sometimes diffusion of the food occurs between the patches (in that case per day 10% of the food is divided among all patches). Sometimes the leaving rate is proportional to the inverse of the food concentration in the patch (leaving rate of patch i is q/X_i).
Unless otherwise stated the parameters have the following values:

a=3; b=2/3; K=3; L_b=0.16; L_j=0.5; I=10; u=30; c=3.

Measures of Spatio-Temporal Irregularity

A.V. Holden[1], *J. Brindley*[2], *and R.M. Everson*[2]

[1]Departments of Physiology, The University of Leeds, Leeds LS29JT, UK
[2]Applied Mathematical Studies, The University of Leeds, Leeds LS29JT, UK, and
Centre for Nonlinear Studies, The University of Leeds, Leeds LS29JT, UK

ABSTRACT

Methods for identifying and characterising periodicity, quasi-periodicity and chaos in experimental time series are well developed. Many experimental measures are samples from spatially extended processes, and measures for quantifying such spatial and spatio-temporal fields are, in contrast, poorly developed. A preliminary proposal is to mimic visual analysis of spatio-temporal patterns by measures that represent computations performed in the visual system.

INTRODUCTION

Although the fact that simple systems can generate complicated behaviour has been known since the time of Poincaré, the current interest in chaotic dynamics in the biological, chemical and physical sciences was to a large extent triggered by May's 1976 review article [1]. Here chaotic behaviour in the logistic map was used to illustrate how a very simple nonlinear difference equation can generate chaos. A naive interpretation of this map is that each successive iteration represents the annual breeding population of an insect species for which there is no overlap of successive generations. Although this quadratic mapping, $x_{n+1} = a x_n (1 - x_n)$, does illustrate a possible mechanism for irregular fluctuations in such a population, the values of a, $3.57 < a < 4$, that correspond to chaotic behaviour, are much higher than the corresponding maximum per capita rates of increase seen in actual insect systems with discrete non-overlapping generations [2].

This quantitative discrepancy may be avoided by interpreting the map as a Poincaré section: the procedure for identifying chaos is to reconstruct vector phase portraits embedded in a high dimensional Euclidean space from actual time series, say by the method of delays, and then to project down to a few significant directions using singular value decomposition [3,4]. The appearance of structure in the resulting phase portrait suggests an attractor, which may be identified as chaotic if it has a positive Lyapunov exponent. The attractor may be quantified by estimating the spectrum of Lyapunov exponents, its various fractal dimensions, entropies and informations [5].

These methods are appropriate for a closed system that may be represented by a system of ordinary differential equations: however, populations changes occur in spatially extensive, perhaps open systems. We are interested in quantifying spatio-temporal patterns and irregularity in modelled and experimental systems from a range of disciplines (fluid flow, neuronal structure and activity) and in this paper we outline some approaches to measure spatio-temporal irregularity that we are considering.

DATA FORMAT

The data we are interested in has a natural Euclidean spatial metric: perhaps a static structure in three spatial dimensions (dendritic arborisation of a neurone) or a planar spatio-temporal pattern. Some spatial data in ecodynamics does not have such a natural metric e.g. epidemiological records from different towns.

The planar spatio-temporal data may be thought of as a temporal sequence of digitized images: it may in fact be a digitized video-sequence, or a series of computer-generated frames. The experimental video-image can be a direct recording of a changing spatial pattern (e.g the wave- and cloud- like patterns on the skin surface of a cuttlefish) where the pattern, but not its medium, is moving [6]. Alternatively, both the pattern and its supporting medium might be in motion - the spatio-temporal patterns of fluid flows are often studied by a quasi-Lagrangian approach using "flow visualisation" methods, in which the motion of some secondary indicator of the fluid motion is actually observed. The indicator may be a suspension of small particles (e.g. a dye) whose precise dynamical response to the actual fluid motion is itself a poorly understood phenomenon. In Taylor-Couette flow, for example, the particle velocities comprise a relatively small meridional component superposed on a large cicumferential velocity; a photograph of a particular meridional cross section has a strong analogy with a stroboscopic map of a periodically forced dynamical system [7]. Ecodynamic analogies to these two examples would be the spatio-temporal patterning of an epidemic, and monitoring the population of migratory species by following tagged samples.

Since we are interested in irregularities that may show sensitivity to initial conditions a high resolution is required: however even with modest resolution (say 256 by 256 pixels, with 64 grey scale levels) the storage requirements are of the order of megabytes/frame. Thus the initial technical problems are more to do with efficient image representation and manipulation, rather than extensive numerical computation: the requirements are for a graphics workstation rather than data links to a mainframe or supercomputer. The aim is to replace each frame of the image sequence by a few measures, whose evolution can then be followed, and so the first step involves a massive data reduction.

SOME MEASURES

[1]. Time series analysis.
Statistics from time series analysis - statistical moments, autocorrelation function, power spectra - readily generalise to the plane, and the advantage of such measures is that maximal likelihood estimators, and their confidence limits, are well known. However, although in time series analysis these methods have proved powerful in distinguishing periodicity from "stochasticity", they have been of little use in identifying chaos, apart from distinguishing quasi-periodicity from chaos. We regard the estimation of these measures as a necessary but not very informative procedure.

[2]. Direct computation.
Reconstruction of a vector phase portrait from a scalar time series, containing no explicit spatial information, is usually accomplished by the method of delays. Spatio-temporal records generally provide an embarassingly large vector time series. For example, a typical digitisation of a video frame yields a vector with 256 x 256 = 65536 components. However, one generally expects that these components are not all independent. Indeed, the objective is usually to describe the dynamics in terms of a low dimensional attractor, involving a few

variables only. The singular value decomposition [8] provides a convenient means of reducing the dimensionality by projecting on to new basis vectors. This technique of locating significant directions is related to factor analysis and the Karhunen-Loeve expansion. Having reduced the dimensionality of the time series, dimensions, entropies and Lyapunov exponents may then be estimated in the conventional manner.

[3]. Fractal dimension of dendritic structures.
Some slowly growing spatial structures, such as the dendritic arborisation of a neurone, or viscous fingering, may have a fractal dimension that is constant with time, or slowly changing: for dendritic structures that grow at their tips the branch pattern presents a history of the growth process. The fractal dimension D of a spherically symmetric, dendritic structure with thin branches may be estimated by the number of branches n within a radius r: n(r) $r^{(D-1)}$. Diffusion-limited aggregation in three spatial dimensions has D = 2.5: what is of interest is that the estimated D for some mammalian neurones are close to 2.5. For asymmetric neurones the estimated D are different in different directions, and the estimated value of D provides a means of classifying the neurones [9]. These methods are also being applied to the tufted dendritic structure of invertebrate neurones. Applications to ecodynamics seem limited: coral growth might be an example.

[4]. Markov random fields.
A possible model for an irregular spatial "pattern" is a Markov random field, of single values $X_{i,j}$ taken at lattice point i,j in the plane. By analogy with a Markov chain, a pattern will involve local dependencies between the "neighbouring" $X_{i,j}$, where the concept of neighbourhood needs to be defined (von Neumann, Moore; first-, second-,... order). The conditional probabilites provide a means of specification and analysis of these spatial dependencies. This was originally proposed in an ecological context [10], and has recently been applied to multi-unit recordings from neurones [11].

[5]. Mutual information
An image at a particular time may be regarded as a probability density p(x,y), the value assumed by a pixel with coordinates (x,y) is then p(x,y). An information theoretic measure of the complexity, or "bumpiness", of this density is given by the mutual information

between the coordinates (x,y):

$$I(x;y) = \iint p(x,y) \log [p(x,y)/p(x)p(y)]\, dx\, dy. \quad (1)$$

Here p(x) and p(y) denote the marginal probability densities. Scenes with little variation yield small I whereas complex scenes have a high mutual information. Higher dimensional analogues of the mutual information, the redundancy, may be defined for use with patterns in three spatial dimensions.

[6]. Visual identification

Given the uncertainty of choice of a suitable measure, and even if there is a single measure that will be of use in quantifying all spatio-temporal irregularities, visual comparison of modelled and experimental patterns is a useful, but not very objective, means of assessing a model. One interesting approach is to attempt to mimic the operations performed by the visual system in pattern analysis and recognition: perhaps measures that may readily be implemented by neural nets ("neuro-computers") that simulate retinal mechanisms may be of value. Here, it is encouraging to note that Markus et al [12] in measuring the order-disorder transition between spiral waves, and convectively disturbed spiral waves in the Belousov-Zhabotinsky reaction, found that minimization of the autodifference function was the most useful measure. Evaluation of the autodifference function is computationally slow, but only involves summations and subtractions, and so is readily implementable by a neural net.

COMMENTS

Previous paragraphs have dealt with measures to summarise the complexity of a spatial pattern by a single statistic, whose temporal variation may then be analysed. However, when dealing with spatially extended systems, it must be recognised that dimensions and entropies, the quantities traditionally used to characterise chaotic dynamical systems, are extensive, and that many systems are open; that is, information flows across the system's boundaries. Grassberger [13] has made preliminary investigations into spatially extended, open systems. We examine the relevance of this to characterising experimental spatio-temporal patterns.

There is increasing evidence [14] that the dimension, measuring the number of degrees of freedom, and entropy of a system grow in proportion to its volume. One thus writes

$$\delta = \lim_{V \to \infty} \frac{D(V)}{V} \quad \text{and} \quad \eta = \lim_{V \to \infty} \frac{h(V)}{V} \qquad (2)$$

where D(V) and h(V) are the dimension and entropy of a system of volume V and δ, η are the dimension and entropy densities. They measure the number of digits required to specify unit volume of the system to a given length scale and the rate of information production per unit volume. Large, but finite, dynamical systems may show an effective dimension that increases as the volume from which a sample is taken is increased until the sample volume approaches the system volume and the overall dimension is measured.

Writing a spatial pattern in a volume V at a time t as a vector $X_V(t)$, the measures described above summarise the complexity of the pattern in a statistic $C(X_V(t))$ allowing an analysis of the temporal variation of C using conventional methods. An alternative approach is to directly compute the dimension D(V) of the trajectory $\underline{X}_V(t)$, regarding $\underline{X}_V(t)$ as an element of phase space. The dimension density might then be estimated by increasing the measurement volume and using equation (2). It is not yet clear that these two approaches yield the same result and further work on simulations and experimental data is in hand.

Systems with open boundaries, such as a measurement volume, allow flow of information across these boundaries. It is not yet clear how this information scales with length, but Grassberger suggests that it may dominate the production of information from small length scales. The information flow will often be anisotropic, such as flow in a pipe, or the migration of a species. Instead of measuring the entropy density, one will measure the rate of information flow through the boundaries of the measurement volume. These techniques have yet to be evaluated on a variety of systems, but it is important to bear in mind that spatially extended and open systems are fundamentally different from closed systems.

REFERENCES

[1] May, R.: Simple mathematical models with very complicated dynamics. Nature **261** 459-467 (1976)

[2] Schaffer, W.M. and Kot, M.: Differential systems in ecology and epidemiology. in Chaos ed. A.V.Holden. Manchester and Princeton

University Presses (1986)
[3] Broomhead, D.S. and King, G.P: Extracting qualitative dynamics from experimental data. Physica **20** D 217-236 (1986)
[4] Everson, R.M.: Quantification of chaos from periodically forced squid axons. In Chaos in Biological Systems ed. H.Degn, A.V.Hilden and L.F.Olsen. Plenum: New York (1987).
[5] Meyer-Kress,G. (ed) Dimensions and Entropies in Chaotic Systems Springer Verlag: Berlin (1986)
[6] Holden, A.V. and Matsumoto, G.: Patterned and irregular activity in exitable media. In: Structure, coherence and Chaos ed. P.L.Christiansen and R.D.Parmentier. Manchester Univ. Press 1988.
[7] Brindley, J. and Mobbs, F.: An experimental investigation of transition to turbulence in Taylor Couette flow using digital image processing. In Proceedings of European Turbulence Conference, Lyon Springer-Verlag: Berlin (1987)
[8] Broomhead, D.S. and King, G.P.: In Nonlinear Phenomena and Chaos ed. S.Sarkar Adam Hilger: Bristol (1985)
[9] Schierwagen, A.K.: Dendritic branching patterns. In Chaos in Biological Systems ed. H.Degn, A.V.Holden and L.F.Olsen. Plenum: New York (1986)
[10] Besag, J. : Spatial interaction and the statistical analysis of lattice systems. J.Roy.Stat. Soc. **36B** 192-227 (1974)
[11] Boriyuk, G.H., Boriyuk, R.M., Kirillov,A.B., Kovalenko, E.I and Kryukov,V.I.: New methods for the analysis of neuronal activity. Acad. Nauka CCP, Puschino (1986) (in Russian).
[12] Markus, M.,Muller,S.C.,Plesser, T., and Hess, B.: Characterization of order and disorder in spatial patterns. In Chaos in Biological Systems ed. H.Degn, A.V.Holden and L.F.Olsen. Plenum: New York (1987)
[13] Grassberger, P. Information content and predictability of lumped and distributed dynamical systems. University of Wuppertal preprint WU B 87-8 (1987)
[14] Ruelle, D.: Large volume limit of the distribution of characteristic exponents in turbulence. Commun. Math. Phys. **87** 287-302 (1982)
Foias, C., Manley, O.P., Temam,R. and Treve, M.T.: Asymptotic analysis of the Navier-Stokes equations. Physica **9D** 157-188 (1983)
Nicolaenko,B.: Some mathematical aspects of flame chaos and flame multiplicity Physica **20D** 109-121 (1986)
Manneville, P., in Macroscopic Modelling of Turbulent Flows. Lecture Notes in Physics **230** p 319 Springer: New York (1985)

Index of Contributors

Research Reports in Physics

The categories of camera-ready manuscripts (e.g., written in TEX; preferably hard plus soft copy) considered for publication in the **Research Reports** include:

1. Reports of meetings of particular interest that are devoted to a single topic (provided that the camera-ready manuscript is received within four weeks of the meeting's close!).
2. Preliminary drafts of original papers and monographs.
3. Seminar notes on topics of current interest.
4. Reviews of new fields.

Should a manuscript appear better suited to another series, consent will be sought from the author for its transfer to the other series.

Research Reports in Physics are divided into numerous subseries, e.g., nonlinear dynamics or nuclear and particle physics. Besides covering material of general interest, the series provides an opening for topics that are too specialized or controversial to be published within the traditional context. The implied small print runs make a consistent price structure impossible and will sometimes have to presuppose a financial contribution from the author (or a sponsor). In particular, in the case of proceedings the organizers are expected to place a bulk order and/or provide some funding.

Within **Research Reports** the timeliness of a manuscript is more important than its form, which may be unfinished or tentative. Thus in some instances, proofs may be merely outlined and results presented that will be published in full elsewhere later. Since the manuscripts are directly reproduced, the responsibility for form and content is mainly the author's, implying that special care has to be taken in the preparation of the manuscripts.

MIX
Papier aus verantwortungsvollen Quellen
Paper from responsible sources
FSC® C105338

If you have any concerns about our products,
you can contact us on
ProductSafety@springernature.com

In case Publisher is established outside the EU,
the EU authorized representative is:
Springer Nature Customer Service Center GmbH
Europaplatz 3, 69115 Heidelberg, Germany

Printed by Libri Plureos GmbH
in Hamburg, Germany